Gemmology

Related titles

The Diamond Formula
A. Barnard

Gems (sixth edition)
M. O'Donoghue

Synthetic, Imitation and Treated Gemstones
M. O'Donoghue

Artificial Gemstones
Michael O'Donoghue

Gem Identification Made Easy
Antoinette L. Matlins & A. C. Bonanno

Gemmologists' Compendium
Robert Webster. Rev. by E. Allan Jobbins

Identification of Gemstones
Michael O'Donoghue, Louise Joyner

The Spectroscope and Gemmology
Basil Anderson & James Payne. Ed. by R. Keith Mitchell

Front cover: Amethyst and diamond necklace on amethyst crystal (Courtesy of P.J. Watson)

Gemmology

Third edition

P.G. Read

NAG Press

First published in Great Britain in 1991

Second edition 1999

Third edition published in 2005 by NAG Press, an imprint of
The Crowood Press Ltd, Ramsbury, Marlborough Wiltshire SN8 2HR

enquiries@crowood.com

www.crowood.com

This impression 2022

British Library Cataloguing-in-Publication Data
A catalogue record for this book is available from the British Library.

ISBN 978 0 7198 0361 1

Printed and bound in India by Parksons Graphics Pvt Ltd

Contents

Preface

My *Beginner's Guide to Gemmology* was first published in 1980, and although this was only intended as an introduction to the subject, it was apparently used with some success as a textbook by many students. For several years I had acted as tutor to students taking the Gemmological Association's correspondence courses, and this experience prompted me in 1991 to produce a more expansive and up-to-date volume in the original hardback edition of *Gemmology*.

Both the first and second editions of the book were written with the aim of providing a readable account of the relatively modern science of gemstones as well as forming a text to assist students preparing for the Gemmological Association's Preliminary and Diploma examinations.

A similar format has been retained in this third edition, but more emphasis has been placed on the use of the book as a work of reference. To this end the index has been enlarged, and cross-links between chapters have been increased. The replacement of the original Preliminary course examination by the new Foundation course examination (and its associated practical endorsement requirement) is covered in Appendix E – Examination notes.

In line with the increased practical content of the Foundation course, more emphasis has been placed on the identification of natural gemstones, their simulants and their synthetic counterparts. The text in many of the chapters has been revised and updated. In particular, new HPHT diamond enhancement methods and beryllium lattice diffusion techniques for corundum are covered as is the low-pressure high-temperature CVD synthesis of thick film diamond. Finally, a new chapter has been added to provide a concise guide to practical gemstone identification.

Peter G. Read
Bournemouth, Dorset
2005

Acknowledgements

In this third edition I would again like to acknowledge my debt to those three pioneering gemmologists Basil Anderson, Alec Farn and Robert Webster all of whom I was fortunate enough to know, and whose lectures, books and articles have strongly influenced this present volume, particularly in its emphasis on practical gem identification.

I am also grateful to Vivian Watson of P.J. Watson Ltd who once again has provided the front cover illustration and some of the colour plates, to Lorne Stather who made sure I had a wide choice of digital pictures to illustrate the appropriate sections of text, and to Ian Mercer for keeping me up to date with the Gem-A's educational activities. My thanks must also go to my publishers, Robert Hale Limited.

Finally, I must thank my wife Joan for all her patience and encouragement during this project.

Chapter 1

Introduction

The evolution of the science of gemmology

The science of gemmology is concerned with the study of the technical aspects of gemstones and gem materials and the use of these aspects for identification purposes. For well over 2000 years, philosophers and scientists have been captivated by the beauty and the enigma of gems, and down the years have left records of their observations on these ornamental products of nature. One of the first gemstone books in English was written by Thomas Nichols as long ago as 1652*, but it was only in the last half of the nineteenth century that the science of gemmology began to emerge as a specialized offshoot of an already well-established branch of science, mineralogy.

Highlights of the last 170 years

In view of the important part that gemmology plays today in the identification of modern synthetic gems, it is perhaps appropriate that we go back around 170 years in time and begin this summary of the highpoints in gemmological history with the first attempt at gemstone synthesis. In 1837, the French chemist Marc Gaudin managed to grow some small crystals of ruby by melting together potassium aluminium sulphate and potassium chromate. This was in the period when there was much interest in reproducing the growth of crystalline substances, and when the first experiments were being made to dissolve the constituents in a solvent 'flux' of lower melting point.

In view of the importance of spectroscopy in present-day gemmology, it is also appropriate to mention the letter written by Sir Arthur Church in 1866 to the learned English periodical, *The Intellectual Observer*. In this he describes his experiments with an early spectroscope and his discovery of absorption bands in the spectrum of Ceylon zircons and almandine garnets. However, it was not until 1932 that a comprehensive study of gemstone spectra for identification purposes was to be undertaken by Basil Anderson.

Some five years after Church's letter appeared in print, the South African diamond rush was in full spate, and 5000 diggers were reported to be working along the banks of the Vaal, Modder and Orange rivers. In 1873, the primitive mining town that had sprung up around the site of the De Beers farm was formally named Kimberley after the British Secretary for the Colonies, the Earl of Kimberley.

In 1877, the French chemist Edmond Frémy manufactured the first synthetic rubies of commercial quality. These crystals were grown in a large porcelain crucible containing a lead oxide flux in which was dissolved alumina powder mixed with a trace

---

* *A Lapidary; the history of precious stones*, Cambridge University Press.

of a chromic salt. However, the resulting crystals were small and expensive to produce and were therefore no threat to natural rubies. The following year was marked by the discovery and identification of a new gem variety, green demantoid garnet.

One of many disturbing events in the jewellery trade occurred in 1885, when a quantity of relatively large 'Geneva' rubies appeared on the market. Initially, the presence of bubbles within the rubies proved them to be synthetic. Later, however, it was thought that the stones had been produced by fusing together smaller fragments of natural ruby, and for this reason they were then called 'reconstructed' rubies. More recently, analysis of surviving specimens coupled with attempts to reproduce the reconstruction process has shown that such a fusion of small stones could not possibly have produced transparent rubies, and that the Geneva ruby was probably created by the multi-step melting of a mound of ruby powder in a flame.

Meanwhile, in South Africa, Cecil Rhodes and Barney Barnato had finally agreed to amalgamate their holdings, and a controlling company, De Beers Consolidated Mines Ltd, was incorporated in 1888. The name was taken from the De Beers brothers' farm which had become the site of the famous 'Big Hole' of Kimberley. In the same year the first successful synthesis of gem quality emerald crystals was achieved by the French chemists Hautefeuille and Perrey using a flux process.

The pace of gemstone synthesis began to quicken towards the turn of the century. In 1891, the French scientist Vernueil, a former assistant to Frémy, was perfecting the furnace he had designed for the production of synthetic corundum. Over 100 years later, furnaces of this type were to be producing in excess of 1000 million carats of synthetic corundum per annum worldwide.

The year 1902 saw the discovery and documentation of the pink kunzite variety of spodumene, and three years later the 3106 carat Cullinan diamond was prised out of a sidewall in the opencast workings of the Premier mine near Pretoria in South Africa. In the same year in England, Dr Herbert Smith produced his first refractometer (Figure 1.1), providing gemmologists at last with an instrument specifically designed for measuring the refractive index of gemstones (this was followed in 1907 by a larger brass version). By 1910, the first synthetic rubies produced by the Verneuil method appeared on the market, although ironically, these could not be identified as synthetic by means of the refractometer.

Figure 1.1 Two early gem refractometers: the Herbert Smith model (left); the Tully version (right)

One event that did more than any other to establish gemmology as a serious science was the formation of the Gemmological Association of Great Britain, which began its life in 1908 as the Education Committee of the National Association of Goldsmiths.

The Association held its first membership examinations in 1913, just three years after synthetic Verneuil rubies appeared on the market. World War I broke out in 1914, and it was not possible to reinstate the Association's examinations for another eight years.

The next milestone in the development of the young science of gemmology occurred in 1925, when Basil Anderson, fresh from university with degrees in chemistry and mineralogy, was engaged by the Diamond, Pearl and Precious Stone Section of the London Chamber of Commerce to set up a pearl testing laboratory in Hatton Garden. The urgent need for such a laboratory arose from the rapid growth of the Japanese cultured pearl industry and the problems that jewellers were experiencing in distinguishing native pearls from the cultured product. The growing importance of the refractometer to gemstone identification is indicated by the introduction in 1925 of yet another version. This new model (Figure 1.1), designed by the famous jeweller–gemmologist B.J. Tully, used a rotatable hemisphere of glass.

The following year, the first synthetic spinels were produced by the Verneuil method, and an endoscope pearl tester (Figure 1.2) was installed in Basil Anderson's new Hatton Garden Laboratory. This equipment, brought over from France, made it possible to test over 200 pearls an hour. By 1928, when C.J. Payne joined the laboratory, nearly 50 000 pearls were being examined each year. The laboratory moved to new premises in 1928, and an X-ray unit was installed so that undrilled pearls could be tested using diffraction techniques. So began gemmology's practical service to the jewellery trade, first with the identification of pearls (well over 4 million were tested in the laboratory which was merged much later with the Gemmological Association) and then with the detection of the new Verneuil synthetic rubies and spinels.

Figure 1.2 The endoscope pearl tester in use at the Chamber of Commerce's Gem Testing Laboratory in London's Hatton Garden

Another milestone event occurred in 1931. Robert Shipley, who had been awarded the British Gemmological Association's Diploma in 1929 and who had then pioneered his own gemmological correspondence course in the USA, founded the Gemological Institute of America. In the mid-1930s he was joined by his eldest son, Robert Shipley Jr, who helped develop a series of gem testing equipment which included a gem microscope, a diamond colorimeter, a refractometer and a polariscope.

In 1934, Robert Shipley Sr founded the American Gem Society as a professional body of leading jewellers. Volume 1, No. 1 of *Gems & Gemology*, the journal of the Gemological Institute of America, appeared in 1935 and in the same year the American Gem Society held its first examinations.

During the depression years of the 1930s, Basil Anderson in the UK made good use of the downturn in trade to carry out research into various gem testing/identification techniques. In particular he and C.J. Payne were able to study and record details of the absorption spectra of many gems (these were later to be published in Robert Webster's *The Gemmologist's Compendium* and in Dr Herbert Smith's 1940 edition of *Gemstones*), and to develop an experimental blende version of the Tully refractometer. During their refractometer work they also formulated a new contact liquid (a solution of sulphur and tetraiodoethylene in di-iodomethane) which until more recently was to become the accepted refractometer contact liquid and used throughout the gemmological world.

Following unsuccessful attempts by the Rayner Optical Company, manufacturer of the Tully refractometer, to fabricate the hemispherical prism in blende rather than glass, a truncated prism version was developed and fitted into a small instrument of Rayner's own design. The change from a hemisphere to a prism-shaped refractometer 'table' made it possible to make a less expensive standard glass model, and this design became the basis for all future Rayner refractometers (and for overseas 'clones'!). The blende refractometer was followed by a diamond and then a spinel version.

Other important research work by Anderson and Payne during the comparative leisure of the mid-1930s included a reassessment of the majority of gem constants. Several stable and relatively safe heavy liquids for the determination of specific gravity were also established. In 1933, Basil Anderson took over the Chelsea Polytechnic classes in gemmology. One of the students in his first Diploma class was Robert Webster. During this period, an emerald filter (called the 'Chelsea' filter) was developed jointly by the London laboratory and the gemmology students (Figure 1.3).

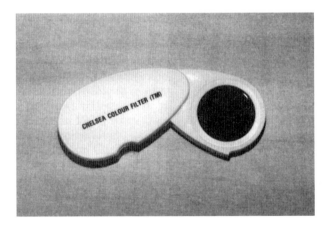

Figure 1.3 The Chelsea emerald filter designed to distinguish between emeralds and their simulants

Early in 1935, news appeared in the London press of a synthetic gemstone having 'all the qualities of diamond' and capable of 'deceiving 99 per cent of the experts'. Today the story has an all-too-familiar ring, but back in the 1930s, this new product caused quite a stir. The simulant behind the scare story was colourless synthetic spinel, which was thought to have been manufactured in Germany.

The same news caused similar consternation when it appeared in the North American press. There was further dismay in the trade on the announcement of the successful synthesis of diamond in gem qualities and sizes by a Mr Jourado, a self-styled gem expert. The Jourado stone was identified as a spinel by Anderson, and reports by him of the stone's characteristics appeared in several of the leading gemmological journals and jewellery magazines.

Another significant event that occurred in 1935 was the pilot-scale synthesis of emerald by the German firm I.G. Farbenindustrie, which had developed a new flux-melt process. Although many samples were produced, the advent of World War II interrupted the company's work. The 'Igmarald' synthetic emerald was never launched commercially, and production was finally discontinued in 1942.

Because of its high value, the synthesis of emerald became the aim of several laboratories, and in 1940 the American chemist Carroll Chatham also succeeded in growing gem-quality crystals. Although the method of manufacture was kept secret, the Chatham emeralds were close enough in character to the German Igmarald to indicate that they were grown by a flux process. (In a move to avoid the use of the word 'synthetic', Chatham finally obtained permission in 1963 from the US Federal Trade Commission to market his product as 'Chatham created emerald').

After World War II, Anderson and Payne were joined in the Hatton Garden laboratory by Robert Webster and Alec Farn, and the work of gemstone identification continued. During the following years, Robert Webster was to carry out pioneering work on the use of ultraviolet light in the identification of gem minerals, and Alec Farn was to become, among other things, the UK's leading expert on pearl testing.

Soon the Hatton Garden laboratory was working at full capacity again as parcels of rubies and sapphires containing up to 10% synthetic stones started to arrive in London. In 1946, more than 100 000 stones were tested in the laboratory, and a year later its problems were further increased by the successful production of star rubies and sapphires by the Linde Division of the Union Carbide Corporation of America.

Synthetic rutile, the first of a series of man-made diamond simulants, appeared in 1948 and was marketed under the trade names 'Rainbow Gem' and 'Titania'. In 1951, a new rare gem species was confirmed and named 'Taaffeite' after its discoverer, Count Taafe. X-ray and chemical analysis were used to verify its principal constituents as beryllium, magnesium and aluminium (except for double refraction, taaffeite closely resembles spinel). Although still a rare species, a few taaffeites have since been found in Sri Lanka.

In the 1950s, strontium titanate was introduced as yet another diamond simulant under the trade names 'Fabulite' and 'Diagem'. Unlike the earlier simulant, synthetic rutile, there appeared to be no counterpart for this material in nature. During this period, progress was also being made in the field of diamond synthesis. Although failing to have their work verified at the time by an independent investigator, the ASEA group of Sweden claimed to have developed a repeatable process in 1951. As they were unaware of any other company working on diamond synthesis, ASEA kept their process secret while they worked on improving the size and quality of their product. They were also unaware of the great industrial importance of grit-size synthetic diamonds. ASEA only revealed their earlier breakthrough after the General Electric group in the USA announced its own successful synthesis of diamond grit in 1955. Four years later, the De Beers Diamond Research Laboratories in Johannesburg also succeeded in synthesizing industrial diamonds (Figure 1.4). Since then many countries have developed this capability including Russia, Japan and the People's Republic of China.

A brown gemstone from Sri Lanka, formerly classified as a peridot, was identified in 1954 as a new mineral species and named sinhalite after its country of origin. In 1957, yet another new gem species was identified and named 'Painite' after its discoverer, A.C.D. Pain.

Figure 1.4 An example of industrial synthetic diamond grits. (Courtesy of De Beers)

The first issue of the *Australian Gemmologist*, the official journal of the Gemmological Association of Australia, appeared in 1958. It carried a report that the British Atomic Energy Research Establishment at Harwell was undertaking the commercial irradiation of diamonds (as a means of enhancing their colour). It also contained news of Australia's first pearl culturing venture using the mollusc *Pinctada maxima*.

The first experimental reflectance instrument for the identification of gemstones was developed in 1959 by L.C. Trumper, who was awarded a Research Diploma by the Gemmological Association of Great Britain for his thesis on the measurement of refractive index by reflection. His instrument took the form of an optical comparator in which the intensity of reflection from a gem's surface was visually matched against a calibrated and manually adjustable source of illumination. In the same year, Caroll Chatham in the USA began marketing synthetic rubies produced by a type of flux-melt process. A new type of synthetic emerald, first called 'Emerita' and then 'Symerald', was produced by the Austrian chemist J. Lechleitner in 1960. He used the hydrothermal method to deposit a thin coat of synthetic emerald on to a faceted beryl gem of poor colour. The Lechleitner emerald process was later acquired by the Linde Company of America, who subsequently marketed the product as the 'Linde' synthetic emerald.

In 1963, Pierre Gilson in France marketed his Gilson synthetic emerald which he manufactured using an improved version of the flux-melt process originally developed by I.G. Farbenindustrie. A year later, Lechleitner produced hydrothermally grown emeralds from seed crystals, and this was followed in 1965 by a similar product from Linde.

During this period of intensive work on emerald synthesis, a new thorium-rich radioactive gem mineral was found in the alluvial gravels of Sri Lanka. The metamict gem was named 'Ekanite' after its discoverer, F.D. Ekanayake.

In 1967, deposits of gem-quality transparent blue zoisite were discovered in Tanzania and given the name 'Tanzanite'. Two years later, another two diamond simulants having no counterpart in nature were introduced. The first of these was yttrium aluminium garnet (YAG), which was marketed as 'Diamonaire' and 'Diamonique', and the second was lithium niobate, marketed as 'Linobate'.

In 1970, synthetic gem-quality carat-size diamonds were grown under laboratory conditions by General Electric of America, but the resulting tabular crystals were not economically viable as a commercial product. In 1971, Russian research workers announced that they too had synthesized gem-quality diamond crystals but they also decided that their product was too costly to market.

The publication of Dr Eduard Gübelin's book *Internal World of Gemstones* in 1974 set new standards in the microphotography of gem inclusions (a worthy successor,

Photoatlas of Inclusions in Gemstones was published in 1986 as the result of collaboration between Dr Gübelin and J. Koivula of the GIA – a second edition appeared in 1992).

The rare alexandrite variety of chrysoberyl was successfully synthesized in 1975 by means of both the crystal-pulling and flux-melt processes. This was followed by the introduction of Gilson synthetic opals, which produced yet another identification problem for the jeweller and gemmologist.

The 1970s and 1980s saw the marketing of an increasing number of sophisticated new synthetics (including the Kashan, Knischka, Ramaura and Seiko rubies, the Cresent Vert, Regency, Biron/Pool and Lennix emeralds, and flux-melt Chatham sapphires). Identification challenges also appeared in the form of new diamond simulants. Gadolinium gallium garnet (GGG) was marketed in 1973, and cubic zirconium oxide (CZ), which virtually superseded all previous diamond simulants, appeared in 1976.

During this period, the need for help with the identification and appraisal of gems at the jeweller level resulted in a proliferation of new gem test equipment. Perhaps the most frequently used of these is the reflectance meter and the thermal conductance tester which have been developed mainly for the detection of diamond and its many simulants. The first commercial reflectance meters were marketed in 1975 (Figure 1.5), and the first commercial thermal tester appeared in 1978 (Figure 1.6). Because of the useful complementary features of these two types of test instrument, dual versions were introduced in 1984.

Figure 1.5 (left) The Sarasota Instruments 'Gemeter 75' – one of the first reflectance meters
Figure 1.6 (right) The Ceres 'Diamondprobe' – the first commercial thermal tester for diamonds and diamond simulants

In 1986, Sumitomo Electric Industries of Japan announced that it was able to produce economically viable gem-quality synthetic diamonds in sizes up to 2 carats. Although these stones were deep yellow in colour and were marketed for industrial applications only, at least one faceted Sumitomo synthetic diamond was identified in Hatton Garden two years later! In 1987, the De Beers Diamond Research Laboratory in Johannesburg, South Africa, revealed that it had also developed a method of growing commercial quantities of gem-quality synthetic diamonds. Using a high-pressure flux vessel, crystals had been grown up to 11 carats in size for specialized industrial applications. In the same year a report by the Soviet news agency Tass claimed that Soviet scientists had manufactured synthetic diamonds weighing 3 kilograms. However, it was subsequently revealed that these crystals were cubic zirconium oxide – the misquote had occurred because the Russian word for simulant and synthetic is the same!

In March 1987, timed to coincide with their centenary celebrations, De Beers announced the recovery (in 1986) of a 599 carat diamond from the Premier mine (Figure 1.7). The same year saw the introduction of the world's first commercial computer program, *Gemdata*, for gemstone identification (developed by the author).

Figure 1.7 (right) The 599 carat 'Centenary' diamond recovered in 1986 from the Premier mine in South Africa: (above) The 273 carat 'Centenary' diamond cut from the rough. (Both courtesy of De Beers)

In 1990 The Gemmological Association of Great Britain ended its long partnership with the National Association of Goldsmiths and moved into separate premises in the Hatton Garden area of London. During the same year, the Gemmological Association merged with the London Gem Testing Laboratory to become the Gemmological Association and Gem Testing Laboratory of Great Britain (GAGTL). In 2002, the GAGTL changed its logo to Gem-A.

By 1993, yellow gem-quality synthetic diamonds were being grown commercially in Russia for use in the jewellery trade. In 1996, De Beers' DTC Research Centre in the UK announced that they had developed two instruments 'DiamondSure' and 'DiamondView' specifically for the identification of synthetic and natural diamonds (Figure 1.8). In the same year, a new diamond simulant, synthetic moissanite (silicon carbide), was introduced by C3 Incorporated of the USA. By 2003 diamonds were being synthesized by chemical vapour deposition (CVD), a low-pressure high-temperature process using hydrogen and methane starter gases in a microwave-induced plasma.

During recent years, while most simulants used as imitations of natural gem materials can be identified by the use of the refractometer, the hand lens/microscope and the spectroscope (as discussed in Chapters 9, 11 and 13), identification problems associated with the growing range of sophisticated synthetic gems have become much more acute. These have been partly solved by increased reliance on the detection of diagnostic inclusions and growth characteristics using the hand lens and microscope (subjects covered in Chapter 16). Where this fails, use must be made of a range of

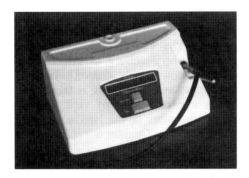 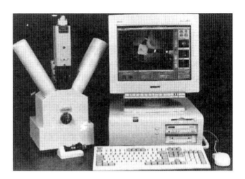

Figure 1.8 The 'DiamondSure' (left) identifies natural and synthetic diamonds by detecting the presence of the 415 nm absorption line in the natural stone. If this test is inconclusive, the 'DiamondView' (right) is used to reveal the diagnostic fluorescent growth patterns characteristic of natural/synthetic diamonds (see also Chapter 16). (Courtesy of De Beers)

hi-tech laboratory instruments including ultraviolet-visible-infrared spectrophotometers, electron microprobes, electron microscopes and cathodoluminescence equipment (also described in Chapter 16).

During the 1980s, 1990s and into the present century, there has also been an increase in the introduction of new gem enhancement techniques, from the heat treatments of corundum, the irradiation of yellow sapphire and blue topaz, the glass filling of fractures in diamond, ruby and sapphire, to the laser drilling of diamond and, in 1999, the colour improvement of brown Type II natural diamonds by HPHT annealing techniques. High-temperature surface-diffusion and deep-diffusion of pale colour corundums is achieved by the addition of transition element oxides, and more recently the lattice diffusion of beryllium additives has produced orange and yellow sapphires but has also created new identification challenges. All of these enhancement techniques are described in Chapter 14.

As we have seen in this brief review of selected highlights from the last 170 years in the evolution of the science of gemmology, there is an urgent and continuing need for the professional gemmologist to discover ways and means to identify new synthetics and detect new enhancement treatments as they are introduced. This has become what is perhaps the most exciting and challenging aspect of gemmology, and one that has resulted in the development of a wide range of gem test equipment and detection techniques.

The essential qualities of a gem material

So far in this chapter we have attempted to convey some idea of the historical background to the science of gemmology. Now it becomes relevant to consider the qualities that make gem materials suitable for use in jewellery. The first and most obvious of these qualities is *beauty*.

Unlike a gem's more tangible properties, its beauty cannot easily be quantified as it depends in the main on subjective factors to do with its appearance. If the stone is a transparent coloured gem, the depth of colour and degree of transparency will be the prime factors. However, in the case of a gem such as a diamond, beauty will be determined by its brilliancy, fire, optical purity and, in general, the absence of any body colour. With precious opal, the quality of its iridescent play of colour will be the deciding factor.

Rarity is another quality which must be present in some degree in all gemstones worthy of the name (coloured glass, while it may be quite beautiful in terms of hue and transparency, is by no means rare). Unlike beauty, however, the rarity of a gem can be affected by factors such as supply and demand, and by both fashion and the scarcity of the source material. Amber and pearls have become popular again and therefore more expensive, while amethyst was once a rare and expensive gemstone until the discovery of the extensive Brazilian sources in the eighteenth century.

Unusual optical properties add to a gem's rarity factor as with alexandrite and cat's-eye chrysoberyl. Diamonds are expensive, but as the total world production of all types of rough diamonds in 1999 was in excess of 110 million carats, the cost of the finished product is hardly attributable to its rarity. In this case, De Beers' influence on the supply side of the supply/demand equation, coupled with the economics of the mining, polishing and marketing of the gem, plays a dominant role.

The third essential quality which must be present in a gem before it can be considered suitable for use in jewellery is its *durability*. This is a more practical quality than either beauty or rarity, but without it a gemstone would not be able to survive either the day-to-day wear and tear experienced by a piece of jewellery, or the chemical attack from pollutants in the atmosphere, and would soon lose its surface polish.

Durability, which includes the property of hardness and toughness, is therefore a most important quality in a gemstone from the wearer's point of view. Equally important is the influence of a gem's hardness on the work of the lapidary and the diamond polisher, and this aspect will be discussed fully in Chapters 5 and 6.

Organic and inorganic gems

Jewellery has, from the earliest times, included gem materials which have an organic as well as a mineral content, and because of this the science of gemmology today covers not only mineralogy, geology, optics and chemistry, but also overlaps into the fields of zoology, biology and botany. Among the gem materials used in jewellery, the largest group is that of the mineral kingdom. The first part of this book therefore deals principally with the characteristics of gemstones having a mineral origin. Gems having an organic origin, such as ivory, bone, pearl, coral, tortoiseshell, jet, and amber are covered separately in Chapter 18, which also describes the methods of distinguishing them from their simulants.

Chapter 2

The geological origin, occurrence and locality of gemstones

The Earth's structure

The Earth is the third innermost planet of the star we call the Sun. Its structure, as indicated by studying seismic wave records of earthquakes, is made up of a central core, a mantle and a crust (Figure 2.1). The core has a diameter of around 4300 miles (7000 kilometres) and is thought to be composed of two parts. The outer core is about 1300 miles (2100 kilometres) thick and appears to consist of a liquid rock 'magma'. Beneath this lies the solid inner core having a diameter of around 1700 miles (2700 kilometres) and is believed to consist of very dense rocks composed of 80% iron and varying amounts of nickel, silicon and cobalt.

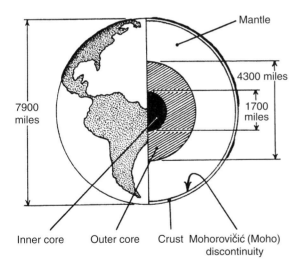

Inner core   Outer core   Crust Mohorovičić (Moho) discontinuity

Figure 2.1 Simplified sketch showing the Earth's structure (comprising the core, mantle and crust)

Between the core and the crust is the mantle, an intermediate semi-molten zone having a thickness of around 1800 miles (2900 kilometres). Above the mantle floats the solidified crust of the continents whose thickness varies from a mere 3 miles (5 kilometres) under the oceans to 45-mile (72-kilometre) deep 'roots' thrusting down beneath the higher mountains. Of these three structural zones it is the Earth's crust which is of prime interest to us, as this is the birthplace of our gem minerals and its geology is the key to their formation.

Minerals and rocks

At this point it is relevant to define exactly what is meant scientifically by the terms 'mineral' and 'rock'. A mineral is a substance which has been formed within the Earth by the forces of inorganic (i.e. non-living) nature. It is also a homogeneous, or uniform, substance with a definite atomic structure, and has a chemical formula and a set of physical and optical characteristics which are constant throughout its bulk.

In mineralogy, there are several thousand listed minerals, but only about sixty of these have the necessary qualities of beauty, rarity and durability to make them suitable for use as gem materials. Within this select group of some sixty gem materials there are several metallic minerals. These include gold, silver, platinum and the platinum group metals rhodium and iridium (rhodium is used as a bright protective coating on silver, and iridium is often alloyed with gold or platinum). These 'precious' metals share the distinction with diamond of being chemical elements in their own right, instead of being compounds of elements as are all other gem materials (the concept of elements and compounds will be discussed in more detail in Chapter 3).

Although minerals are mined from the Earth's crust in various states of purity, the bulk of this crust is made up of various mixtures of minerals, and these mixtures or aggregates are called rocks. Granite, for example, is composed of a mixture of feldspar, quartz and mica. With a few exceptions such as lapis lazuli (which contains a mixture of lazurite, sodalite, calcite and pyrite, and can therefore be classified as a rock), the majority of gemstones are composed of just one mineral.

The formation of rocks in the Earth's crust

Sedimentary rocks

Returning to the Earth's structure, the crust itself consists of three distinct layers (Figure 2.2). The upper and thinnest layer is composed mainly of the fine deposits of sand, grit and clay eroded by the action of rain, wind and flowing water from the ancient pre-existing rocks in the middle layer of the crust, and compressed to form layers of sandstone or limestone. Because of the mechanism of their formation, these top layer rocks are called *sedimentary*.

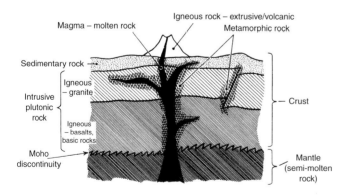

Figure 2.2 Sketch showing the relative positions of sedimentary, igneous and metamorphic rocks in the Earth's crust

Igneous rocks

The middle layer of the crust is formed by the solidification of molten magma. It is composed of rocks which are described as *igneous* (meaning 'fiery'), and these are mainly granite. The lower layer also contains igneous rocks. These are darker and denser, and consist mainly of basalts, and other basic and ultra basic rocks low in silica content and rich in iron and magnesium. The igneous rocks which solidified in the middle and lower layers are called *intrusive*, *plutonic* or *abyssal* rocks, while those which erupted through the upper layer and were formed by the rapid cooling of magma at the surface of the crust are called *extrusive* or *volcanic* (e.g. lava).

Metamorphic rocks

Some igneous and sedimentary rocks in the various layers of the crust have undergone changes as the result of heat and pressure, and these are called *metamorphic* rocks. Marble is a metamorphic rock which has been produced in this way from limestone. Other examples of a metamorphic reaction is where molten magmas were forced into cooler rocks, and where the temperatures and pressures caused by the large-scale shearing and crushing of rocks produced similar changes.

Mineral groups, species and varieties

In mineralogy, there are over 2000 different minerals. In order to subdivide these into manageable numbers they are split up initially into a series of *groups*, each of which contains those minerals which have similar features or characteristics. In gemmology, with its more limited number of gem minerals, there are only a few which have enough in common to qualify under this heading. These are the feldspar, garnet and tourmaline groups.

Gem minerals are further subdivided into *species*, which have their own individual chemical composition and characteristics, and *varieties* of species, which differ from each other only in colour or general appearance. Starting with a group such as garnet, we can draw a family tree branching out through species such as andradite, grossular and spessartite, to their colour varieties.

It is easy to become confused by the alternative names a gemstone may have under the headings of group, species and variety. To help resolve any confusion, those gem-stones which have several legitimate names under two or more of these categories are listed in Table 2.1.

The origin of gems

The *origin* of a gem is its place of genesis in the Earth's crust or mantle. Except for organic substances such as amber and jet, sedimentary rocks contain no primary gem material. However, if the pre-existing weathered rocks contained heavier minerals (i.e. gem minerals), these may have been washed out and swept away to form secondary or *alluvial* deposits. These gemstones often end up as water-worn pebbles, and can be seen in the gem gravels of Brazil, Myanmar (Burma) and Sri Lanka. Opal's origin begins with low-temperature silica-bearing water which solidifies as thin veins of silica gel material in the cracks and fissures of rocks.

Most of our important gem minerals, such as feldspar and quartz, tourmaline, beryl, topaz and zircon, originated in intrusive or plutonic igneous rocks whose slower rate of cooling in the middle and lower parts of the crust made it possible for quite large crys-tals to form from the original molten residues (Figure 2.3). As the temperature of the magma began to drop, minerals separated out in a process known as fractional crystal-lization. The feldspars were the first to solidify, and having plenty of space they pro-duced large well-shaped crystals. As the magma continued to cool, other minerals

Table 2.1

Group name	Species name	Variety name or description
	Beryl	Emerald, aquamarine, morganite (pink), heliodor (yellow), goshenite (colourless)
	Chrysoberyl	Chrysoberyl (yellow, greenish-yellow), alexandrite (red in tungsten light, green in daylight), cymophane (greenish-yellow cat's-eye)
	Corundum	Ruby, sapphire (blue, violet, green, yellow, pink, orange, colourless)
	Orthoclase	Adularia (colourless), moonstone (yellow, colourless with iridescence), orthoclase (yellow)
	Microcline	Amazonite (green)
	Sanidine	(Colourless to brownish – rare)
Feldspar	Plagioclase	Labradorite (multi-colour sheen and yellow), oligoclase (yellow), sunstone or aventurine feldspar (bronze or green spangled), albite moonstone (white, green, fawn or brownish pink with blue iridescence)
	Almandine	(Purplish red)
	Pyrope	(Blood red)
Garnet	Grossular	Hessonite (orange-brown, green, pink), massive grossular or hydrogrossular (jade green), tsavolite (green)
	Andradite	Demantoid (green), yellow andradite, melanite (black)
	Spessartite	Orange, yellow, flame red
	Uvarovite	Emerald green
	Opal	White opal and black opal (both with iridescence), water opal (colourless or brownish yellow with iridescence), Mexican fire opal (orange, sometimes with iridescence), potch or common opal (whitish without iridescence), hyalite (colourless glass-like without iridescence)
	Quartz	Amethyst, citrine (yellow), rose quartz, rock crystal (colourless), aventurine quartz (green, blue or brown with mica spangles), tiger's-eye (yellow-brown), hawk's-eye (blue-green), jasper (red, brown), quartz cat's-eye (light green or brown)
	Chalcedony (cryptocrystalline) (quartz)	Chalcedony (blue, grey unbanded), agate (curved concentric bands), cornelian (red), chrysoprase (green), onyx/sardonyx (brown and white with straight banding)
	Achroite	Colourless
Tourmaline	Schorl	Black
	Indicolite	Blue
	Other varieties	Red, pink, green, yellow and brown

crystallized out. Of these, quartz was one of the last to solidify, and, as it had much less room than the others to grow, was not always able to produce such large and well-defined crystals. Many of the intrusive gem–bearing rocks formed as coarse-grained granites called pegmatites. Geodes also originated in igneous rocks in which quartz and other minerals have been precipitated as crystals in almost spherical cavities formed by chemical-rich molten or aqueous residues trapped in the magma (Figure 2.4).

The chemical reactions generated in metamorphic rocks when molten magma was forced into cooler rocks created the gemstone varieties of emerald, alexandrite, ruby and sapphire. Other gem minerals were formed as a result of the large-scale shearing and crushing of rocks. Examples of these are garnet, andalusite, serpentine, nephrite and jadeite.

Figure 2.3 From top left to bottom right: single crystals of feldspar, quartz, tourmaline, beryl, zircon and topaz formed in intrusive or plutonic rocks

Figure 2.4 The sawn and polished half-section of a geode. Rapid cooling produced a thick outer layer of microcrystals, while a more leisurely fall in temperature allowed larger crystals to form towards the centre

Diamonds (Figure 2.5) differ from the rest of the gem minerals in that they were formed somewhere in the region between the lower part of the Earth's crust and the beginning of the mantle. This is the transitional zone at the base of the crust. It is known as the *Mohorovičić Discontinuity* or Moho (see Figures 2.1, 2.2), and is named after the Yugoslavian professor who discovered it. The current theory is that diamonds crystallized at least 70 miles (120 kilometres) beneath the Earth's surface from carbon (in the form of carbon dioxide or methane) at very high temperatures and pressures. The diamond-bearing magma was then forced up to the Earth's surface by explosive gas pressures in a volcanic-type eruption. The magma eventually cooled and solidified to form the present-day kimberlite pipes which make up the bulk of the world's primary source of diamonds.

The tops of these pipes are thought to have originally extended above the surface as cone-shaped hills or even as mountains. Over hundreds of millions of years these kimberlite hills were eroded away by the weathering action of wind and rain into

Figure 2.5 Rough diamonds. From left to right: a triangular twinned 'macle'; a 'shape' (distorted octahedron); a 'stone' (octahedron); and a 'cleavage' (broken crystal). (Courtesy of De Beers)

low-lying hillocks or 'kopjes' (a South African word pronounced as 'copy'). The diamonds contained in the eroded top section of the pipes were washed away to form alluvial deposits along river beds and, in the case of South West Africa, along the marine terraces of the Namibian coastal strip.

The manner in which gemstones were created in nature can therefore be related to the sedimentary, igneous or metamorphic processes of rock formation as well as the chemical content of molten and aqueous residues.

Gem occurrences

The *occurrence* of a gem is the geological environment in which it is found or mined (e.g. gem gravels, gem-bearing veins or a diamond pipe). Many gemstones are found at the site where they were originally formed, and this type of deposit is of particular interest to the mineralogist as it provides evidence of the method of gemstone formation. *Alluvial* gem deposits, on the other hand, are the result of gemstones which have been carried from their place of formation either by weathering agents such as wind or rain, or by rivers. Evidence of the distances travelled by gems found in alluvial deposits can be seen in their abraded surfaces (e.g. water-worn topaz pebbles and the rounded profiles of diamond crystals from the Namibian coastline). Sometimes, gems that were released from rocks by weathering have been deposited with little or no transportation or concentration from the site of the parent rocks. These residual deposits are described as *eluvial*.

The major gem localities

The *locality* of a gem deposit is the country or area in which it is found. Some areas of the Earth seem to have been more blessed than others by geological conditions favourable to the formation of gemstone deposits. Although they are only relatively small land masses, the islands of Sri Lanka (formerly Ceylon) and the Malagasy Republic (originally Madagascar) have been especially fortunate in this respect. Sri Lanka is host to a wide range of gem varieties including ruby, sapphire, the more common varieties of garnet, chrysoberyl (alexandrite), quartz, moonstone, spinel, topaz, zircon, tourmaline, andalusite and sinhalite. The main gems missing from this impressive list are emerald and diamond. The Malagasy Republic has aquamarine and the pink variety of beryl (but not emerald), many of the varieties of quartz and garnet, plus topaz, tourmaline, orthoclase feldspar and the kunzite variety of spodumene.

Not far from Sri Lanka are the rich sapphire and ruby deposits of Myanmar (Burma) and Thailand (Siam), while further south-east on the vast continent of Australia are the diamond, sapphire and opal fields which are major sources of these gems. In South America there are the famous emerald deposits of Columbia and Brazil, the latter also being a prolific source of amethyst, aquamarine and topaz. The North American continent, among many varieties, contains important sources of tourmaline, while moving due east again across the vast stretches of Russia (the first alexandrite deposits were discovered here in Siberia) we find malachite, tourmaline, the demantoid variety of garnet, and many other gems including all the varieties of quartz and the heliodor, aquamarine and emerald varieties of beryl.

Other commercially important deposits of emerald are mined in India, Pakistan, Zimbabwe and the Republic of South Africa, which is also well known for its tiger's-eye variety of quartz and is one of the world's major sources of diamond. Among the Republic's famous diamond mines are those at Kimberley (site of the famous 'Big Hole') and Pretoria (the 3106 carat Cullinan diamond was discovered in the Premier mine 20 miles (32 kilometres) north-east of the city). There are other important diamond mines in southern Africa (including Namibia, Botswana and Angola) (see Figure 2.6). Equally important are the diamond mines in the Siberian plateau of the Confederation of Independent States (CIS, formerly the USSR), in the People's Republic of China and in Australia. Although historically India was the first source of diamonds, followed by Brazil, both of these countries are now only minor sources of the gem.

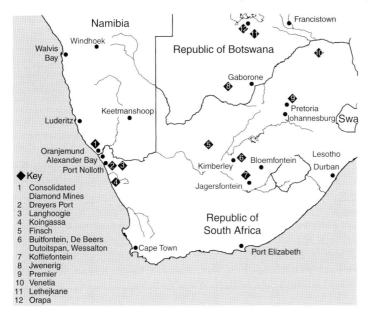

Figure 2.6 The main diamond mines of Namibia, Botswana and South Africa

In the Argyle mine in Western Australia, the transport rock magma has a different composition to the kimberlite in which diamonds are found in southern Africa and is called *lamproite*. Another relatively new source of diamonds is being mined in the Canadian Arctic territories at Lac de Gras. Numerous kimberlite pipes have been found in the area, many of them beneath lakes, and by 1998, the Ekarti diamond mine was in full production. In Angola, another new mine, the Catoca, also began production in the late 1990s (its development was funded by the Russian ARS Company).

For more information on the localities of the major deposits of organic and inorganic gems, reference should be made to the gem profiles in Appendices B and C.

Mining techniques

The recovery of gems from the Earth's crust varies from simple manually-intensive methods to highly mechanized operations. The mining of the alluvial gem deposits in Myanmar, Sri Lanka, Thailand and Brazil consists of sinking round or square section pits of up to 50 feet (15 metres) deep until a layer of gem gravel is reached (Figure 2.7, left). The gravel (known as 'byon' in Myanmar and 'illam' in Sri Lanka) is dug out and hauled or winched to the surface where the gem content, which is more dense than the sand and silt, is separated out by a rotary-motion hand-washing operation using a shallow woven basket (Figure 2.7, right). The sand and silt filters out through the fine mesh of the basket, and the lighter gravel content is spun out over the basket's rim, leaving the heaver gem content to settle at the bottom. This concentrate (called 'dullam' in Sri Lanka) is then tipped out and picked over for its gem content.

Figure 2.7 (left) Gem mining near Ratnapura, Sri Lanka – a shaft sunk to recover the layers of alluvial gem gravel: (right) Using a woven basket to wash and concentrate the recovered gem gravels

These alluvial deposits are often discovered in the area of flood plains or along the courses of ancient dried-up river beds. Sometimes the gem-bearing gravels are found nearer the surface in the beds of existing rivers, and coffer dams and suction-dredging techniques are employed to recover the stones. Where there is a bend in a river, the water flow on the inside of the bend will slow and at this point denser minerals such as gem materials and precious metals will sink and form a placer deposit (Figure 2.8). If the river meanders, forming an 'ox-bow' (the U-shaped collar of an ox-yoke), a channel may be cut from bank to bank to divert the river. The by-passed section of the river is then dammed at both ends and pumped dry to enable the recovery of gem gravels (Figure 2.9).

Where the gem deposit is nearer the surface and is spread over a large area, as in the case of some topaz deposits in Brazil, a more mechanized mining operation is employed. Wide stretches of top soil are removed by a drag-line excavator which is then used to gather in the underlying gem-bearing soil (Figure 2.10). The silt is washed out by the use of sieves and high-pressure water jets (Figure 2.11), and the topaz crystals recovered by hand picking.

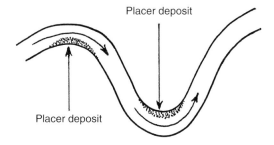

Placer deposit

Placer deposit

Figure 2.8 Dense minerals are deposited where river water slows around the inside of a bend

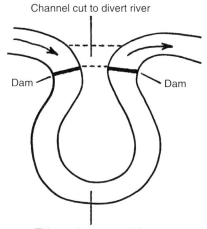

Channel cut to divert river

Dam   Dam

This section pumped dry and gem gravels recovered

Figure 2.9 Recovering gem gravels from an 'ox-bow' river meander

Figure 2.10 Drag-line removal of clay topsoil and excavation of gem deposits at the Capo do Lana topaz mine in Brazil

Figure 2.11 Using a high-pressure water jet to separate topaz crystals from a mixture of mineralized clay and kaolin at the Capo do Lana mine in Brazil

In the case of emerald and opal, the gem-bearing veins may traverse a hillside, and these are exposed by bulldozing horizontal terraces in steps along the slope of the hill. At one major South Australian site, opal mining is aptly described by the aboriginal place name of *CooberPedy* meaning 'man in a hole'. Even the miners' homes are underground in this hot and arid climate. In the emerald mines of Colombia and South Africa tunnel mining has been employed to reach less accessible pockets of gem crystals. The more mechanized the mining technique, the more important is the need for an adequate supply of water, and where a natural supply is not available this shortcoming is often rectified by the construction of a reservoir.

The methods employed in diamond mining range from the simple alluvial river bed operations just described (Figure 2.12), where the stones are separated from the gravels by means of a panning operation similar to that used when panning for gold, to highly mechanized shaft mining. Many of the diamond mine operations are concerned with pipe deposits, and for the first few years of their life these mines are worked as open-cast sites. As the top layers of the pipe are excavated using bulldozers and blasting to recover the diamond-bearing rock, so the country rock surrounding the top of the pipe has to be cut away to maintain a spiral roadway access to the working level.

Figure 2.12 Panning for diamonds on the bank of the Gbobora River in Sierra Leone. (Courtesy of De Beers)

Figure 2.13 Removing the overburden of sand to recover diamond-rich gravels along the desert coastline of Namibia in South-West Africa. (Courtesy of De Beers)

A point is reached eventually when it becomes more economic to sink a vertical shaft alongside the pipe and to mine it from underground tunnels cut into the pipe. The mined rock is often crushed underground before it is brought to the surface for further treatment. As diamond is one of the heavier gem minerals, many of the separation techniques concentrate the diamond content of the crushed rock by using high-density liquid slurries and centrifuge devices.

Diamond is not easily wettable and because of this it tends to stick to grease. This characteristic is exploited in yet another method of extraction which uses vibrating grease tables or grease belts to separate the diamond crystals from the crushed rock. A more effective method employed at the newer mines is X-ray separation, which is based on the fact that all diamonds fluoresce under X-rays, but the majority of unwanted materials in the crushed rock do not.

One other highly mechanized type of diamond mining of quite a different character is that employed along the coastal desert strip north of the mouth of the Orange river in Namibia. Here the diamonds are recovered by using drag-lines and huge earth-moving vehicles to remove up to 30 feet (9 metres) of sand to reach the diamond-rich gravels lying on ancient marine terraces (Figure 2.13). In an even more dramatic mining operation large sections of the shoreline are dammed at low tide to form paddocks from which diamonds normally lying beneath the sea are recovered. There are also off-shore recovery operations along the Namibian coast ranging from divers working from small powerboats to larger vessels using vacuum pipes to suck up diamond gravels from selected areas of the seabed.

Chapter 3

The chemical composition of gemstones

Atoms, elements, molecules and compounds

Philosophers and scientists of the ancient world believed that the world was composed of varying proportions of the four elements – earth, water, fire and air. This theory was comfortingly simple, and lasted well into the seventeenth century when scientists in Western Europe began to reveal by systematic experiment and analysis the existence of not four but 90 elements which make up the fabric of our planet.

As scientific methods developed, further experiments over the years confirmed that all matter, whether it is solid, gaseous or liquid, is composed of these elements either singly or in combination. Perhaps the most startling scientific revelation of all was the discovery that the elements are themselves composed of *atoms*, the basic chemical building blocks of the universe.

We now know that chemical elements are homogeneous substances which are composed entirely of the same type of atom, and each of these constituent atoms is the smallest division of that element which still retains the properties of the element.

The original concept of atoms as the indivisible building blocks of the universe was in turn shattered when scientists were able to show that atoms themselves are made up of a number of even smaller particles, some of which are still being discovered, and all of which have quite different properties to the element formed by the whole atom (even the concept that all matter consists ultimately of two basic nuclear building blocks – quarks and gluons – may yet be replaced by the 'superstring' theory!).

For the purposes of gemmology, however, we only need to be concerned with the three most important atomic particles, the *proton*, the *neutron* and the *electron*. (Atomic particles can, according to the quantum theory, be regarded as energy waves and vice versa. The photon, for example, is a unit quantity of electro-magnetic radiation in the form of light, and can be produced by electron interactions and by electron transitions within the atom. The photon is referred to in Chapters 8 and 12).

The structure of the atom can be visualized as a miniature solar system, with the neutrons and protons taking the place of the sun at the centre or nucleus, and the electrons forming the orbital planets (see Figure 3.1 – for simplicity the electron orbits are shown lying in the same plane). The proton is a particle possessing a single positive charge, but the neutron, which has the same mass as the proton, is electrically neutral. The electron, whose mass is much smaller than that of the proton and neutron, has a single negative charge. Atoms are normally electrically neutral, having equal numbers of protons and electrons.

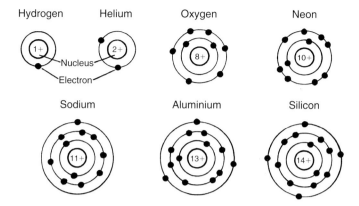

Figure 3.1 The Danish physicist Niels Bohr's representation of the atom consists of a central nucleus containing a number of positively charged protons (e.g. one in the case of hydrogen or 14 in the case of silicon) surrounded by an equal number of negatively charged orbital electrons. The charges on the protons and electrons balance out, and the atom is normally neutral

The 90 naturally occurring elements in the Earth's crust range from hydrogen, which is the lightest, to uranium which is the heaviest. In their normal state all of these elements, with the exception of hydrogen, have both protons and neutrons in their nuclei. Those elements containing neutrons in excess of their normal number are called *isotopes*. On the rare occasion when the nucleus of hydrogen contains a neutron, it is called *deuterium*, which is the isotope of hydrogen.

The number of protons in an element's nucleus is called its *atomic number* (referred to as Z; e.g. for oxygen Z = 8). The *atomic weight* of an element is its mass compared with that of an atom of oxygen (which is assigned a value of 16.0 atomic mass units, or 16.0 u for short). From this the atomic weight of hydrogen is 1.0079 u, and that of carbon is 12.011 u. When atoms combine together they form a new entity which is called a *molecule*. The molecule is the smallest part of a substance which can have a separate stable existence. Groups of molecules formed by the combination of different elements are called *compounds*. Even the atoms of an individual element often become more stable when they join together to form a molecule. For example, oxygen and hydrogen atoms are more stable when they join together in pairs, and these elements are termed *diatomic* (O_2 and H_2). The atoms of many elements such as helium (He) are quite stable in the single state, and by definition they exist as both atom and molecule in this condition. Such elements are called *monatomic*.

The distinction between a substance which is produced by mechanically mixing together two or more elements, and one in which these elements have combined together chemically to form a compound is also important to the understanding of minerals. A substance produced by simply mixing elements together will retain something of the properties of the individual elements, and the proportion of the constituents is unimportant.

When a compound of two or more elements is formed, however, this involves a more fundamental change (i.e. a chemical reaction) which is usually accompanied by the emission or absorption of heat. The properties of the compound produced in this way are always quite different to those of the individual elements, and the proportions of these elements are usually in a strict relationship governed by the resulting compound.

For example, two diatomic molecules of hydrogen can be combined with one diatomic molecule of oxygen to produce the compound water, whose properties are dramatically different from those of the individual gases. The chemical reaction which occurs between hydrogen and oxygen can be expressed in the following simple equation (here and in more complicated reactions between elements and compounds, the equation must balance in that it must show the same number of atoms of each substance on each side):

$$2H_2 + O_2 = 2H_2O$$

The number in front of the chemical symbols indicates the number of molecules, while the subscripts indicate the number of atoms in each molecule.

Of the 90 naturally occurring elements in the Earth's crust, only eight of these account for the bulk of the crust. Of these eight elements, oxygen and silicon are the dominant pair, and it is these two elements which combine together with the remaining six elements, aluminium, iron, calcium, sodium, potassium and magnesium, to form silicate minerals such as the aluminium-silicate feldspars. Any remaining silicon combines with oxygen to form silica (SiO_2). Silica is found worldwide as quartz and makes up around 12% of the Earth's crust and upper mantle.

For convenience, all elements are identified by letter symbols. Table 3.1 lists the symbols of the eight principal elements we have referred to so far. A complete list of

Table 3.1

Element	Symbol	Valency	Atomic number (Z)	Atomic weight
Aluminium	Al	3	13	26.98
Calcium	Ca	2	20	40.08
Iron	Fe	2 and 3	26	55.85
Magnesium	Mg	2	12	24.31
Oxygen	O	2	8	16.00
Potassium	K	1	19	39.10
Silicon	Si	4	14	28.09
Sodium	Na	1	11	22.99

all the 90 naturally occurring elements and their symbols, atomic numbers and atomic weights appears in Appendix H, which also includes 11 man-made elements all of which are isotopes and (with the exception of technetium and francium) are heavier than the heaviest naturally occurring element, uranium.

Valency

The electron orbits surrounding the atom's nucleus can be thought of as concentric 'shells', each shell representing an energy level. In some atoms such as aluminium there are several electron energy levels (this concept will become particularly relevant in Chapter 12 when we consider the mechanisms of luminescence). The outer electron shell of an atom is called the *valence* shell.

Elements form compounds by gaining, giving up or sharing electrons with each other in order to produce a stable arrangement of two or eight electrons in their outer shells. The number of electrons which are gained, donated or shared in this way is called the *valency* of the element.

An example of the practical effect of valency can be seen in the compound which forms aluminium oxide (i.e. the mineral corundum, Al_2O_3, whose varieties are ruby and sapphire). Aluminium has three electrons in its outer shell, and oxygen has six. For the two elements to combine as a stable compound, it is necessary for them to end up with eight electrons in their outer shell. Each oxygen atom requires an extra two electrons to become stable in a combination. This situation can only be resolved by a combination of two aluminium atoms (providing a total of six electrons) and three oxygen atoms (requiring a total of six electrons for stability). As oxygen is a diatomic molecule, the balanced chemical reaction equation looks like this:

$$4Al + 3O_2 = 2Al_2O_3$$

Aluminium, with its three outer electrons, is said to have a valency of three, (i.e. it is *trivalent*) and oxygen, with its need to find two extra electrons is said to have a valency of two (i.e. it is *divalent*).

However, with some elements, such as iron and copper, both the outer shells and the inner shells can take part in a chemical reaction to form compounds. As a result, these elements possess more than one possible valency. Iron, when it behaves as a divalent element, forms *ferrous* compounds; as a trivalent element, with a valency of three, it forms *ferric* compounds (this distinction will become relevant in Chapter 14 when the heat treatment of gems is discussed). Copper, with a valency of one, forms *cuprous* compounds; with a valency of two it forms *cupric* compounds.

To complete the valency picture it should be mentioned that there are several elements which already possess stable valency shells containing either two or eight electrons (i.e. they have a zero valency). These are virtually inert elements in that they do not easily form compounds with other elements. Examples are the so-called 'noble' gases helium (two electrons), and neon, argon, krypton, xenon and radon (eight electrons). Gold, silver, platinum and the platinum group metals are also termed 'noble' (although they do not have stable valency shells) because they are relatively resistant to chemical action and, except for silver, do not tarnish in air or water.

Bonding

The chemical joining together of elements into a compound is called *bonding*, and the forces which hold the elements together are called *bonds*. There are two types of bond which are relevant to the study of gemmology. One of these, called *ionic bonding*, is found in salts, and the other, *covalent bonding*, is found in almost all crystals that are neither salts nor metals.

A simple example of ionic bonding is that which occurs between sodium metal and chlorine gas to form sodium chloride (i.e. common salt; see Figure 3.2). The balanced chemical reaction equation is as follows:

$$2Na + Cl_2 = 2NaCl$$

In this form of bonding, one element donates one or more electrons, while the other acquires the appropriate number according to its valency. By donating its single outer electron, the sodium atom becomes unbalanced electrically with an overall positive

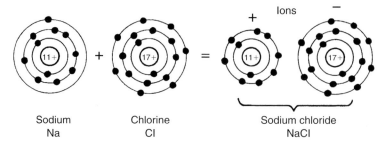

Sodium   Chlorine   Sodium chloride
  Na   Cl   NaCl

Figure 3.2 The chemical combination of sodium and chlorine to form sodium chloride is an example of ionic bonding in which the sodium atom donates an electron to the chlorine atom. This unbalances the atomic charges and results in a positively charged sodium ion and a negatively charged chlorine ion, the opposite charges producing the ionic bonding force of attraction

charge (Na^+). Atoms which are no longer electrically neutral are known as *ions*. A positive ion is called a *cation*.

In acquiring an extra electron to add to its existing seven, the chlorine atom becomes negatively charged (Cl^-). This type of ion is called an *anion*. As exactly the same number of electrons have been both donated and acquired, the overall electrical charge again becomes neutral. However, it is the electrical force of attraction that exists between the positive cation and negative anion ions in the compound that produces the bond between them. Corundum (Al_2O_3), hematite (Fe_2O_3), spinel ($MgO.Al_2O_3$) and chrysoberyl ($BeAl_2O_4$) are examples of gem minerals which have ionic bonding between their atoms, as do most metallic salt minerals.

In *covalent bonding*, electrons are shared by an interlinking of the valency shells of the combining atoms to produce an electrically stable configuration. A simple example of this type of bonding is illustrated in Figure 3.3 where carbon dioxide is formed by the combination of one carbon atom with two atoms of oxygen:

$$C + O_2 = CO_2$$

In forming the gas carbon dioxide, the two oxygen atoms each share two of the carbon electrons, while the carbon atom shares two electrons from each of the oxygen atoms. With this type of bonding there is no donating or acquiring of electrons, and as a result no ions are produced. The bonding force in this case is the need for both elements to produce a stable arrangement of eight orbital electrons in their valency shells.

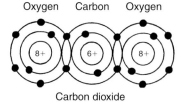

Oxygen   Carbon   Oxygen

Carbon dioxide

Figure 3.3 The chemical combination of two oxygen atoms with one carbon atom to produce carbon dioxide is an example of covalent bonding. In their outer valency shells, the two oxygen atoms share two electrons with the carbon atom, and the carbon atom shares two electrons with each oxygen atom. Unlike ionic bonding, the individual atoms remain electrically neutral. Here the bonding force is the need for all three atoms to achieve eight electrons in their outer valency shells

Diamond, which is composed of the element carbon, is a prime example of covalent bonding and, unlike carbon dioxide, has a more immediate link with the study of gemmology. Carbon has two crystalline forms, a property called *polymorphism*. When it crystallizes as diamond its atoms all share two electrons with each neighbouring atom in the crystal lattice to achieve the ideal eight-electron outer shell. In diamond, all carbon-carbon bond distances are equal and the resulting structure forms the strongest and hardest substance known. In contrast to this, carbon can also crystallize as graphite to produce one of the softest and most slippery substances known.

The reason for the dramatic differences in the two crystalline forms of carbon is that in graphite only three of the four carbon bonds are covalent bonds. The fourth bond is very much weaker and is known as the *van der Waals'* bond. This results in the formation of large, very strong sheets of carbon 'chicken wire' which are only weakly bonded together and which easily cleave or slip one against the other.

Like diamond, quartz is another gem mineral formed by covalent bonding. In general, however, minerals seldom contain bonds which are solely ionic or covalent, and there is often a transition between both types of bond.

Oxygen, for example, which combines more strongly with silicon than it does with any other element, has Si-O bonds which are about 50% ionic and 50% covalent in most silicate crystal structures. The tetrahedral SiO_4 unit bonds strongly to metallic elements such as iron and calcium to form a range of silicate materials, such as peridot and garnet.

As electronics and electrics play an increasing role in gem test instruments, it is relevant to mention yet another form of bonding. This is called *metallic bonding* and occurs in most metals. The metallic atoms exist as positive ions, but are surrounded by a cloud of valence electrons which are not bound to any particular atom's valency shell. This comparative freedom allows the electrons to wander randomly between the atoms, while the bulk of the material remains electrically neutral. When a potential difference (i.e. a voltage) is applied across the metal, these electrons move through the metal under the influence of the applied electric field to form an electric current. This type of bonding makes metals good *conductors* of both electricity and heat.

While metals are good electrical conductors, the strong attachment of valence electrons to their atoms in other materials results in a very poor conduction of electricity. Materials of this kind are classed as electrical *non-conductors* or *insulators*, and are usually poor conductors of heat. The majority of diamonds are unique among minerals in this respect as they are virtually non-conductors of electricity and yet can conduct heat better than most metals.

In between the conductors and non-conductors of electricity there are the semiconductors such as carbon, germanium, silicon and one particular type of diamond. With these materials it is necessary to apply a larger electrical field than that required in metals to persuade valence electrons to break out of their shells and travel through the material as a current.

The relationship between chemical composition and durability

In Chapter 1, mention was made of the essential qualities that make a mineral suitable for use as a gem. Among these is durability – a quality which determines whether a gem material can withstand the everyday abrasive wear and tear (and the chemical attack from pollutants and perspiration!) that are experienced by jewellery.

With the exception of diamond, which is the most durable of all the minerals and one which exists as an element (carbon), all of our gemstones are compounds of elements. For gemmological purposes, these can be divided into four main groups of oxides, carbonates, phosphates and silicates:

Gem oxides consist of a metal combined with oxygen
These are generally hard and resistant to chemical attack

Examples are:	Chrysoberyl	$BeAl_2O_4$
	Corundum	Al_2O_3
	Quartz and chalcedony	SiO_2
	Spinel	$MgO.Al_2O_3$

Gem carbonates are formed by the action of carbonic acid on metals
These are soft and easily attacked by acids

Examples are:	Calcite	$CaCO_3$
	Malachite	$Cu_2(OH)_2CO_3$
	Rhodochrosite	$MnCO_3$

Gem phosphates are produced by the action of phosphoric acid on metals
These are soft and not very resistant to acid attack

| Examples are: | Apatite | $Ca_5(F,Cl)(PO_4)_3$ |
| | Turquoise | A complex hydrated phosphate of copper and aluminium |

Gem silicates result from the action of silicic acids on metals
These are hard and very durable. They represent the majority of gems

Examples are:	Beryl	$Be_3Al_2(SiO_3)_6$
	Feldspar group gems	Aluminium silicates in combination with calcium, potassium or sodium
	Garnet group gems	Silicates of various combinations of magnesium, manganese, iron, calcium, aluminium and chromium
	Jadeite	$NaAl(SiO_3)_2$
	Nephrite	Complex calcium, magnesium or iron silicate
	Peridot	$(Mg,Fe)_2SiO_4$
	Rhodonite	$MnSiO_3$
	Topaz	$Al_2(F,OH)_2SiO_4$
	Tourmaline	A complex borosilicate of aluminium and iron
	Zircon	$ZrSiO_4$

Chapter 4

Crystallography

Students of gemmology often find their first encounter with the subject of crystallography to be rather daunting, even though only a superficial knowledge of the subject is required for vocational purposes. Sadly, many students simply learn by rote the dry bones of the subject sufficient to pass an examination. However, an understanding of the basics of crystallography is as important a tool to the study of gemstones as mathematics is to engineering. In this chapter, emphasis will be placed on the practical aspects of crystallography as a means of fleshing out the dry bones of the subject.

Crystalline and non-crystalline materials

All solid matter is composed of material which is either *crystalline* or *non-crystalline**, or a mixture of these two states. Like many specialized terms, the word crystalline is derived from the Greek *krystallos* which means ice. The term gradually came to be used for any substance that had the clarity and transparency of ice, and is now also applied to crystals that are neither colourless nor transparent.

In a non-crystalline substance, the atoms and molecules are positioned randomly throughout the material, and are not aligned in any special order or pattern. Because of this, a non-crystalline material can never develop any naturally occurring characteristic shape. In the molten rock-forming magma, tetrahedral SiO_4 units (see under 'Bonding' in Chapter 3) join together and also link up with metal ions to form islands or long chains. This produces some orderly structure within the molten liquid and results in a viscous fluid which does not flow easily. When sudden cooling occurs, there is not enough time for the islands and chains to become reorganized into a large-scale crystalline structure. The solidified material then has an internal structure having only tiny scattered regions of crystalline order.

Glass is a material of this type containing frameworks of disordered silicon tetrahedra. Because of the lack of overall crystal structure, glass cannot exhibit any of the crystalline characteristics such as external shape, or cleavage and pleochroism (subjects which are covered in Chapters 5 and 10). Other non-crystalline materials include amber and jet, both of which have an organic origin.

The majority of minerals are wholly crystalline substances whose atoms and molecules are arranged in an ordered and symmetrical three-dimensional pattern or lattice. In most instances, this underlying symmetrical crystal structure makes itself visible in the external shape of the mineral specimen. An important exception to this is opal

---

* Also called *amorphous* – an obsolete term no longer used in modern crystallography.

which, although a mineral, is non-crystalline. In contrast, there are innumerable non-mineral substances such as sugar, naphthalene and the man-made diamond simulants YAG and GGG that can occur as crystals.

Among gem minerals there are also a few which, although they are crystalline, do not have an identifiable external profile. Such minerals are called *massive*, a term which does not necessarily refer to their size or weight, but more to their lack of any character-istic shape. A common example of a massive crystalline gem mineral is the pink variety of quartz (i.e. rose quartz). Although mainly massive in external form, rose quartz does, however, very occasionally occur in the characteristic shape of a quartz crystal (other minerals which occur in the massive form are *polycrystalline* materials such as jadeite and nephrite, and *microcrystalline* materials such as agate and chrysoprase – see later in this chapter).

Perhaps the most important feature of a crystalline material, and one that is missing from non-crystalline substances, is that its physical characteristics and properties vary with the orientation of the crystal. With a non-crystalline material, its properties are the same no matter what the direction of measurement or viewing, but in a crystalline substance they are related to the high degree of orderliness of its constituent atoms and molecules.

Diamond's cleavage and hardness are two good examples of directional-dependant properties. Diamond can be *cleaved* or split only in the directions parallel to its octa-hedral crystal faces (Figure 4.1). Diamond's hardness is also directional-dependant and because of this the stone is much easier to saw and polish in one direction than in others. This is an important factor that has to be taken into account by both the diamond sawyer and polisher who must avoid the cubic and octahedral planes of maximum hardness when fashioning this most durable of all the gem materials.

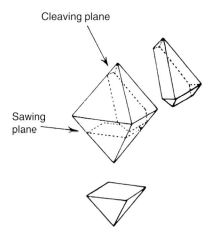

Cleaving plane

Sawing plane

Figure 4.1 Two examples of directional-dependant properties in diamond. The cleavage planes are parallel to the octahedral faces, which are also planes of near maximum hardness. The easiest direction for sawing is at right-angles to the dodecahedral plane, which is also the plane of minimum hardness

As well as cleavage and hardness, optical properties such as colour can also vary with the direction of viewing in many crystalline materials. This can become a con-trolling factor in non-diamond coloured gems such as ruby, sapphire and tourmaline when the lapidary is deciding how the rough stone should be cut to bring out the best colour in the finished product.

So far we have seen how a basic knowledge of crystallography is used by the dia-mond cleaver, who may be able to provide an alternative to a lengthy sawing process by splitting the stone in two along a cleavage plane. The diamond polisher who is deciding where to place his facets on the stone also uses this knowledge to avoid the directions of maximum hardness.

In contrast to the use of cleavage by the diamond cutter, the lapidary must do his or her best to avoid polishing a gem facet parallel to a cleavage plane as this could result either in an uneven facet surface, or in the gem cleaving in two during the polishing operation. The lapidary must also use his or her knowledge of the gem's directional optical properties to bring out its colour to best effect.

Before going on to discuss the basic principles of crystallography, it is relevant to form some idea of the way in which atoms are arranged in a crystalline mineral, and the underlying forces which govern this arrangement.

The atomic structure of a crystal

In 1669, Nicolas Steno, a Danish anatomist, made the fundamental crystallographic discovery that wherever a quartz rock crystal was found, its corresponding inter-facial angles were always the same regardless of the shape or size of the crystal. Subsequently, it was discovered that the angles between corresponding faces of other mineral crystals were also unvarying and specific to each individual mineral. As a result, Steno's observations proved to have a universal application.

Over 100 years later sufficient proof of this had been gathered to enable the Abbé Haüy, who is regarded as the 'father of crystallography', to state in mathematical terms that each substance has its own unique characteristic form. From this he deduced that the rigid external conformity shown by crystals probably indicated that their internal structure was also orderly and regular. He further suggested that crystals grow and produce these characteristic external shapes by the 'assembly' of tiny units or building blocks. His concept of minute elementary blocks or cells is illustrated in Figure 4.2 where a particular assembly of small cubes displays the habit of the rhombic dodecahedron (compare this with the rhombic dodecahedron in Figure 4.6).

This building block theory can be substantiated in a very practical way by taking a cleavage rhombohedron of the Iceland spar variety of calcite (Figure 4.3). This very

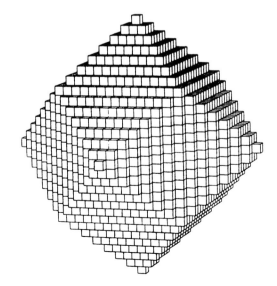

Figure 4.2 A stack of cubes, drawn in perspective by means of computer graphics, shows how unit cells in the cubic crystal system can be assembled to display the habit of the rhombic dodecahedron. (After Abbé Haüy)

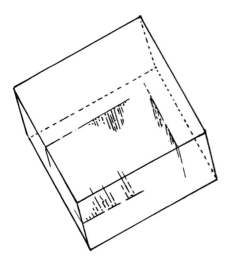

Figure 4.3 A rhombohedron of the Iceland spa variety of calcite showing incipient cleavage in three directions

easily cleaves into even smaller rhombohedra, each of which is an identical copy of its parent. Even if one of these is crushed to powder, the individual particles will all be found to be minute rhombohedra when viewed under the microscope. If it were possible to continue this process of dismantling the calcite crystal down to the molecular level, we would finally arrive at calcite's basic building block.

This basic building block, or *unit cell* as it is called, is the smallest atomic or molecular lattice structure which still retains all the characteristics of the mineral. Its shape is determined by the chemical composition of the mineral and the forces of valency and bonding that were described in the previous chapter.

Finally, if the two-dimensional plan profiles of all the unit cells of all the mineral crystals could be visualized, these would consist entirely of three-, four- and six-sided shapes. The reason for this is that these are the only possible profiles which can link together without leaving gaps in the crystal lattice. Such linking is not possible with a five-sided profile and is the reason why there is no mention of a five-fold axis of symmetry later in this chapter in the section dealing with the elements of symmetry.

The term 'growth' has already been used several times when discussing the formation of minerals, and it is relevant here to describe briefly the process of crystal nucleation and growth so as to make a distinction between this concept and that of biological growth!

If a chemical compound, such as common salt, is dissolved in water, and the water is allowed to evaporate, a point will be reached eventually when there is only just enough water to hold the salt in solution. When this occurs, the salt solution is said to be *saturated*, and any further evaporation will result in salt being expelled from the solution as growing crystals. This process will happen more quickly if the salt is first dissolved to saturation level in hot water (most solids are more soluble in hot water than in cold) and the water allowed to cool rather than evaporate. In this case, a temperature will be reached at which there is too much salt for the water to hold in solution. Once again crystals will start to grow, their size being dependant upon the rate of cooling.

The parallel to these processes can be seen in the natural growth of mineral crystals from molten or aqueous solutions in the Earth. In rocks which cooled slowly from the molten condition, large well-shaped crystals were formed in trapped molten solutions or from aqueous solutions under high pressures. Where rapid cooling took place, only very small crystals had time to grow and these often formed as microcrystalline masses. These and the much larger well-formed crystals are found in cavities known geologically as *geodes, druses* or *vugs*.

Classification of crystals by symmetry

In crystallography, the concept of *symmetry* as applied to the crystal structure is of prime importance. As we will see shortly, crystals are classified into seven *systems* which range from the most symmetrical, the *cubic* or *isometric* system, to the least symmetrical, the *triclinic* system.

For most practical gemmological work it is only necessary to be able to recognize the habits of the various gem crystals and to be aware of their individual optical and physical characteristics. However, the student of gemmology should also be familiar with the convention of imaginary crystallographic *axes* which are used to describe the idealized shape of a crystal. Of equal importance are the three *elements of symmetry*, as they include *axes of symmetry* which alone define the seven crystal systems.

The *crystallographic axes* act as a frame of reference for identifying the positions of the crystal faces. They achieve this by means of their mutual angles and relative lengths. They are best thought of as imaginary lines which run through the centres or junctions of crystal faces to meet at a point within the ideal crystal called the 'origin'. The three *elements of symmetry* can be defined as follows:

Axis of symmetry

An imaginary line positioned so that when the crystal is rotated around it, the characteristic profile of the crystal appears two, three, four or six times during each complete rotation as indicated in Figure 4.4. (There are usually several possible axes of symmetry in a crystal; these are described as two-fold, three-fold, four-fold or six-fold axes. Alternatively, they may sometimes be labelled as digonal, trigonal, tetragonal or hexagonal axes of symmetry):

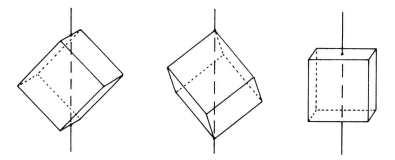

Figure 4.4 From left to right, examples of two-fold, three-fold and four-fold axes of symmetry in a cubic crystal. There are four three-fold axes of symmetry (see centre sketch), one through each pair of opposite corners of the cube, and these are the characteristic axes which determine the cubic system

Plane of symmetry

This is an imaginary plane though a crystal which divides it into two mirror-image halves (see Figure 4.5). A cube has nine such planes.

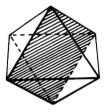 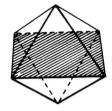

Figure 4.5 Showing two of the planes of symmetry in an octahedron

Centre of symmetry

A crystal is said to possess a centre of symmetry when identical faces and edges occur on exactly opposite sides of a central point.

The seven crystal systems and their elements of symmetry

As mentioned earlier, crystals can be grouped into seven systems. These can be further subdivided into 32 *classes*. The subdivision is based on the different degrees of symmetry (as specified by the elements of symmetry) of the crystals within each system. The seven crystal systems themselves are classified in terms of the number of their crystal axes, their relative lengths and the angles between them.

The elements of symmetry included in the following descriptions of the seven crystal systems refer to the *highest* level of symmetry in each case (it should be remembered that crystals do form with *lower* levels of symmetry and may sometimes have no centre of symmetry at all). The systems are divided into three groups as indicated, and the characteristic *axes of symmetry*, which determine the system to which a crystal belongs, are printed in italics.

The cubic system (Group 1, Figure 4.6)

Crystals in this system have the highest order of symmetry and are also called isometric. The cubic system has three crystal axes (a_1, a_2, a_3), all of which are of equal length ($a_1 = a_2 = a_3$) and intersect each other at right angles (90°).

Axes of symmetry: 13 (six two-fold, *four three-fold*, three four-fold)
Planes of symmetry: 9
Centre of symmetry: 1
Common forms: cube, eight-sided octahedron, twelve-sided dodecahedron
Examples: diamond, garnet, spinel, fluorite

The tetragonal system (Group 2, Figure 4.7)

This has three crystal axes. The two lateral ones are of equal length ($a_1 = a_2$) and at right angles (90°) to each other. The third (principal or *c*) axis is at right angles (90°) to the plane of the other two and is shorter or longer than them ($c \neq a_1$)

Axes of symmetry: 5 (four two-fold, *one four-fold*)
Planes of symmetry: 5
Centre of symmetry: 1
Common forms: four-sided prism with square cross-section
Examples: zircon, scapolite

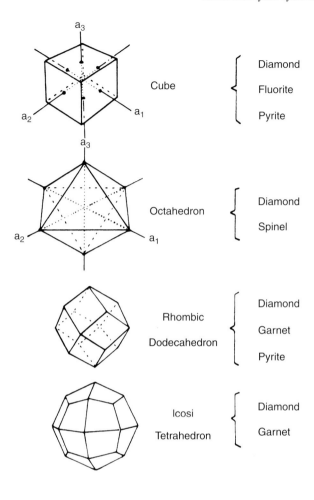

Cube

{ Diamond

Fluorite

Pyrite

Octahedron

{ Diamond

Spinel

Rhombic
Dodecahedron

{ Diamond

Garnet

Pyrite

Icosi
Tetrahedron

{ Diamond

Garnet

Figure 4.6 The cubic crystal system, also showing some of the habits of gemstones in the cubic system

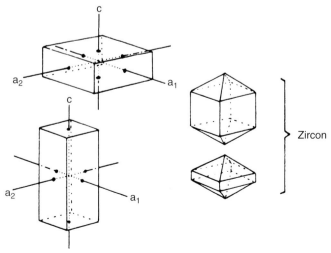

Zircon

Figure 4.7 The tetragonal crystal system

The hexagonal system (Group 2, Figure 4.8)

Contains four crystal axes, the three lateral ones are of equal length ($a_1 = a_2 = a_3$) and intersect each other at $60°$ in the same plane. The fourth (or principal) c axis is at right angles to the other three and is usually longer.

Axis of symmetry: 7 (six two-fold, *one six-fold*)
Planes of symmetry: 7
Centre of symmetry: 1
Common forms: six-sided prism
Examples: beryl, apatite

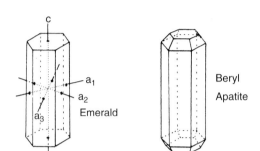

Figure 4.8 The hexagonal crystal system, showing a second-order prism (in a first-order prism the lateral axes intercept the junction of the prism faces)

The trigonal (or rhombohedral) system (Group 2, Figure 4.9)

This system (sometimes treated as a subdivision of the hexagonal system) has four crystal axes which are arranged in the same manner as in the hexagonal system. The symmetry of the trigonal system is, however, lower than that of the hexagonal system.

Axes of symmetry: 4 (three two-fold, *one three-fold*)
Planes of symmetry: 3
Centre of symmetry: 1
Common forms: three-sided prism, rhombohedron
Examples: calcite, corundum, quartz, tourmaline

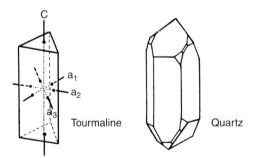

Figure 4.9 The trigonal crystal system

The orthorhombic system (Group 3, Figure 4.10)

This system has three crystal axes, all at right angles ($90°$) to each other and all having different lengths ($a \neq b \neq c$). The principal or c axis is the longest, and of the remaining two lateral axes, the longer b is known as the *macro* axis and the shorter a is called the *brachy* axis.

Axes of symmetry: (*three two-fold*)
Planes of symmetry: 3
Centre of symmetry: 1
Common forms: rectangular prism (prism with cross-section of playing card 'diamond'), bipyramid comprising two four-sided pyramids joined at the base
Examples: topaz, peridot, chrysoberyl, andalusite, sinhalite, zoisite

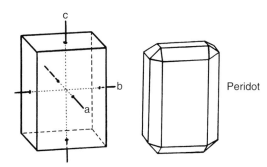

Peridot

Figure 4.10 The orthorhombic crystal system

The monoclinic system (Group 3, Figure 4.11)

There are three crystal axes in this system, all of unequal lengths (a ≠ b ≠ c). The *b* axis, known as the *ortho* axis, is at right angles to the plane of the other two which cut each other obliquely. The longest of these is the *c* axis, and the one inclined to it (at an angle other than 90°) is called the *a* or *clino* axis.

Axes of symmetry: (*one two-fold*)
Planes of symmetry: 1
Centre of symmetry: 1
Common forms: prisms and pinacoids
Examples: orthoclase feldspar (moonstone), diopside

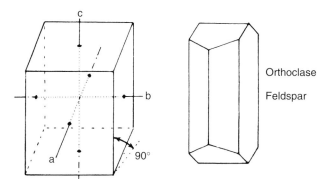

Orthoclase

Feldspar

Figure 4.11 The monoclinic crystal system

The triclinic system (Group 3, Figure 4.12)

This system has three crystal axes, all of unequal lengths (a ≠ b ≠ c) and all inclined to each other at angles other than 90°. The longer lateral axis is called the *macro*, and the shorter is called the *brachy* as in the orthorhombic system.

Axes of symmetry: *none*
Planes of symmetry: none
Centre of symmetry: 1
Common forms: prism (tilted sideways and backwards) with pinacoids
Examples: plagioclase feldspar, microcline feldspar (amazonite), rhodonite, turquoise (usually in microcrystalline aggregates)

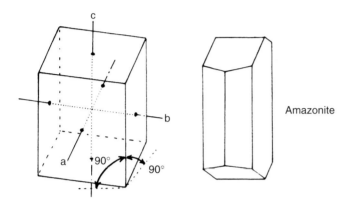

Figure 4.12 The triclinic crystal system

Crystal forms

A crystal form is composed of a group of crystal faces which are similarly related to the crystal axes. A form made up entirely of identical interchangeable faces is called a *closed* form (e.g. a cube or an octahedron). A form which is only completed by the addition of other forms is called an *open* form. An open form cannot exist on its own and must be completed by the addition of other open forms which act as suitable terminations to complete the crystal shape (see Figure 4.13).

A tetragonal *prism*, for example, is an open form whose top and bottom can be terminated by a pinacoid or by four-sided pyramids, the latter producing the closed form of a zircon crystal.

The *pinacoid* open form (which consists of a pair of crystal faces which are parallel to two crystal axes and are cut by the third) can occur in several positions. When it

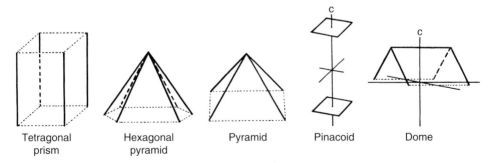

Figure 4.13 Examples of open forms

terminates a prism, as in the case of the emerald crystal in Figure 4.8, its faces are parallel to the crystal's lateral axes and it is called a *basal* pinacoid. Another open form is the *dome*, which is often found as the terminating face on a topaz prism. The dome is defined (in a somewhat complicated way) as a form whose two faces intersect the vertical *c* axis of a crystal and one lateral axis, but are parallel to the other lateral axis (see Figure 4.13).

Crystal habits and their use in identification

The shape in which a mineral usually crystallizes is referred to as its *habit*. From the sketches illustrating the seven crystal systems it can be seen that minerals belonging to the same system can often have very different habits, despite the similarity of their internal crystal structures.

These differences in habit can be the result of combinations of more than one form. Other differences may be due to the variety of terminations which some crystals adopt. Some of the more bizarre shapes are caused by the parallel growth or interpenetration of two or more crystals.

The many differences in crystal habit can be confusing at first encounter, as it is often difficult to associate a habit with the idealized shape of one of the seven systems. This very factor can sometimes be an advantage, however, as it can be used to identify a mineral simply by the individuality of its habit.

The specialized terms used to describe crystal habits are shown in Table 4.1.

Table 4.1

Habit	Description	Example
Acicular	Slender needle-like crystals	Rutile/tourmaline in quartz
Bipyramidal	Two pyramids joined at base	Sapphire
Botryoidal or globular	Resembling a bunch of grapes, hemispherical masses	Chalcedony, hematite
Columnar or fibrous	Series of very slender prisms	Kyanite, calcite
Dendritic	A branching or tree-like feature or inclusion	Moss agate
Dodecahedral	A 12-sided dodecahedron	Garnet
Euhedral	Crystal with well-formed shape	Diamond, spinel, quartz
Mamillary	Rounded intersecting contours	Malachite
Massive	Without external crystal shape	Rose quartz
Octahedral	An eight-sided octahedron	Diamond, spinel
Prismatic	A crystal whose main faces are parallel to the *c* axis	Tourmaline, quartz, zircon
Scalenohedron	A 6-sided bipyramid with unequal sides	Sapphire
Striations	Growth lines on surface of crystal	Tourmaline, quartz, pyrites
Tabular	Short stubby crystal, usually prismatic	Ruby

Sometimes the shape of a crystal is due to influences which modify its normal habit. When quartz crystallizes inside wood, for example, it can replace the wood fibres to produce 'petrified wood'. Quartz can also replace the fibres in crocidolite (asbestos) and produce pseudo-crocidolite which is more popularly known as 'tiger's-eye'. Other modifications due to heat, pressure and chemical processes can also modify the way in which crystals form. The following terms are used to describe these modified shapes:

Enantiomorph: a crystal which has mirror image habits and optical characteristics, occurring in both right- and left-handed forms (e.g. quartz – see Figure 4.14).

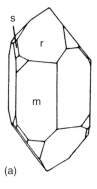
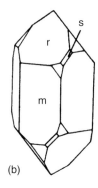

Figure 4.14 Showing the property of enantiomorphism in quartz crystals. In the left-handed crystal (a) the trigonal pyramid 's' is situated in the upper left-hand corner of the prism face 'm' beneath the positive rhombohedral 'r'. In the right-handed crystal (b) the opposite occurs

Hemihedral: a crystal which has only half of the full number of faces of its symmetry class.

Hemimorphic: a crystal which has differing forms at the opposite ends of its axis of symmetry.

Isomorphic: minerals exhibit isomorphism when they have identical external forms but differ chemically (e.g. the garnet group).

Isomorphous replacement: The replacement of one element in a mineral by another. While the same form and crystal structure is retained, this may cause wide variations in the mineral's physical properties (e.g. the garnet group).

Lamellar: A crystalline structure composed of straight or curved layers often due to intermittent growth or twinning.

Polymorph: Minerals which differ in shape but have the same internal composition (e.g. diamond and graphite; andalusite, kyanite and sillimanite).

Pseudomorph: a mineral which has adopted an external form other than its normal habit by copying, for example, the shape of a pre-existing crystal or organic structure.

Examples of the more common gem minerals and their crystal habits are given in Table 4.2.

Twinned crystals

A twinned crystal is one which consists of two or more individual crystals which have grown together in a crystallographic relationship to produce a symmetrical shape. Twinning, which is very common with quartz crystals, usually occurs in one of two forms: contact twins and interpenetrant twins.

Contact twins (Figure 4.15) occur when the twin-halves of a crystal have grown with one half rotated through 180° to the other half. Spinel and diamond often occur in this form. With diamond a contact twin is called a *macle*. In some minerals repeated twinning of this type (known as 'polysynthetic' twinning) produces a lamellar structure consisting of thin plates of alternate orientation. This often results in a symmetrical habit uncharacteristic of the gem's crystal system (e.g. the pseudo-hexagonal twinning of chrysoberyl called *trilling*). It also produces planes of false cleavage known as 'parting' planes. Contact twins can usually be identified by the re-entrant angles between them (see chrysoberyl trilling in Figure 4.15).

Table 4.2

Gemstone	Crystal system	Habit
Apatite	Hexagonal	Six-sided prism, often terminated with pyramid (also tabular prism and massive)
Beryl (emerald, aquamarine)	Hexagonal	Six-sided prism, often striated vertically; terminations are rare
Calcite	Trigonal	Rhombic prisms, scalenohedra and six-sided prisms
Chalcedony (agate)	Trigonal (microcrystalline)	Massive, botryoidal, mamillary, nodules, geodes
Chrysoberyl	Orthorhombic	Prismatic crystals; triple contact twins forming 'hexagon'
Corundum (ruby, sapphire)	Trigonal	Sapphire: tapering barrel-shaped bipyramid; ruby: tabular hexagonal prism
Diamond	Cubic	Octahedron, dodecahedron, icositetrahedron (cubes rare), contact twin (macle)
Feldspar (orthoclase)	Monoclinic	Crystals of both types resemble each other in habit;
Feldspar (microcline and plagioclase)	Triclinic	both are prismatic and often blocky with wedge-shaped faces
Fluorspar	Cubic	Cube, interpenetrant cubes and octahedral crystals (naturally occurring octahedral are rare, but the cube cleaves readily into this form)
Garnet	Cubic	Dodecahedron, icositetrahedron (and combinations of both)
Jadeite	Monoclinic (polycrystalline)	Massive
Nephrite	Monoclinic (polycrystalline)	Massive
Peridot	Orthorhombic	Prismatic
Pyrite	Cubic	Cube, dodecahedron (also massive and granular forms)
Quartz	Trigonal	Six-sided horizontally striated prism with rhombohedral terminations
Rhodochrosite	Trigonal	Massive
Rhodonite	Triclinic	Tabular and massive
Rutile	Tetragonal	Four-sided prism with pyramidal terminations (also acicular and massive granular)
Scapolite	Tetragonal	Four-sided prism (also massive)
Spinel	Cubic	Octahedron and spinel twin (contact twin)
Topaz	Orthorhombic	Flattened four-sided prism with pyramidal or dome termination (prism faces often vertically striated)
Tourmaline	Trigonal	Triangular prism (generally with rounded faces, heavily striated along length)
Turquoise	Triclinic (microcrystalline)	Massive
Zircon	Tetragonal	Four-sided prism with bipyramidal terminations

Interpenetrant twins (Figure 4.15) consist of two or more crystals which have grown in proximity and have penetrated each other with a direct relationship between their axes. Common examples can be seen in the interpenetrant cubes of fluorite, and in staurolite twins.

While not producing a twinned crystal, mention should be made here of *parallel growth* as this is often mistaken for twinning. In parallel growth all the edges and faces of one crystal are parallel to those of its neighbour. In twinned growth the adjoining crystals are not parallel to each other, but their faces have a symmetrical orientation.

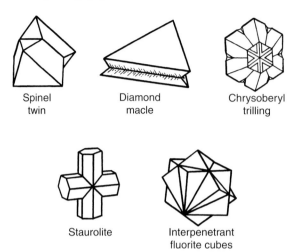

Spinel
twin

Diamond
macle

Chrysoberyl
trilling

Staurolite

Interpenetrant
fluorite cubes

Figure 4.15 Top: three examples of
contact twins. Bottom: two examples of
interpenetrant twins

Polycrystalline and microcrystalline minerals

Some minerals such as jadeite and nephrite are described as *polycrystalline*. They consist of aggregates of randomly orientated small crystals or crystalline fibres which can be seen under magnification (or sometimes by eye alone). In addition there are many minerals which are composed of an aggregate of very much smaller crystals or crystalline fibres. The size of the crystals in these minerals is so minute that they *cannot* be detected by eye or even with the aid of a standard optical microscope. Such materials are described as *microcrystalline* or *cryptocrystalline* (from the Greek word 'kruptos' meaning hidden). Gemstone examples include the chalcedony varieties and turquoise.

Because of their internal structure, polycrystalline and microcrystalline minerals are massive in their habit. Most polycrystalline or microcrystalline gems are also semi-transparent or opaque, and the random orientation of their crystals or crystalline fibres results in optical properties which are significantly different from normal crystalline minerals (these will be discussed in Chapters 9 and 10). *A word of warning here*: although all polycrystalline and microcrystalline materials are massive because of their internal structure, there are several massive gem minerals that are neither polycrystalline nor microcrystalline (e.g. rose quartz, rhodonite and rhodochrosite).

Metamict minerals

There are some minerals which have experienced natural alpha-particle irradiation (either from their surroundings in the Earth or from radioactive impurities or constituents). As a consequence, the crystalline structures of these minerals have become damaged to a point where they are virtually non-crystalline. Such minerals are described as metamict. In 'low' zircon, this metamict state has been caused by the presence within the gemstone of small amounts of radioactive uranium and thorium. The rare gem mineral ekanite, which was discovered in the Sri Lankan gem gravels during the 1960s, is also a metamict material and contains enough thorium to activate a Geiger radiation counter!

Chapter 5

Cleavage, parting and fracture

The mechanism of cleavage

The cleavage property of a gemstone is of particular interest to the lapidary and the diamond cutter. Like other properties possessed by minerals, cleavage is a directional feature and can only exist in crystalline substances.

Cleavage occurs in a gemstone as a well-defined plane of weak atomic bonding which allows the stone to be split in two leaving reasonably flat surfaces. These cleavage planes are always parallel to a crystal face in a perfectly formed single crystal. In some gemstones which have cleavage properties, however, the crystal faces may not always be present as a convenient guide to the direction of the cleavage plane. Furthermore, if the stone happens to be microcrystalline or polycrystalline this will inhibit any cleavage property which might otherwise have been present in a single-crystal specimen of the mineral.

In those gemstones that possess the property of cleavage, the cleavage planes are the result of the atoms lying parallel to these planes being more closely linked together than the atoms between the planes. The bonding force along the planes of atoms is therefore much stronger than between the planes.

As examples of cleavage we can look at the atomic structure of graphite and diamond, the two crystalline polymorphs of carbon (polymorphs are minerals which differ in shape but have the same composition). The valency of carbon is four, and to form a crystalline lattice it needs to link with four other atoms of carbon. With diamond, the five carbon atoms are linked together by four covalent bonds (as described in Chapter 3) in what is known as tetrahedral bonding. Each atom is at the geometric centre of four others which form the corners of a tetrahedron (Figure 5.1). When these tetrahedra link together into a three-dimensional diamond lattice, the result is a series of layers of 'puckered' hexagonal rings which represent the slightly rippled octahedral cleavage planes of diamond (Figure 5.2). Although the length of each bond is the same (0.154 nm), there are more bonds along the cleavage planes than between them.

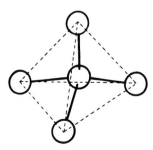

Figure 5.1 The tetrahedral bonds of the carbon atoms in diamond. The dotted outline indicates the four triangular faces of a tetrahedron, although it is probable that a true tetrahedral diamond does not exist as an individual crystal

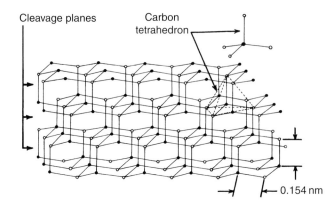

Cleavage planes   Carbon tetrahedron

—← 0.154 nm

Figure 5.2 Sketch showing the lattice structure formed by the covalent atomic bonds in diamond, and three parallel cleavage planes. For clarity, the central atom of the carbon tetrahedron is shown black, and the remaining four atoms are shown white

In graphite, these hexagonal layers are flat with shorter (and therefore stronger) covalent bonds between the atoms in each layer (0.142 nm). However, the fourth set of bonds (known as 'Van der Waals' bonds) between the layers of atoms are longer (0.335 nm) and therefore much weaker (Figure 5.3). As a result, the layers cleave apart from each other so easily that this characteristic makes graphite useful as a lubricant.

Cleavage can sometimes be of help in identifying a gemstone. An indication that a stone possesses cleavage is often revealed by the interference colours developed within the stone or on its surface. This is the result of flaws or cracks caused by incipient cleavages. Such rainbow-like areas of colour can often be seen in fluorite, topaz and calcite crystals.

When such indications are seen, it is necessary for identification purposes to be aware of the directions of cleavage for those gems which possess this property. These directions of cleavage are related to the crystal faces, and are described as *octahedral* when they are parallel to the octahedral faces or planes of a cubic crystal, *prismatic* when they are parallel to the faces of the prism, *basal* when they are parallel to a pinacoid, or *rhombohedral* when they are parallel to the rhombohedral faces. An example of a gemstone with octahedral cleavage is diamond (Figure 5.4). Spodumene has prismatic cleavage, topaz has basal cleavage, and calcite has rhombohedral cleavage in three planes (see Figure 3.4 in Chapter 3).

It is also important to know the quality of a gemstone's cleavage surface. This is generally described as being *perfect* (as in the case of diamond, fluorite and topaz), *good* (as with feldspar), *distinct* (andalusite, sphene), and *poor* (corundum, quartz). Equally important is the ease with which a gem material cleaves, and this can be described as *easy*, *moderate* or *difficult*. Diamond for example has perfect cleavage, but only cleaves with difficulty. Fluorite and calcite have perfect cleavage and are also easy to cleave. Perfect cleavage can be recognized by the presence of a series of very shallow steps where splitting has occurred along different but parallel layers of atoms. Descriptions of gemstone cleavages in terms of quality and direction are included in the gem profiles in Appendix C at the rear of this book.

Lapidary problems caused by cleavage planes

When polishing a gemstone which has perfect or good cleavage, the lapidary must ensure that none of the facets lie exactly parallel to a cleavage plane. If an attempt is made to polish the stone at an angle of less than 5° to a cleavage plane, this will result

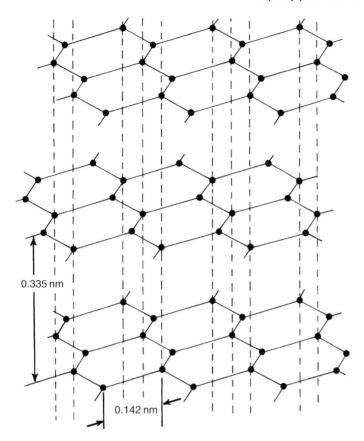

0.335 nm

0.142 nm

Figure 5.3 Sketch showing the 'chicken wire' lattice structure in graphite formed by the hexagonal bonds between the carbon atoms. Only three of the four carbon bonds are covalent; the longer broken lines between the 0.335 nm cleavage layers indicate the weaker Van der Waals' bonds

One of four directions of
octahedral cleavage

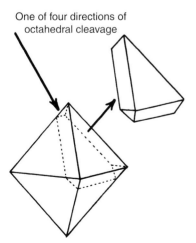

Figure 5.4 The cleavage planes in diamond are parallel to the octahedral faces

in an uneven finish due to the polishing lap lifting up cleavage sections from the stone's surface (see Figure 19.17).

When handling a stone which has easy cleavage, the presence of incipient cleavage cracks must also be checked as these could cause the stone to split in two during the polishing operations.

The use of cleavage in the fashioning of diamond

In a diamond-cutting factory, the person who decides how a particular diamond should be worked in order to obtain the best 'yield' from the stone may use diamond's perfect cleavage property to remove a flawed section of the crystal. He or she may also decide that the cleavage planes are such that they are suitable for splitting a large irregularly shaped stone into suitably smaller sections.

Both of these decisions will be made because of the much longer, and therefore more expensive, alternative of sawing the diamond. As will be described more fully in Chapter 19, the cleaving of diamond is a specialized craft. Before a diamond can be cleaved, the direction of the cleavage plane must be identified and a groove (called a *kerf*) cut along this plane in the surface of the stone. The cleaver's blade is then inserted in this groove, the blade is given a sharp tap, and provided that the kerf has been correctly positioned, the diamond will split cleanly in two. However, on the rare occasional that an error of judgement is made, the stone may break into several pieces – an occurrence which, by way of geological forces (or simply rough handling), can also happen to other minerals. Such a breakage or fracture has its own characteristics which are described later in this chapter.

Parting (false or pseudo-cleavage)

Some minerals which do not possess cleavage properties can be divided in two along a plane of weakness rather than along a plane related to the stone's crystal faces. Such a direction of weakness is called a parting plane. Parting is usually caused by a form of twinning. In the case of corundum and labradorite, parting occurs along directions of lamellar twinning. A significant difference between parting and cleavage is that the former only occurs at discrete intervals along the twinned crystal, while cleavage planes are separated only by atomic layer intervals. Neither cleavage nor parting plane surfaces are as flat as the surface of a well-polished facet.

Fracture and its use as an aid to identification

The way in which a mineral breaks (other than by cleavage or parting) is known as fracture. Unlike cleavage and parting, fracture occurs in a random direction, often as the result of a sharp impact. It is typical of non-crystalline materials such as glass, but can also occur in those crystalline materials which do not have significant planes of cleavage or weakness (e.g. quartz and beryl). Fracture can also happen occasionally with other crystals, even (as indicated earlier) with diamond.

Sometimes the surface contour of a fracture can be distinctive enough to constitute a useful identifying feature. In the initial examination of a gemstone it is often rewarding to inspect it under a hand lens for signs of characteristic fracture damage. Typical types of fracture, together with the gem materials in which they occur, are as follows:

Conchoidal: This is a shell-like fracture consisting of a series of scalloped rings similar to those seen in a seashell. Conchoidal fractures occur in quartz and the garnets. It is particularly characteristic of glass (Figure 5.5).

Figure 5.5 Conchoidal fracture in goldstone, a glass simulant of the feldspar gem sunstone

Hackly or splintery: This takes the form of long fibrous splinters and can be seen in nephrite, jadeite and ivory.

Smooth or even: This type of fracture, although not perfectly flat, shows no signs of significant irregularities. Examples of smooth fractures can often be seen in rough diamonds.

Chapter 6

Hardness

The durability of a gemstone

In Chapter 1, mention was made of the essential qualities of a gem material. These qualities included *beauty*, *rarity* and *hardness*. While the beauty and rarity of a gemstone may be the prime factors in its selection for use in jewellery, its hardness, which is part of its overall 'durability', must also be taken into consideration.

The reason for this is that the gem must be capable of withstanding the wear and tear to which jewellery is subjected in everyday use. While a ring-mounted gemstone may have to endure the occasional hard knock, it is the presence in airborne dust of microscopic particles of silica or quartz that is the real test of a gem's durability. These particles are in continual contact with the gem, and if the stone is relatively soft they will gradually abrade its polished surface. For this reason, the hardness of a material is defined in gemmology and mineralogy as the ability of that material to resist abrasion.

This ability of a gemstone to withstand abrasion is not due to the 'hardness' characteristic of the atoms that make up its structure, but rather to the strength of the bonds between these atoms when they combine together. Diamond, with atomic covalent bonds only 0.154 nm in length, has a compact structure which makes it the hardest of all the minerals.

In addition to hardness, durability also encompasses 'toughness', which gives a gemstone the ability to resist fracture and 'stability', a quality which enables a material to resist alterations due to heat, light or chemical attack. Some materials, because of their structure, are inherently tougher than others even if they are not as hard. This toughness is exhibited by microcrystalline and polycrystalline materials such as jadeite, nephrite and agate.

Despite its hardness and durability diamond is not, however, indestructible. In the early days of prospecting in South Africa, very few people were experienced enough to be able to distinguish diamond from other stones. Many prospectors based their test for diamond on the belief that it was the only mineral that could survive a blow from a sledge-hammer! (This was one of the tests advocated by the Roman philosopher Pliny some 2000 years ago.) Because of this belief, many fine stones were destroyed by this most destructive of all hardness tests.

The hardness of a gem also has an effect on its appearance. Hard gemstone materials generally take on a brighter polish than soft ones. Their facets are also flatter and their facet edges sharper than softer gems whose edges have a more rounded or moulded appearance under the hand lens. Because of these factors, a polished diamond can often be visually distinguished from an imitation (other identifying tests are covered later in Chapter 17).

Mohs' scale of comparative hardness

The hardness of a material can be more precisely described as its ability to resist abrasion when a pointed fragment of another substance is drawn across it, using a pressure less than that which would cause cleavage or fracture. The scale of hardness used in gemmology and mineralogy was devised by the German mineralogist Friedrich Mohs in 1822. The scale is a purely comparative one in that its divisions are not equally spaced in terms of hardness intervals. It is based on the principle that any substance on the scale having a given hardness number will scratch another one having a lower number, and will in turn be scratched by one having a higher number.

Friedrich Mohs chose as his standards ten fairly common minerals each having a distinctive hardness and each of which could be easily obtained in a high state of purity. That they could be easily obtained was important if the scale was to be of universal use, and the fact that the minerals were available in a high state of purity ensured that their hardness factor was reasonably constant and independent of the source of supply.

The ten minerals that were selected for the Mohs scale are as follows, and are numbered from one to ten in ascending order of hardness:

1. Talc
2. Gypsum
3. Calcite
4. Fluorspar
5. Apatite
6. Orthoclase feldspar
7. Quartz
8. Topaz
9. Corundum
10. Diamond

Referring to the list, it can be seen why gems having a lower hardness than 7 are vulnerable to abrasion from quartz dust particles. For example, a ring-mounted cabochon of malachite, with a Mohs hardness of 4, may need to have its surface repolished after several years of use.

One of the important facts to remember about the Mohs scale is that it is a *relative* and not a *linear* one (i.e. the difference in hardness between each successive number is not the same). For example, the difference in hardness between corundum at 9 and diamond at 10 is greater than the difference between talc, the softest mineral on the scale, and corundum.

As mentioned earlier, hardness is not the sole arbiter of a gemstone's durability. Although of less importance to a gemstone than hardness or toughness, its degree of *brittleness* may also affect its durability. Zircon, for example, has a hardness of around 7 to 7.5, but is brittle, and can suffer from a type of chipping known as 'paper wear' when several stones are kept loose together in a stone paper. The two jade minerals nephrite and jadeite, on the other hand, have a considerable degree of toughness despite their hardness values of 6 and 7 respectively. This is because they are composed not of single crystals, but of a mass of microscopic interlocking fibres or crystals. In prehistoric times, it was this toughness that made jade preferable to the more brittle flint and volcanic glass for Neolithic man's tools and weapons.

Hardness tests (using hardness pencils and plates)

Although a gemstone's hardness as measured on the Mohs' scale can be a useful identification constant, its use for this purpose in gemmology is limited. This is because a hardness test is, in general, a *destructive* one as it usually results in a permanent mark on the gemstone being tested. Another reason for its limited usefulness in identification is that many minerals have similar values of hardness.

As there are often more useful confirmatory tests that can be made on a gemstone, a hardness test should be considered only as a last resort, and even then should be carried out with great care and restricted to the smallest possible scratch on the least noticeable part of the gem, such as its girdle. The hardness test is probably one of the more dubious heritages from the earlier days of gemmology before other more reliable test techniques were developed.

Having made the point about the generally destructive nature of the scratch test, it is only fair to mention that there are a few circumstances where it can be justified. With a large carving, for example, where other tests may not be practicable, it might be permissible to make a scratch on the base without detracting from the carvings' appearance. Another case where a hardness test can be justified is when testing for diamond. Diamond is the only gemstone that can scratch corundum, so in this case the test is confirmatory (and, with care, not damaging to the diamond). Hardness scratch tests can also be made on uncut mineral specimens, particularly where the size or condition of the specimen rules out other tests.

Special *hardness pencils* are manufactured for use in scratch tests (Figure 6.1.), and these consist of wood/metal holders in which are cemented pointed fragments of a selection of the minerals used as standards in the Mohs' scale. The hardness number of each mineral is usually indicated on the ends of the pencils.

Figure 6.1 A set of nine hardness pencils from a boxed set. (Gem-A Instruments Ltd)

When using hardness pencils, the gemstone under test should be scratched (preferably in the region of the girdle where a mark will be least visible) starting with the softest pencil and working up the scale until one is reached that just leaves a visible scratch. The hardness of the gem will be somewhere between the hardness of this pencil and the preceding one. When inspecting the scratch (using a hand lens) it should first be wiped to make sure that it is a genuine scratch and not a line of powder from the test point.

A safer alternative to the rather risky use of hardness pencils is the 'reverse' hardness test. With this test, a facet edge (or preferably the girdle) of the stone being checked is drawn across the flat surface of a series of plates cut from minerals of known hardness. Starting with the softest plate, tests are made moving up the hardness scale until a plate is reached which cannot be scratched. The stone's hardness will then be somewhere between that of the unscratched plate and the previous one.

A set of *hardness plates* made from orthoclase (6), quartz (7), synthetic spinel (8) and synthetic corundum (9) can be ordered from a lapidary, and will be adequate for most hardness tests. A polished section of a synthetic corundum boule can also provide a useful confirmatory test for diamond.

The popular belief that a diamond can be identified by its ability to scratch glass ignores the fact that there are several simulants of diamond (e.g. synthetic moissanite, synthetic corundum, synthetic spinel, topaz, YAG and even quartz) which will also scratch glass. However, if a scratch test is made on glass using diamond and then repeated with one of the harder diamond simulants, the mark made by diamond is certainly deeper, and the diamond 'bites' into the glass more readily than the softer material.

It is quite instructive to make a series of scratch tests on a glass microscope slide using various pieces of mineral and applying the same pressure for each test. While diamond makes a deep scratch with ease, those made using corundum and topaz or synthetic spinel become progressively less deep. When a hardness value is reached which is only one Mohs number higher than that of the glass (whose hardness is usually about 5.5), it becomes quite difficult to start the scratch without applying extra pressure. As a very simple non-destructive way of estimating a gem's hardness, a glass microscope slide can form a useful reference guide.

Finally, a word of warning about one of the older tests used to differentiate between soft 'paste' (i.e. man-made glass) stones and genuine ones. This involved drawing the point of a fine file across the girdle of the gemstone on the assumption that anything scratched by the file must be glass. However, as a steel file has a hardness of around 6.5, there are several valuable gems including opal and lapis lazuli which would suffer damage from such a test.

Directional hardness

Some crystalline materials possess *directional hardness*. Kyanite, for example, has a hardness of 4 in one direction and at right angles to this a hardness of 7 (it can be easily scratched with the point of a needle in a direction parallel to its prism, but resists scratching across the prism (see Figure 6.2e,f). Diamond has an even more marked directional hardness; the diagonals on the cube plane (Figure 6.2a) are the hardest directions, while the planes parallel to the octahedral faces (Figure 6.2b) are the next hardest, followed by the axial directions on the cube plane (Figure 6.2c). The softest direction is in the dodecahedral plane (Figure 6.2d) parallel to the axis of the crystal. The difference in hardness between the diagonal cube directions and the dodecahedral planes in diamond can be as great as a hundred to one.

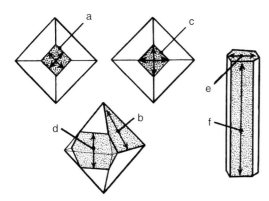

Figure 6.2 The three octahedral sketches on the left show the hardest (a) to the softest (d) directions in diamond. The prism sketch illustrates the hardest basal plane (e) and softest prismatic plane (f) in kyanite

Changes in crystal growth which can produce twinning (as in a diamond macle) will further complicate matters in a mineral possessing directional hardness, as problems can arise when sawing or polishing across twinning boundaries. All of these factors are related to the crystallography of gemstones, and affect the work of both the lapidary and the diamond polisher who must be aware of the best directions for sawing and polishing a gemstone.

The fact that diamond, the hardest of all minerals, can be cut at all depends on its directional hardness property. Diamond dust, which is used as the cutting abrasive when sawing or polishing diamonds, consists of minute randomly positioned particles of diamond. It is this random orientation of the diamond dust particles which ensures that some of their cutting edges will be in the direction of maximum hardness when they come into contact with the surface being polished or sawn. Even then, the diamond polisher has to arrange that none of the planes of the facets he or she is cutting lies in the plane of maximum hardness. If necessary the angle of polishing can be offset by a small amount so as to obtain a working differential between the directional hardness of the diamond being cut and that of the diamond dust.

Polishing the jade minerals usually results in a characteristic 'orange peel' surface when using abrasives having a hardness of 9 and below. This is due to the directional hardness of the material which causes undercutting of the softer granular or fibrous areas of the gemstone and results in a mottled finish. This directional hardness problem in nephrite, jadeite and some other gem materials can be reduced by using diamond dust as the abrasive. The increased speed of cutting achieved with diamond dust (and the increasingly competitive price of synthetic diamond abrasive) is also making its use by the lapidary more widespread.

Engineering hardness tests

Most engineering hardness tests are designed for use on metals. They are based generally on the force required to press a profiled indenter into the surface of the metal, and on the diameter or penetration of the resulting indentation. Brinell, Vickers, Rockwell and Knoop are examples of this type of test. Unfortunately, although these tests produce results which are linear in relationship to each other, they are not really suitable for use on many minerals because of the possibility of cleavage damage. Most of them are also unsuitable for use on diamond. However, Dr C.A. Brookes of Exeter University developed an indentation test using a five-sided diamond indenter. This gave the following average indentation hardnesses (see Table 6.1).

Table 6.1

	Indentation hardness	Mohs'
Diamond	10 000	10
Corundum	2200	9
YAG	1730	8
Spinel	1700	8
Strontium titanate	530	5.5

Abrasion tests which measure the *wear factor* of materials, and are therefore more akin to the Mohs' scratch test, have also been made on gem materials. Dr E.M. Wilkes of Oxford University used a miniature grinding wheel impregnated with diamond dust. The wheel was applied to the test sample (in the direction of maximum hardness)

Table 6.2

	Wear factor	Mohs'
Diamond	1	10
Corundum	5000	9
YAG	2500	8
Spinel	20 000	8
Strontium titanate	250 000	5.5

for a fixed time period to check the wear factor of various materials. The wear was quantified as the amount of material abraded from the sample (i.e. the weight lost) compared to diamond (results shown in Table 6.2).

The test results, which transposed the values for corundum and YAG and made spinel 'softer' than YAG, showed that there was no simple relationship between the Wilkes 'wear factor' and the Mohs' scratch test.

The influence of hardness on mining techniques

Diamond's unique hardness has resulted in the formation of a polishing industry quite separate from that of the rest of the non-diamond gems. This extreme hardness is also reflected in many of the mining methods used to extract diamond from its primary sources in the ground. Unlike many other types of gemstone mining, the toughness of the diamond crystal allows rock drills and explosives to be used both above ground and underground in the extraction of the diamond-bearing rock. This toughness also allows the use of crushers and mills to break up the mined rock and release the entrapped diamonds.

In the mining of other minerals, which are relatively fragile compared with diamond, explosives are used only very sparingly to open up seams of gem-bearing rock containing, for example, beryl or even corundum. Gem-bearing seams are more likely to be exposed by mechanical excavation, and will then be worked by hand using pick-axe and hammer to ease out the crystals. The removal of any rock matrix from the stones will also be done manually rather than by machine because of the much softer and often shock-sensitive gem material.

Chapter 7

Specific gravity, density and relative density

The *specific gravity* (SG) of a gemstone is a constant which can be of great help in iden-tifying unmounted stones. The various methods of determining a gemstone's SG form one of a series of important test techniques used in modern gemmology.

Definitions

The SG of a substance is the ratio of its weight to the weight of an identical volume of pure water (at standard atmospheric pressure and at 4°C – the temperature at which water is most dense). Because SG is expressed as a ratio, no units of measurement are neces-sary. The SG of diamond, for example, is 3.52, which indicates that diamond is 3.52 times heavier than an identical volume of water. Water is chosen as the standard for defin-ing SG as it is both stable and universally available. By definition, the SG of water is 1.0.

Note: For all normal gemmological purposes, neither temperature nor atmospheric pressure variations introduce any significant errors in the estimation of SG.

The *density* of a substance is defined as its weight per unit volume, and is measured not as a ratio but in units of weight and volume. The international SI units chosen for density measurement are the kilogram and the cubic metre (previously the units were the gram and the cubic centimetre, i.e. gm/cc). Kilograms per cubic metre are expressed mathematically as $kg\,m^{-3}$. Using these units, the density of diamond is $3520\,kg\,m^{-3}$ (in older textbooks this is shown as $3.52\,g/cc$, and is numerically the same as diamond's SG).

Relative density is a term sometimes used in connection with liquids, and like SG is the ratio between the weight of the liquid and the weight of an identical volume of water at 4°C. However, when referring to heavy liquids in this chapter and elsewhere in this book use will be made of the term 'SG'.

Archimedes' principle and the measurement of SG

Formulated by the Greek scientist, mathematician and philosopher over 2000 years ago, Archimedes' principle states that a body immersed in a fluid experiences an upward force equal to the weight of the fluid it displaces. It was this concept which Archimedes developed to provide the first foolproof method of testing gold for purity of content, and which has also become the basis for SG measurements in gemmological work.

As the value of SG is a reliable constant for the majority of gemstones, it can be a useful aid in the identification of an unknown (and unmounted) specimen. Although a gemstone cannot normally be identified by its SG alone, together with other measurements this can help in narrowing down the range of possibilities.

The SG of a substance depends in part on the atomic weights of its constituent elements, and in part on the compactness of the structure formed by these elements. Diamond, for example, with its light but compactly arranged atoms of carbon, has a higher SG than quartz, which consists of heavier but less densely packed atoms of silica and oxygen.

When handling gemstones it soon becomes evident that among stones of a similar size, some have a greater 'heft', i.e. they feel appreciably heavier. Zircon, size for size, is twice as heavy as opal. This method of sensing a gem's SG is of course very crude, but is sometimes of value when making a quick subjective assessment of a stone.

As a gem's SG depends on the ratio between its weight and the weight of a similar volume of water, its measurement would seem at first sight to be quite a problem. Finding the stone's precise weight is not difficult, but unless the specimen happens to be a perfect cube or similar regular shape, the calculation of its volume (necessary in order to find the weight of a similar volume of water) could be quite complicated mathematically.

Measurement of SG by displacement

Archimedes faced this same problem. His solution was to make the specimen under test (a gold artifact in his case) displace its own volume of water into a measuring vessel. To achieve this he designed an apparatus called the *eureka can* (Figure 7.1). This type of device is still in use today, and consists of a metal container fitted with an overflow pipe.

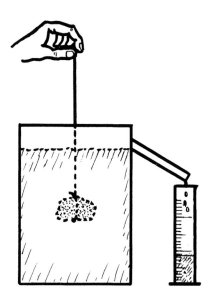

Figure 7.1 The eureka can is used to measure the SG of an object by means of its volumetric displacement of water. For medium-sized to large objects, the suspending cord has a negligible effect on the accuracy of results

To measure the SG of a specimen, the eureka can is first filled with water until it just begins to flow out of the pipe. When the level of the water has settled and the overflow has ceased, an empty beaker is placed under the pipe, and the specimen is gently lowered into the can until it is totally immersed. The volume of water displaced into the beaker then represents the volume of the specimen. The displaced water can either be weighed, or its volume measured in a vessel graduated in cubic centimetres (1 cubic centimetre of water weighs exactly 1 gram at 4°C).

The SG of the specimen is then calculated by dividing the weight of the specimen by the volume (or weight) of the displaced water:

$$SG = \frac{\text{weight of gemstone (in grams)}}{\text{volume of displaced water (in cc)}}$$

or

$$SG = \frac{\text{weight of gemstone (in grams)}}{\text{weight of displaced water (in grams)}}$$

Although the displacement method of determining specific gravity using an eureka can is suitable for large objects such as carvings or mineral specimens, it is not accurate enough for use with gemstones because of their comparatively small volume. For more precise work on small gemstones, use is made of an SG bottle (Figure 7.2). The bottle, also called a *pycnometer*, consists of a small glass phial having a ground-glass stopper through which runs a capillary channel.

Figure 7.2 An SG bottle (also called a pycnometer) can be used to measure a small gemstone's SG by the displacement method. It is designed, however, for the precise determination of the SG of liquids (see end of chapter)

The SG bottle is more often used for the precise determination of the SG of heavy liquids (for this reason the internal volume of the phial, measured at 20°C, is usually engraved on its exterior). Details of its use for this purpose are given at the end of this chapter under the heading 'Precise measurement of SG using heavy liquids'.

When using the SG bottle to measure the displacement of a gemstone, the specimen is first weighed in the normal way (*W1*), and then the empty and dry SG bottle and stopper is weighed (*W2*). Next, the bottle is completely filled with water, and the stopper inserted, forcing surplus water out of the top of the stopper through its capillary channel. The outside of the bottle is then dried (removing any excess water from the top of the stopper without withdrawing water from the capillary channel), and the bottle reweighed (*W3*).

Finally, the stopper is removed, the gemstone is inserted in the bottle and the stopper reinserted (the water which now overflows through the capillary channel is equal to the volume of the gemstone). After the previous drying operation has been repeated, the bottle is again weighed (*W4*).

The weight of water originally filling the bottle = *W3* − *W2* = *A*. The weight of water filling the bottle with gem inside = *W4* − *W2* − *W1* = *B*. Therefore the weight of water lost from bottle when gem introduced = *A* − *B*:

$$\text{SG of gem} = \frac{\text{weight of gem}}{\text{weight of displaced water}} = \frac{W1}{A - B}$$

The following alternative methods, all based on Archimedes' principle, are frequently used in gemmological work.

Hydrostatic methods of SG measurement

As these methods depend on the precise measurement of weight (see Appendix J), with small specimens they should only be attempted on a high-sensitivity balance of the analytical type. Even with small specimens, hydrostatic weighing is not practicable for gemstones weighing less than 2 carats, as the margin of error (due to the limitation of balance sensitivity) increases as the weight of the specimen decreases. However, with very large objects such as carvings, the volume of water displaced is large enough to provide reasonably accurate results even when using a spring balance.

The hydrostatic weighing technique is based on the principle that an object immersed in a fluid experiences an upward force (buoyancy or loss of weight) equal to the weight of the displaced fluid. The method consists of weighing the gemstone in air, and then again when it is *totally* immersed in pure water. The three methods to be described cover the use of two-pan, single-pan and spring balances.

Two-pan balance method (Figure 7.3)

In addition to a two-pan beam balance, a few simple accessories are required comprising a glass beaker, a metal or wooden platform for positioning the beaker above one of

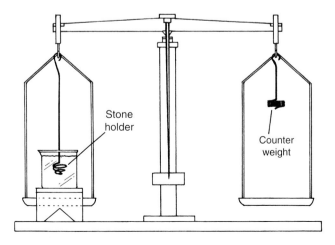

Stone holder

Counter weight

Figure 7.3 Diagram showing the use of a two-pan beam balance for the hydrostatic measurement of SG

the balance pans, a spiral wire stone holder for suspending the gem specimen in the water, and a counterpoise to offset the weight of the wire stone holder (Figure 7.4). Both the stone holder and the counterpoise can be made from 20–22 swg tinned copper wire.

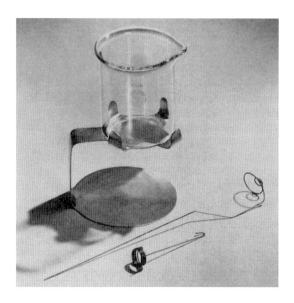

Figure 7.4 A kit of accessories for the measurement of SG by hydrostatic weighing

First, the beaker is filled with purified or distilled water and placed on the platform over the weighpan. The wire stone holder is attached to the weighpan hanger on the balance by a length of fine-gauge mono-filament nylon (not nylon-braided glass yarn or woven cord as these will absorb water by capillary action). The cord length is then adjusted so that the stone holder remains immersed in the water throughout the vertical swing of the balance. Alternatively, the suspending cord can be dispensed with and the wire forming the stone holder extended so that it can be attached directly to the weighpan hanger; however, this may reduce the sensitivity of the balance by affecting its free swing.

Finally, to simplify the SG calculations, the counterpoise is suspended from the other weighpan hanger and its weight adjusted to balance the weight of the empty stone holder when it is totally immersed in the water.

To determine the SG of a specimen gemstone, it is first weighed in the normal way in air (A), and then weighed again suspended in water by the wire stone holder (B). The stone's SG can then be calculated from the following simple formula:

$$SG = \frac{\text{weight of gemstone}}{\text{loss of weight in water}} = \frac{A}{A - B}$$

Although SG measurements, by definition, should be made at 4°C, errors introduced by using water at room temperature will mainly affect the third decimal place in the results, and can therefore be disregarded for normal gemmological work. Of more importance are the possible errors introduced by air bubbles adhering to the gemstone, and by the surface tension of the water, the latter causing a friction-like drag on the wire support as it moves through the water.

Bubbles can be removed by thoroughly wetting the gemstone before immersion. Any residual bubbles can then be carefully removed from the stone and the stone holder wire using a camel-hair paintbrush (not, as once described in an exam answer, by a *camel's hair brush!*). The use of distilled or boiled water will help to prevent bubbles forming from air contained in the water. Surface tension effects can be reduced by mixing a drop of detergent with the water.

Single-pan balance methods

There are two methods of hydrostatic weighing which can be used with a single pan balance. One of these is shown in Figure 7.5, and also uses the *upthrust* or loss of weight experienced by a totally immersed gemstone (the stone holder in the illustration consists of a perforated metal pan of known immersed weight). The procedure and SG calculations are exactly the same as for the two-pan method previously described except for an allowance made for the immersed weight of the stone holder (which can alternatively be tared, i.e. adjusted to zero, if the balance is an electronic one).

Figure 7.5 A set of SG accessories for a single-pan balance. The beaker standing on the metal platform contains a thermometer and an immersed perforated stone-holder pan. (Sauter)

The second method uses the *downthrust* produced when a gemstone, suspended from an independent base rather than from the balance hanger, is inserted into a beaker of water placed on the weighpan of the balance (Figure 7.6). Although the downthrust is exactly equal to the upthrust in the two-pan method, the procedure and SG calculations are a little different.

The weight of the gemstone in air is measured in the usual way ($W1$). The wire cage stone holder is suspended from an independent support and arranged so that it is totally immersed in a beaker two-thirds full of water placed on the balance weighpan. The stone holder wire is marked at the point where it enters the water so that it can be reinserted later to precisely the same point. The weight of the beaker and water is then measured with the stone holder immersed at this setting ($W2$).

Next, the stone is carefully inserted into the immersed cage (without spilling any water) and the height of the cage readjusted so that the mark previously made on the

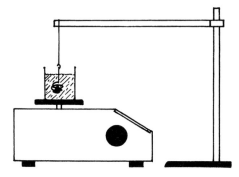

Figure 7.6 An alternative single-pan method of hydrostatic weighing uses an independent suspension of the specimen to measure the downthrust produced by its displacement of water

wire again coincides with the surface level of the water. At this point the stone should be checked to see that it is fully immersed (any bubbles on the stone can be removed as described for the two-pan method). Finally, the weight of the beaker with the stone immersed is measured (*W3*).

The SG of the gemstone is calculated as follows:

Weight of equivalent volume of water $= W3 - W2 = A$

$$SG = \frac{W1}{A}$$

Spring balance method

For larger gem specimens, such as carvings, where the difference between the weight in air and the immersed weight are proportionally larger, the less sensitive spring balance (Figure 7.7) can be used to provide reasonably accurate results. With care, the difference between the specific gravities of the jade minerals (nephrite = 3.0,

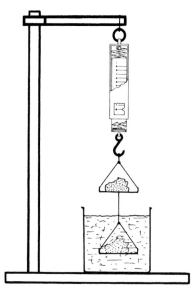

Figure 7.7 Sketch showing a spring balance being used to check the in-air weight and the in-water loss of weight of a carving or large mineral specimen

jadeite = 3.3) can be detected (weight in air divided by *loss* of weight in water) and the type of jade identified using a spring balance and a suitably large container of water.

Before leaving the subject of hydrostatic weighing, mention must be made of specialized balances designed to give a direct readout of a specimens' SG. Figure 7.8 shows one version of this type of balance. The lower weighpan is immersed in a beaker of water, and the gemstone to be checked is first placed in the upper (in-air) weighpan. Next, the beam is balanced by adding weights to the hanger at the right-hand (0) end of the beam scale. The weighing is then repeated with the gemstone transferred to the lower immersed pan, and the beam is brought back into balance again by sliding the weight hanger towards the central fulcrum of the beam. The SG of the gemstone is then read directly from the beam's calibrated scale at the weight hanger suspension point.

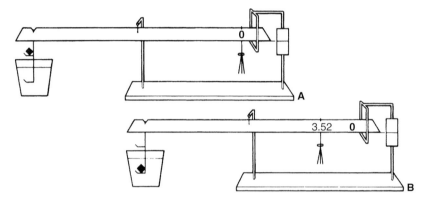

Figure 7.8 A direct-reading SG balance. Two gemstone holders are provided at the left end of the beam for the in-air and in-water weighings (the lower one is immersed in the beaker of water during both weighings). The counterweight is set to the '0' mark for the in-air weighing and is loaded (by means of hook-on weights) for balance equilibrium (A). For the in-water weighing, the counterweight is moved to the left to restore equilibrium (B), its new position on the scale indicating the gemstone's SG. (Hanneman)

Approximation of SG using heavy liquids

This method, although less precise than the hydrostatic weighing method, has the advantage that it is fast and that there is no bottom size limit to the specimens. In its simplest form it consists of four liquids having specific gravities between 2.65 and 3.32. The gemstone under test is immersed in each liquid in turn, and its SG lies between that of the liquid in which it just sinks and that in which it floats on the surface.

The principle in use here is again that of Archimedes, and if the gemstone being tested has, by chance, exactly the same specific gravity as the liquid in which it is immersed, it will experience an upward force exactly equal to its own weight. As this condition can only be met if it is totally immersed, it will float freely within the liquid.

Before going on to describe the liquids and the test procedure in detail, it should be mentioned that heavy liquids are unsuitable for testing stones that are porous, such as opal or turquoise, or stones which have surface cracks or flaws, as these may become discoloured by the liquids and/or produce inaccurate results. Plastic materials may soften in the liquids, as may the cement used in composite stones.

Heavy liquids were first put to serious gemmological use in the 1920s by Basil Anderson who was Director of the London Chamber of Commerce's pearl testing laboratory. By blending heavy liquids it was found possible to separate the denser Japanese cultured pearls from the naturally formed variety. Much later the technique was to be used by the laboratory to provide forensic evidence on imitation pearls in a murder trial!

While a variety of heavy liquids were formulated and used by mineralogists in the 1870s and 1880s, few of these are in use today because of their cost and their poisonous characteristics. The most important of the surviving liquids is the organic fluid compound *di-iodomethane*, CH_2I_2 (also called *methylene iodide*). This heavy liquid (with an SG of 3.32) was first suggested by R. Brauns in 1884 and with another organic compound, *bromoform* $CHBr_3$ (SG of 2.86), they have come into general use because of their stability and relative safety. A third heavy liquid, *1-bromonaphthalene*, also known as *monobromonaphthalene* $C_{10}H_7Br$ (with the much lower SG of 1.49), can be used as a dilutant to reduce the SG of bromoform and di-iodomethane. Today, the distinctive smell of these liquids provides an unforgettable background aroma to any gem testing class or laboratory!

A fourth heavy liquid, *Clericis solution* (with an SG of 4.15) is a mixture of thallium malonate and thallium formate in water. The solution was formulated by the Italian chemist Clerici in 1907, and was introduced by Anderson to the gemmology classes at the Chelsea Polytechnic, London, in the late 1930s. Its main attraction was the fact that garnet, spinel, diamond and even corundum floated in the solution which could then be blended with water to the appropriate SG and used to test for any of these gems.

Clerici's solution was later to prove a valuable aid in the accurate determination of a variety of gemstone SGs because its own SG was linearly and precisely related to its refractive index. This meant that if it was blended with water until the specimen under test was suspended, the refractive index of the blended liquid could be measured on a table spectrometer and its SG (and therefore that of the specimen) read from a straight-line graph. At the time, this provided a more accurate means of measurement than the use of an SG bottle.

Clerici's solution, however, is both poisonous and corrosive, and despite its usefulness in gemmological work it proved to be too hazardous for general application. It is still available for specialized laboratory purposes, but is no longer used or recommended for checking gemstone SGs.

The following basic set of heavy liquids (Figure 7.9) is recommended for general test purposes, and can be obtained singly or in boxed sets from suppliers of gemmological equipment (e.g., the Gemmological Association of Great Britain, or the Gemological Institute of America). One such set is illustrated in Figure 7.10, and a similar set from the GIA contains 2.57, 2.62, 2.67, 3.05 and 3.32 liquids. Both of these sets are supplied with two smaller bottles of adjusting liquids (3.32 and 1.50 or 1.56).

1. Bromoform or di-iodomethane diluted with 1-bromonaphthalene to 2.65
2. Di-iodomethane diluted with 1-bromonaphthalene to 3.05
3. Undiluted di-iodomethane, 3.32

Liquid 1 is useful for checking quartz gems, and liquid 2 is blended to the SG of tourmaline. Liquid 3 is useful when separating the heaver gems having an SG above 3.32 or close to this figure. An additional liquid, as suggested by Robert Webster, can be made up when testing for amber and some of its simulants. It consists of a solution

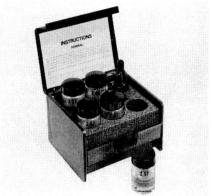

Figure 7.9 A set of heavy liquids used in estimating the SG of a gemstone. The smaller bottle in the centre contains a salt solution for testing amber (see under the heading 'Amber' in Chapter 18)

Figure 7.10 A set of five heavy liquids with two small bottles of adjusting fluids. (Krüss)

made from 10 level teaspoons (50 grams) of common salt dissolved in a half pint (0.28 litre) of water. The resulting solution has a sufficiently high SG (1.13) to cause amber (SG of 1.08) to float and the various plastic and Bakelite simulants to sink (copal resins and polystyrene have an SG lower than 1.13 and cannot therefore be separated from amber by this test).

Several other mixtures of the two basic heavy liquids can be blended as a specific check for various stones: a 2.71 mixture of di-iodomethane and 1-bromonaphthalene can be used to check for varieties of the beryl species; a 3.18 mixture of di-iodomethane and 1-bromonaththlene can be used to check for gemstones such as andalusite, apatite, fluorite and spodumene.

Containers for heavy liquids should be wide-necked for the easy insertion and retrieval of gems, and of clear glass so that the gem can be seen without difficulty during the test (and when attempting to retrieve the stone!). The bottles should have well-fitting glass stoppers rather than plastic ones, as the latter may deteriorate in contact with the liquids.

When mixing heavy liquids to a chosen SG, pour half of the required quantity of the heavier of the two chosen liquids into the bottle. Select a transparent flaw-free sample of the gem whose SG is being matched and place it in the bottle as an indicator. Preferably, the sample should be colourless (e.g. rock crystal rather than the colour varieties of quartz as the latter have SGs slightly above quartz's 2.65). Gemstones having a simple composition also have the most constant SGs – for example, quartz, calcite, corundum, fluorite. Then add the lower SG liquid (slowly at first and finally drop by drop), stirring the mixture between each addition until the indicator gem becomes freely suspended.

Practical points for SG measurement using heavy liquids

1. Hold the stone in tweezers, and insert it just below the surface of the liquid before releasing it. This prevents a small gem, which may have an SG greater than the liquid, from 'floating' on the surface due to surface tension.
2. If the stone sinks in the liquid, observe the rate of sinking. If the stone only sinks slowly, or rises slowly to the surface, its SG will be very close to that of the liquid

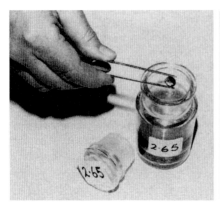

Figure 7.11 Left: An unknown gemstone being inserted in a bottle of heavy liquid whose SG is 2.65. Right: The unknown gemstone suspends freely within the liquid, indicating that its SG is 2.65, and that the stone is most likely one of the quartz varieties

(i.e. the test is very sensitive to small differences in SG). Because of this, a stone rarely suspends freely halfway in the liquid (Figure 7.11), but at best will drift slowly up or down.

3. If the stone immediately bobs to the surface after being released, its SG is significantly less than that of the liquid. It should then be tested in a liquid of lower SG. If the stone sinks rapidly, its SG will be significantly greater than the liquid, and it should be transferred to a liquid of higher SG.

4. Always clean both the gemstone and the tweezers before transferring the stone from one liquid to another. This will prevent the liquids from becoming contaminated.

5. As it is easier to retrieve a gemstone which is floating, start the test using the 3.32 liquid. If the stone floats in this, go to the next highest liquid and if necessary work downwards until the stone sinks. However, if it sinks in the 3.32 liquid, no other SG tests are necessary and the results can be recorded as *SG greater than 3.32* (or SG > 3.32 for short). In the majority of cases, the gem's SG will lie somewhere between two of the liquids. For example, it may sink in the 2.65 liquid and float in the 3.05 liquid. Its SG can then be recorded as being between 2.65 and 3.05, i.e. 2.65 < stone's SG < 3.05.

6. Because the heavy liquids also have high refractive indices, a gem whose refractive index is close to that of the liquid in which it is immersed may 'disappear', a situation which has alarmed many a student! The paler the colour of the gem, the less visible it will be, and retrieval may become difficult. With any retrieval problem, it helps to hold the bottle up to the light at eye level, and if the stone is heavier than the liquid, to tilt the bottle so that the stone can be trapped with the tweezers in the lower corner.

7. Over a period of time, the SG of heavy liquids may change due to differential evaporation. To provide a warning of any change, it is good practice to leave a small piece of indicator mineral in the bottle. A crystal is preferable to a cut stone as it cannot later be confused with the stone being tested.

A word of warning, however. If a 2.65 SG heavy liquid has been blended at a room temperature of, say, 18°C to suspend a quartz indicator, and the test liquid used again on a much colder day, the indicator will be found floating on top. On a

much hotter day it will have sunk to the bottom! This at least gives proof of the sensitivity of the heavy liquid method (as further proof, an indicator that has floated to the surface on a cold day can be made to sink slowly by simply warming the container in the hand).
8. Di-iodomethane and bromoform tend to darken if exposed to daylight, and for this reason they should be kept in the dark when not in use. Copper foil can be used to lighten the liquids and a small piece should be placed in each bottle of heavy liquid.

Precise measurement of SG using heavy liquids

A more accurate way to determine a gemstone's SG is to use a pycnometer or specific gravity bottle. Earlier in this chapter a brief mention was made of use of the SG bottle to measure a gem's SG by displacement (see Figure 7.2). However, the following section describes a more precise method which can be employed even with large stones that are too big for insertion in the bottle. With this method the bottle is used to measure the SG of a heavy liquid which has been carefully blended to a point where it freely suspends the gem under test.

First of all, the empty SG bottle complete with stopper is weighed, and then the *approximate* SG of the stone under test is obtained by use of the standard heavy liquids as described previously. Next, a separate heavy liquid is made up to this approximate SG and carefully adjusted by further blending until the stone becomes freely suspended in the liquid.

The SG bottle is then filled with the blended liquid, its capillary stopper inserted, and the bottle and stopper carefully dried as described earlier. Finally, the bottle is weighed, and the SG of the blended liquid (and therefore that of the gemstone) is calculated by first subtracting the weight of the empty bottle (in grams) from its filled weight. This figure is then divided by the volume of the bottle in cubic centimetres or millilitres (the volume is usually engraved on the bottle) to give the precise SG of the stone.

Sodium polytungstate

Sodium polytungstate ($3Na_2WO_4.9WO_4.H_2O$) was proposed by Hanneman as a safer alternative to some of the organic heavy liquids used for SG determinations. In its dry state this material is a white crystalline powder. When made up as a saturated solution using distilled, deionized or softened water (note: the presence of calcium ions will produce a white precipitate) it has an SG of about 3.10. This can be lowered by the addition of water to 3.05 to test for tourmaline, or to 2.65 to test for quartz.

Because its SG is linearly related to its refractive index (like that of Clerici's solution), if a solution of sodium polytungstate is made up to suspend a gem specimen, then the SG of that solution can be determined by measuring the refractive index of the liquid on a refractometer, and then applying the formula:

SG = 2.80 + 8.43 (refractive index − 1.555)

Note: for refractive indices at exactly 1.555, the SG of the sample is precisely 2.80. For indices above 1.555 the SG becomes more than 2.80, and for indices less than 1.555 the factor within the bracket is negative; when multiplied by 8.43 the product is also negative and the SG becomes less than 2.80.

Safety precautions

The following note, originally printed in the April 1979 issue of the *British Journal of Gemmology*, contains advice from the Health and Safety Executive of the UK Department of Health on the use of gemmological test liquids.

Care should be taken when using ethylene dibromide (a suspect carcinogenic liquid sometimes used for SG determinations and in hydrostatic weighing), or any other heavy liquids used in gemmology, to avoid skin contact or inhalation of vapour. On no account should any of the liquids used by gemmologists for gem testing be swallowed. In case of contact with the skin, liquid should be washed off; if in the eyes, they should be well flushed out with running water; if swallowed, vomiting should be attempted and medical assistance obtained.

As with certain other volatile liquids, it is also advisable to avoid smoking when using heavy liquids. **Gemmological liquids should only be used under well-ventilated laboratory conditions.**

Chapter 8

Colour, lustre and sheen

In Chapter 1, beauty of appearance was given as a prime quality of a gemstone. In a world where colour is one of the dominant visual sensations, it is not surprising that a gemstone's beauty is largely determined by its colour. Although the perception of colour is an everyday experience, and as such is taken for granted, in gemmological studies it is important to understand exactly how the appearance of colour is produced in a gemstone. This in turn provides a valuable means of identification as will be seen in Chapter 11 which deals with spectroscopy.

The electromagnetic spectrum

To gain an understanding of colour, we must first of all take a look at the nature of light itself. Light is a form of energy which is radiated as a wave motion by *electromagnetic waves* (Figure 8.1). These are similar to the ones generated for radio and television broadcasts, but have a very much shorter wavelength. The colour of light is determined by its *wavelength*. The *intensity* or strength of light is proportional to the square of the amplitude of the electromagnetic wave.

The relative positions of both light and radio waves in the electromagnetic spectrum can be seen in the upper section of Figure 8.2, while the lower section shows an expanded view of the visible part of the spectrum. Light waves in the visible spectrum are bounded at the long-wavelength red end by infrared heat waves, and at the violet end by invisible ultraviolet rays.

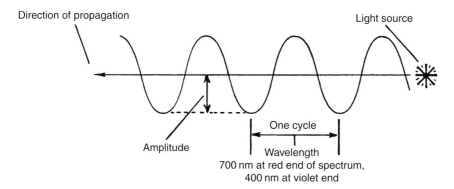

Figure 8.1 The wave nature of light

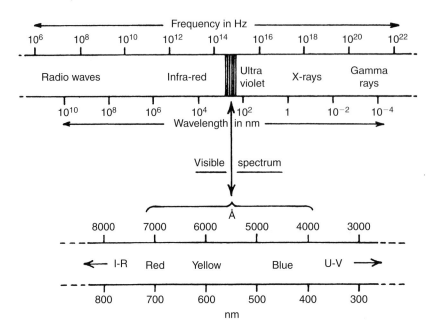

Figure 8.2 The electromagnetic spectrum, with (lower section) an expanded view showing the visible part of the spectrum

The wavelengths of radio and television transmissions are measured in metres and centimetres, but light wavelengths, because they are much shorter, are measured in *nanometres* (or in older textbooks in *ångström* units). A nanometre (nm) is one millionth of a millimetre (10^{-9}m), and is equal in length to ten *ångström* units ($10\,\text{Å}$). Wavelengths, particularly in radio and television, are sometimes expressed as a frequency, which indicates the number of waves (or cycles) passing a fixed reference point per second. As an electromagnetic wave has a velocity or speed of propagation of about 300 000 000 metres per second, the relationship between its wavelength and its frequency can be expressed as follows:

$$\frac{\text{Wavelength}}{\text{(in metres)}} = \frac{\text{velocity}}{\text{frequency}} = \frac{300\,000\,000}{\text{cycles per second}} = \frac{300}{\text{megacycles per second}}$$

$$\text{Wavelength (in nanometres)} = \frac{300 \times 10^9}{\text{megacycles per second}}$$

Note: In the same way that the nanometre has replaced the ångström unit, so has the unit hertz (Hz) replaced cycles per second, and megacycles per second are now referred to as megahertz or MHz.

The photon

So far we have been discussing light in terms of Maxwell's theory of electromagnetic wave propagation, a theory which was eventually verified in 1876 by Herz when he produced and detected radio waves. However, according to Einstein's quantum theory of radiation, light is not a continuous train of waves, but consists of a large number of wavelets or *photons*.

When energy in the form of heat or an electrical charge is absorbed by an atom, this enables some of its electrons to move from their normal low energy orbits into higher energy orbits further away from the nucleus. When they return to their normal low energy orbits again they emit their surplus energy in the form of photons of light (a particular form of this emission, luminescence, will be covered in Chapter 12). A single electron transition of this kind lasts for only one hundred millionth of a second (10^{-8} s), and because of the velocity of light, the resulting photon is emitted as a travelling wavelet some 3 metres long. This radiation of light then ceases until a further electron transition takes place. Continuous light as we perceive it consists of a multitude of these short pulses or photons.

The quantum theory of radiation is strongly supported by the fact that the energy of the light released by electron transitions in the atom corresponds to the difference between the energy levels of the two electron orbits taking part in the transition. The revolution in scientific thought brought about by quantum mechanics meant that the separate concepts of atomic particles and waves of radiation began to merge. The end result is that protons, electrons, neutrons, photons, etc. are now considered to exist both as particles and as waves.

Colour and selective absorption

White light is composed of an approximately equal mixture of all the colours or wavelengths that make up the visible spectrum. When we look at a coloured gemstone in white light, the colour we see is the result of the absorption by the stone of various wavelengths (or bands of wavelengths) in the original white light. In a transparent stone, these wavelengths will be absorbed from the light as it passes through the stone; in an opaque stone, the wavelengths will be absorbed as the light is reflected back from its surface layer. In both cases, the colour appearance of the stone is produced by the combined effect of the unabsorbed parts of the white light spectrum which results in a colour *complementary* to the ones absorbed.

If, for example, the *violet* end of the spectrum is absorbed by the gemstone, the colours in the remaining part of the white light will combine together to produce the complementary colour *yellow* (Figure 8.3). If the wavelengths from yellow through to violet are absorbed, the stone will appear red.

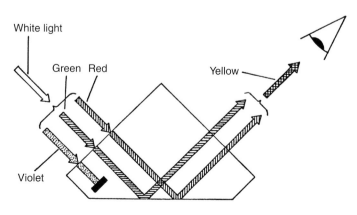

White light

Green Red

Yellow

Violet

Figure 8.3 With a gemstone having an absorption band in the blue/violet end of the spectrum (such as a cape series diamond) the residual red and green part of the white light reflected from the stone gives it a yellow appearance

This suppression of certain wavelengths or colours from the white light illuminating an object is known as *selective absorption*. It can be visually analysed by means of an instrument called a *spectroscope*. The light passing through (or reflected from) a gemstone is directed into the spectroscope where a combination of prisms (or a diffraction grating) spreads the light out into its spectral colours. The various wavelengths which have been absorbed by the stone are visible along the spectrum as a dark band or a series of dark lines or bands. The overall result as viewed through the spectroscope is called an *absorption spectrum*, and is sometimes sufficiently distinctive to identify a gemstone (Figure 8.4). A detailed description of the spectroscope and its use is given in Chapter 11.

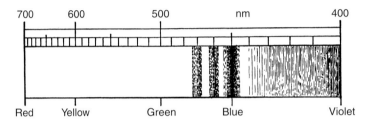

Figure 8.4 An absorption spectrum showing three bands (caused by iron) which are diagnostic for blue sapphire

Allochromatic and idiochromatic gems

The selective absorption of light by most gemstones is caused either by impurities present in the gemstone (such as chromium oxide in ruby, or iron oxide in aquamarine), or by the chemicals in the stone's composition (e.g. copper in turquoise, or manganese in spessartite garnet). Gemstones whose colour is caused by impurities are termed *allochromatic* (i.e. 'other coloured'), while those which owe their colour to their own chemical composition are called *idiochromatic* ('self coloured').

The majority of coloured gemstones are allochromatic, and in a pure state some of these occur as colourless varieties. Examples of allochromatic gem varieties containing no colouring impurity are rock crystal, colourless sapphire and topaz, and the colourless goshenite variety of beryl.

The transition elements

Selective absorption of light in both allochromatic and idiochromatic gemstones is caused mainly by the presence of one or more of eight metallic elements, which are called the *transition elements*. In allochromatic gem minerals, these elements act as the colouring impurities, while in the idiochromatic stones they are an integral part of the mineral's chemical formula. The eight transition elements, in rising order of their atomic weights which range from 22 to 29 (and with examples of the gemstones they colour), are shown in Table 8.1.

Zircon and some colour varieties of topaz, quartz and fluorspar contain no measurable amounts of any transition element. Unlike other stones, their colours can be more easily changed by heat and irradiation (for the cause of colour in these stones, and in diamond, see later in this chapter under 'Colour centres'). At the atomic level, the way

Table 8.1

Titanium	Blue sapphire (with iron), blue zoisite
Vanadium	Grossular garnet (tsavorite), green vanadium beryl, synthetic corundum (alexandrite simulant) some synthetic emeralds, blue/violet sapphire
Chromium	Ruby, emerald *, red spinel, pyrope garnet, chrome grossular garnet, demantoid garnet, uvarovite garnet†, chrome diopside, green jadeite, pink topaz, alexandrite, hiddenite
Manganese	Rhodochrosite†, rhodonite†, spessartite garnet†, rose quartz, morganite variety of beryl, andalusite
Iron	Sapphire, sinhalite†, peridot†, aquamarine, blue and green tourmaline, enstatite, amethyst, almandine garnet†
Cobalt	Synthetic blue and green spinel, synthetic blue quartz (except for a rare blue spinel, cobalt is not found in any natural transparent gemstone)
Nickel	Chrysoprase, synthetic green and yellow sapphires
Copper	Diopside, malachite†, turquoise†, synthetic green sapphire

* In UK and Europe only beryl coloured by chromium can be described as emerald.
† Idiochromatic gemstones.
Note: The above transition elements can be remembered (in the order given above) by the initial letters of the simple mnemonic, 'tints vary chromatically, man I could never copy'.

in which the interaction of electrons in a transition element produces colour is a complex subject bound up with the *crystal field theory*.

Colour-change gemstones

In some instances, the positions of absorption bands produced by transition elements may cause a stone's body colour to change when it is moved from one type of lighting to another. This colour-change is thought by some to be caused by *metamerism*, but because it is seen most dramatically in the rare alexandrite variety of chrysoberyl it is generally known as the alexandrite effect. In this particular gemstone there is a broad absorption band centred about 580 nm in the yellow area of the spectrum. This causes the stone to appear red in the blue-deficient light of an incandescent lamp (i.e. a tungsten filament lamp), and green in the more balanced spectrum of daylight or under a daylight-type fluorescent lamp.

Because of alexandrite's rarity and consequent high price, several simulants have been marketed which attempt to copy this colour-change effect. One of these is a synthetic corundum doped with vanadium. The colour change with this simulant is from an amethyst purple in tungsten light to a pale blue in daylight, which makes it easily distinguishable from the genuine article. A green synthetic spinel simulant has also been produced which approaches more closely to the true alexandrite colours.

In 1973, a true synthetic copy of chrysoberyl was introduced having the correct colour-change of the best quality Siberian alexandrite. Since then several synthetic alexandrites have appeared on the market. Although they are all much more expensive than the synthetic corundum version, they are still only a fraction of the cost of the natural gemstone, and present yet another identification challenge to the gemmologist.

Although alexandrite is perhaps the prime example of a colour-change gemstone, the effect, though rare, does sometimes occur with other natural stones such as corundum, spinel and garnet.

Interference colours

Colour can also be produced in a gemstone by optical effects rather than by its chemical composition. One of these optical effects is caused by the interference between rays reflected from the surface layers of a gemstone (Figure 8.5). If a ray of white light (I) meets a very thin transparent layer, it will be reflected from the top surface of this layer as well as from the lower surface. Both reflected rays (R_1, R_2) will be parallel with each other, but because the one that penetrated the layer has travelled further, it will be out of *phase* (i.e. out of step) with the other one.

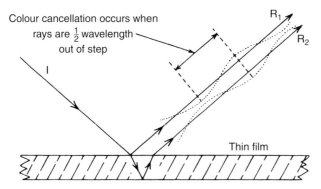

Figure 8.5 Diagram showing how colour is produced in a thin film by the mutual interference between reflected rays. A particular colour is cancelled out when the extra distance travelled by R_2 brings it out of step with R_1 by half a wavelength of that colour. The same colour is reinforced when the extra distance brings R_1 in step or phase with R_2

Depending on the thickness of the layer, at one particular wavelength or colour in the reflected light the two rays will be exactly half a wavelength out of phase with each other, and that colour will be cancelled out. The remaining components in the reflected light then combine to produce the complementary colour (as in selective absorption). At another wavelength the two rays may be exactly in step, or in phase, and this colour will be reinforced in the reflected light. The part that this light interference effect plays in the production of colour in gemstones such as opal, labradorite and moonstone is explained later in this chapter under the heading 'Sheen'.

Dispersion

Dispersion is yet another optical property, possessed in varying degrees by most gemstones, which can generate colour. White light passing through a material possessing dispersion has its individual spectral wavelengths refracted, or bent, by different amounts as they enter and leave the material at an angle other than 90°. The effect of this can be seen most clearly in a glass prism which splits the white light into its spectral colours (Figure 8.6). The violet end of the spectrum is refracted the most, and the red end the least. In a highly dispersive gemstone this results in the production of flashes of coloured light (known as '*fire*') when the gem is moved around under a light source.

The degree of dispersion possessed by a gemstone is generally related to the size of its refractive index (an optical property which will be dealt with in the next chapter) and in gemmology is usually measured as the difference in refractive index of a material at the B and G Fraunhofer wavelengths of 686.7 nm and 430.8 nm. The exception to the relationship between dispersion and refractive index occurs with diamond

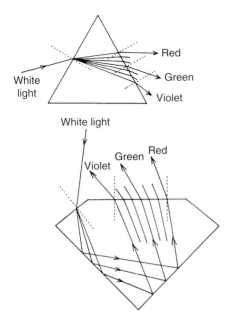

Figure 8.6 White light entering a prism (top) is dispersed into its spectral colours, the individual rays being refracted in varying degrees as they enter the prism and as they leave it. White light is similarly dispersed when it enters a polished gemstone (bottom), and after being totally internally reflected by the gem's pavilion facets the component colours emerge from the crown facets to produce 'fire'

whose high refractive index is only accompanied by a relatively modest degree of dispersion.

Although the prismatic colours produced by dispersion are seen most strikingly in colourless gems, they are also visible in the coloured stones demantoid, garnet and sphene, although masked to some extent by the body colour of these gems.

Colour centres

The colour in some gemstones is produced, or modified, by defects within the crystal lattice. The defect, which can be produced by natural or man-made irradiation, is the result of either an extra electron trapped at a site in the crystal lattice which it would not normally occupy, or an electron missing from a site it would normally occupy (leaving a 'hole'). The extra electron gives rise to an *electron colour centre*, and the missing electron gives rise to a *'hole' colour centre*.

Examples of gemstones whose tints are caused by colour centres include fluorite, quartz and those diamonds whose colour has been artificially altered by irradiation. It is also possible that many of the rare 'fancy coloured' diamonds owe their colour to colour centres produced by natural irradiation in the Earth. Natural zircon colours may be due to the lattice damage caused by irradiation from impurity atoms of uranium and thorium trapped within the gem.

The majority of naturally coloured yellow Cape series diamonds owe their colour to the presence of nitrogen atoms which replace carbon atoms in the crystal lattice. Rough green diamonds usually owe their colour to a thin surface 'skin', but the much rarer naturally coloured diamonds with a uniform green body colour (such as the Dresden Green) owe their fancy colour, as mentioned above, to natural radiation damage. The cause of colour in brown and most pink and mauve diamonds is thought to be due to plastic deformation of the crystals while forming in the Earth. This produces laminations running parallel to the diamond's cleavage planes. The colour of natural blue diamonds is the result of boron atoms similarly replacing carbon atoms. These colours, produced by impurities in the lattice, are the result of electron movement in the lattice rather than within the individual atom (as happens with the transition elements and in colour centre defects). They are explained by the *band theory*, which also covers semiconductor behaviour, but is outside the scope of this book.

Lustre

The lustre of a gemstone is the optical effect created by the reflectivity of the stone's *surface*. Lustre is directly related to the refractive index of a gem material, and although the lustre of some gemstones is visible in the rough, its full potential is usually revealed only when the stone is polished. Because gemstones cover a wide range of refractive indices from 1.43 to 3.32, they also exhibit different degrees of lustre. The descriptive terms in Table 8.2 have come into use to describe the most characteristic of these lustres.

Table 8.2

Metallic	The type of very high lustre associated with metals (e.g. gold, silver, platinum) and seen in hematite and some metallic compounds (e.g. pyrites, galena)
Adamantine	The high surface polish achieved with diamond (zircon and demantoid garnet are classified as 'sub-adamantine')
Vitreous	A glass-like lustre typical of the majority of gemstones (corundum, topaz, quartz)
Resinous	The more subdued polish as seen in amber
Waxy	The almost matt surface typical of turquoise and jadeite
Greasy	The appearance of soapstone and nephrite
Pearly	The lustre seen in mother-of-pearl
Silky	A fibrous lustre typical of satin spar

These adjectives are only intended as relatively broad descriptions of the surface appearance of a polished gemstone. However, in recent years, an instrument called a *reflectance meter* has made it possible to provide a comparative measurement of a stone's lustre and to use this as a means of identification. Details of this technique are given in the next chapter.

Sheen

While lustre is all to do with the *surface* reflectivity of a stone, *sheen* is the optical effect created by light rays reflected from *beneath* the surface of a gemstone. As with lustre, there are several terms (listed below) which are used to describe the various types of sheen exhibited by gemstones.

Chatoyancy

This is the 'cat's-eye', or band of light, effect caused by reflection from parallel groups of fibres, crystals or channels within the stone. In the case of pseudo-crocidolite, or tiger's-eye as it is better known, these channels are the fossilized remains of asbestos fibres which have been replaced by quartz (Figure 8.7). The finer and more highly

Figure 8.7 (left) An enlarged view of the parallel quartz channels just beneath the surface of a polished section of tiger's-eye. The bright chatoyant lines are running roughly at right angles to these channels. (right) A cabochon of tiger's-eye cut to show the chatoyant sheen effect

reflecting the fibres or channels, the brighter the chatoyant 'line'. Chatoyant stones are usually polished as cabochons (whose base is cut parallel to the plane of the fibres) to best reveal this effect. There are many chatoyant minerals (e.g. quartz, tourmaline), but the finest quality cat's-eye stone is the cymophane variety of chrysoberyl.

Asterism

This is the 'star' effect present in some rubies and sapphires (which are polished in the cabochon style to show the effect to best advantage). Like chatoyancy, the effect is due to fine parallel fibres or crystals, but in this case there are three sets of them lying along the crystal's lateral axes and intersecting each other at 60°.

In black star sapphire, the fibres are hematite needles formed parallel to the faces of the second order prism. In all other star corundums, the needles are of rutile formed parallel to the faces of the first-order prism. Some Thai star sapphires may contain both rutile and hematite needles producing a 12-ray star.

Although the best asterism occurs in corundum as a six-pointed star (Figure 8.8), it can also be seen occasionally in rose quartz where it is visible in *transmitted* rather than reflected light (an effect known as *diasterism*; reflected light asterism is called *epiasterism*). In diopside and some garnets it appears as a four-pointed star. In these stones there are only two sets of fibres, and these intersect each other at 90° for garnet, and 73° for diopside. Synthetic star rubies and sapphires have been produced, but with these stones the star effect is sharper and more obviously on the surface of the gem than with the natural stone. The synthetic stones are also a better colour and more transparent.

Iridescence

This is the 'play' of rainbow-coloured light caused by extremely thin layers or regular structures beneath the surface of a gemstone. Like a thin film of oil on water, these

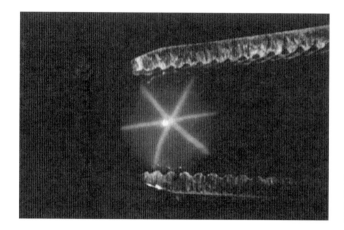

Figure 8.8 A cabochon of
star ruby cut to show the
sheen effect of asterism.
(R.V. Huddlestone)

layers interfere with the reflected light rays, reinforcing some colours and cancelling others (see Figure.8.5).

The effect is seen at its best in *precious* opal. Until the 1960s, the cause of colour in opal was a matter of speculation. Then researchers in the Australian CSIRO Division of Mineralogy and Geochemistry used an electron microscope to investigate the structure of the gemstone. They discovered that opal's play of colour is caused by millions of sub-microscopic spheres of cristobalite (silica gel) which make up the bulk of the stone. These spheres are all the same size (in precious opal) and are arranged in orderly rows and columns (Figure 8.9). Because of their small size, and the symmetry of their arrangement, they colour reflected light by a combination of interference and *diffraction* effects. This latter effect is produced when white light is split up into its

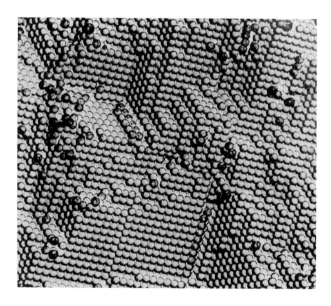

Figure 8.9 The orderly
arrangement of silica gel
(cristobalite) spheres in opal,
as revealed by the electron
microscope at the magnification
of 25 000×

spectral colours by being passed through a narrow aperture (as with the optical grating used in the *diffraction grating spectroscope*, which is described in Chapter 11).

The colours produced by precious opal depend somewhat on the angle of viewing, but mainly on the size of the spheres. An opal containing spheres of 300 nm diameter will reinforce light having a wavelength of up to twice this dimension (i.e. from red to violet), while one with spheres of only 200 nm will only reinforce colours at the blue/violet end of the spectrum. In common or potch opal, the spheres are of random size, and there is very little diffraction or reinforcement of the reflected light. This results in a milky opalescent effect which is almost completely lacking in colour.

Labradorescence

This is a particular form of iridescence which can be seen in the labradorite variety of feldspar and in spectrolite, a beautiful Finnish type of labradorite. In both cases the colour effect (much exploited in carvings – see Figure 8.10) is due to thin flakes of feldspar in the gem's surface layer which are the result of lamellar twinning.

Figure 8.10 Carving of an owl in labradolite using light interference effects caused by lamellar twinning in the material to generate colour on the bird's plumage. (E. Becker)

Adularescence

Also known as 'shiller', this is the bluish sheen seen in the moonstone variety of feldspar. It is yet another form of iridescence, and is also caused by lamellar twinning.

Colour, transparency and identification

The colour of a gemstone also has an effect on its transparency. Deeply coloured stones will pass less light than pale ones. Another factor which will influence transparency is the presence of internal flaws or inclusions (for this reason, chatoyant and star stones are usually not very transparent). In addition, the thicker the stone, the greater the loss of light passing through it. Because of this, a deeply coloured cabochon-cut stone (having no chatoyancy or asterism) is sometimes hollow-cut (i.e. the base is hollowed out to make the stone thinner).

Transparency is an important optical quality of a gemstone which affects both its beauty and its value. The various degrees of transparency, translucency and opacity are defined in Table 8.3.

Table 8.3

Transparent	An object viewed through the stone can be seen clearly (e.g. rock crystal, topaz)
Semi-transparent	The image of an object viewed through the stone will be blurred but still recognizable (e.g. amber, chalcedony)
Translucent	The stone will transmit some light, but objects cannot be seen through it (e.g. chrysoprase, jadeite)
Semi-translucent	Some light can still penetrate the stone, but only through the translucent edges (e.g. aventurine quartz)
Opaque	The stone is sufficiently dense optically to prevent the passage of any light (e.g. malachite, jasper)

Although the colour of a gemstone is probably its most important feature, and certainly has a big influence on its commercial value, it is not often of much use to a gemmologist when it comes to making an identification. There are of course the obvious exceptions, such as the bright grass-green of peridot, the purple of amethyst, the variegated green colour bands in malachite and the orange of fire opal.

With the transparent allochromatic gem minerals beryl, corundum, tourmaline and topaz, however, colour is much less useful as a distinguishing feature, as these gemstones crystallize in many different hues. In these stones, the colour depends entirely on which of the transition elements was present at the time the mineral was forming. In the case of tourmaline, crystals are sometimes found in which the colouring impurity changed during its growth. This results in a prismatic crystal whose colour may change either radially (as in 'watermelon' tourmaline), or lengthwise. As a result, several colours, including blue, green, pink and colourless, may be present in the same crystal.

Because of this unpredictability, it may be difficult on occasion to distinguish between allochromatic gemstones by colour alone. In earlier times, before the chemistry and characteristics of gemstones were fully understood, many stones were classified simply by their colour (red spinel for example was identified as 'balas' ruby). Evidence of this can still be seen in the British Crown Jewels in the Tower of London, where the Black Prince and Timur rubies are, in fact, red spinels.

With opaque gemstones, particularly the idiochromatic species malachite, turquoise, rhodonite and rhodochrosite, colour is a far more reliable identifying feature. Among other opaque gemstones which are easily recognizable by virtue of their colour and surface patterning are aventurine quartz, jasper, amazonite, and the varieties of tiger's-eye and chalcedony.

Chapter 9

Reflection and refraction

As we have seen in the previous chapter, the reflection of light plays an important part in the appearance of a gemstone. However, for identification purposes the most important single item of information about a gemstone is its *refractive index* (RI). This is because the RI of most gemstones is a constant which can be measured to two decimal places (and often estimated to the third decimal place). For this reason, many gems can be distinguished from each other with certainty even when there is very little difference in their RIs (e.g. natural and synthetic spinel; pink topaz and tourmaline).

Snell's laws of reflection and refraction

The Dutch scientist W. Snell was a professor at Leyden University in the seventeenth century. His two laws of reflection are very simple, and because reflected light and reflections are present in our everyday life they may also seem very obvious. However, they form part of the framework necessary for basic optics:

1. The angle of incidence of a light ray striking a flat reflecting surface is equal to its angle of reflection.
2. The incident ray, the reflected ray and the normal (at the point of incidence) all lie in the same plane.

The *incident ray* (Figure 9.1) is the light ray striking the reflecting surface, and its angle is the one made between the ray and a line drawn perpendicular (i.e. at 90°) to the reflecting surface. This perpendicular line is called the *normal*, and acts as a reference

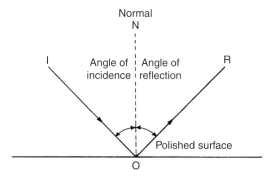

Figure 9.1 Snell's first law of reflection states that the angle of incidence equals the angle of reflection (ION)

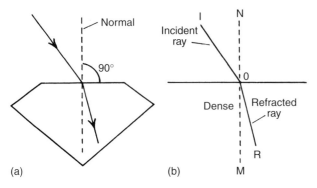

Figure 9.2 (a) The incident light entering a gemstone (at an angle other than 90° to the surface) is refracted towards the normal. Light leaving a gemstone (other than at 90°) will be refracted away from the normal. (b) If air is the less dense medium, the RI of the denser medium is the ratio of the sines of the angles ION and MOR

for the measurement of both the incident ray and the reflected ray angles. In Figure 9.1, angles ION and NOR are equal as stipulated in Snells' first law of reflection.

The *refraction* of a light ray is slightly more complicated, at least mathematically, and for many years it was not fully understood. Then in 1621 Snell discovered the underlying relationship between incident rays and refracted rays and laid the foundation for the subsequent rapid advances in applied optics. He expressed this relationship in his two laws of refraction:

1. When a light ray passes from one medium into another there exists a definite ratio between the sines of the angle of incidence and the angle of refraction. This ratio is dependent only on the two media and the wavelength of the light.
2. The incident ray, the refracted ray and the normal (at the point of incidence) are all in the same plane.

The word 'refraction' simply means angular bending. When a ray of light passes from one medium (such as air) into an optically denser medium (such as a gemstone), at an angle other than 90°, the ray is refracted or bent *towards* the normal (see Figure 9.2(a)). Conversely, when the ray leaves the gemstone and passes into the air, it is refracted *away* from the normal. The greater the difference between the optical densities of the two mediums (or, in the case of a gemstone surrounded by air, the greater the optical density of the gem), the greater will be the amount of refraction.

Refractive index

Using Snells' first law of refraction, we can derive a numerical index from the relationship between the angle of incidence in air and the angle of refraction in the material. Referring to the general case of an incident ray refracted from air into a denser medium (Figure 9.2(b)), the denser medium's refractive index (RI) can be calculated as follows:

$$RI = \frac{\text{Sine of angle ION}}{\text{Sine of angle MOR}}$$

Note: If ION and MOR are made into right-angled triangles by horizontally joining I to N and R to M, then the sine of the angle ION is equal to the ratio of the triangle sides IN/IO, and the sine of MOR is equal to the ratio of the triangular sides RM/RO.

Fortunately there are several more practical ways of arriving at a gemstone's refractive index than measuring angles and looking up sine values in trigonometrical tables!

Although only of academic interest to the gemmologist, there are two additional ways of defining refractive index, both of which again use air as the standard:

1. The ratio of the optical density of the gemstone to that of air.
2. The ratio of the velocity of light in air to the velocity of light in the gemstone.

Snells' first law of refraction also mentions the *wavelength* of light. Because refractive index varies with the wavelength of light, the standard light source chosen for gemmological work is the yellow *monochromatic* light produced by a sodium lamp (if red light is used, the measured refractive index will be slightly lower; for blue light it will be slightly higher). The term 'monochromatic' means that all of the emission energy of the light is confined to a single colour (i.e. a very narrow segment or 'bandwidth' of the spectrum). Sodium light consists of two very closely spaced emission lines whose mean value is 589.3 nm and whose overall bandwidth is only 0.6 nm. Sodium light was originally chosen as the standard because it was easily and cheaply produced by burning common salt in a bunsen flame. Today the sodium vapour lamp is no longer produced commercially for gemmological use because of its cost. However, as will be seen later, it had an important advantage over cheaper non-monochromatic alternatives.

Double refraction

So far we have been considering materials which have only *one* refractive index. Non-crystalline substances such as glass and amber, or gemstones belonging to the cubic crystal system, are in this category, and light entering these materials produces a *single* refracted ray as indicated in Figure 9.2. Such materials are called *isotropic* or *singly refractive*.

However, crystalline materials, including gemstones, belonging to the tetragonal, trigonal, hexagonal, orthorhombic, monoclinic and triclinic systems (i.e. all systems other than cubic) have *two* refractive indices. When a ray of light enters these materials it is split into *two* rays which are polarized at right-angles to each other (see Figure 9.3). These two polarized rays travel through the crystal at different speeds, and as with light of different wavelengths they are refracted by different amounts. Gemstones which produce two polarized rays are called *birefringent* or *anisotropic*. (The subject of polarized light will be dealt with at greater length in the next chapter).

One of the optical effects produced by birefringent materials is *double refraction*, the amount of which, as well as the value of each refractive index, provides valuable information when attempting to identify a stone. Gems with a high double refraction,

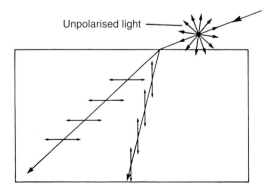

Unpolarised light

Figure 9.3 Unpolarized incident light (vibrating in all directions) is split into two plane polarized rays on entering a doubly refracting material

such as zircon, can often be identified by the obvious doubling of the image of the back facet edges when they are viewed through the main (table) facet with a hand lens.

The refractometer

The instrument in general use for the measurement of refractive index is the *refractometer* (Figure 9.4). This is also called the *critical angle refractometer* or the *TIR* (total internal reflection) *refractometer*, names which indicate its mode of operation.

To understand how this important class of instrument is used to measure refractive indices, we must first look at Figure 9.5. This indicates what happens to an incident ray of light travelling in a dense medium (such as a refractometer prism) when it meets a less dense medium (such as a gemstone) at varying angles of incidence (I_1 to I_5).

Ray I_1, which is inclined at a large angle to the normal, is reflected back into the denser medium from the interface between the two mediums. The reflected ray, R_1, obeys the laws of reflection (angle I_1,0,Nd = angle Nd,0,R_1) with the result that the incident ray I_1 undergoes *total internal reflection* in the denser medium.

As the angle of incidence is reduced, the rays continue to be reflected back into the denser medium (e.g. I_2/R_2). This total internal reflection of rays continues until the *critical angle* of reflection is reached (I_3,0,Nd). At this point the incident ray (I_3) ceases to obey the laws of reflection, and travels along the interface between the two mediums (R_3). As the incident angle is decreased still further, the rays (R_4, R_5) then obey the laws of refraction and pass into the rarer medium where they are refracted away from the normal Nr.

If the denser medium forms a component part of our refractometer (i.e. a glass prism), and the rarer medium is a gemstone, rays of light passing through the dense medium of the prism will be reflected back from the surface of the gemstone over an arc of incident angles *greater* than the critical angle, but will be refracted upwards into the gemstone at angles of incidence *less* than that of the critical angle.

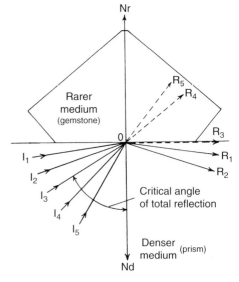

Figure 9.4 A typical critical angle refractometer, complete with built-in monochromator filter. (Gem-A Instruments Ltd)

Figure 9.5 At angles less than the critical angle, rays R_4, R_5 are refracted into the rarer medium

This critical angle is determined by the refractive indices of both the glass prism and the gemstone. As the prism is part of the refractometer, its RI can be taken as constant, and the critical angle thus gives a direct measure of the RI of the gemstone (i.e. the greater the critical angle, the greater the gemstone's refractive index). The relationship between the critical angle and the RI of the two mediums is as follows:

$$\text{Sine of critical angle} = \frac{\text{RI of rarer medium (gemstone)}}{\text{RI of denser medium (refractometer prism)}}$$

$$\text{RI of gemstone} = \text{sine of critical angle} \times \text{RI of refractometer prism}$$

The refractometer is designed optically to use the phenomenon of critical angle to provide direct readouts of refractive indices, but it can only do this if the RI of the gemstone being tested is *less* than that of the refractometer's glass prism (if the gemstone's RI equals or exceeds this value, all the incident rays will be refracted out into the gemstone). In practice, the maximum measurable RI is limited still further by the RI of the *contact fluid* necessary to make good optical contact between the gemstone and the prism (this feature will be discussed later).

The dense glass used in the refractometer is usually made from high lead-oxide content glass with a refractive index around 1.86. In early refractometers the prism was fashioned as a hemisphere, but in the 1930s this was modified to the present-day truncated prism design (see Chapter 1 under 'Highlights of the last 170 years').

The basic construction of the critical angle refractometer is shown in Figure 9.6. Light rays arriving at the interface between the gemstone and the glass prism, and having an angle of incidence less than the critical angle ION, are not reflected into the lens system. However, those rays having a greater angle than ION are reflected back into the lenses and illuminate a scale graduated in RI values (usually from 1.40 to 1.80 in 0.01 increments).

The image of the scale is inverted by a mirror or a prism and then focused with an eyepiece. The end result is viewed as a dark top section of the scale due to the light rays refracted out through the gemstone, and a bright lower section where the rays are reflected back from the gemstone's surface. The horizontal shadow edge between the two sections acts as a cursor or measurement line to indicate the refractive index of the gem on the scale.

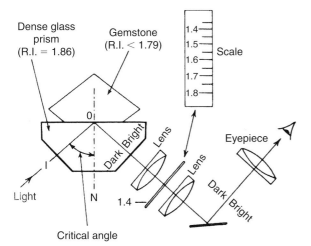

Figure 9.6 Sketch showing the optical design of a modern refractometer

Contact fluid

Because of the difficulty in obtaining a good optical contact between a gemstone's facet and the refractometer prism, use is made of a contact fluid having an RI of 1.79 for standard instruments. A small drop of this fluid is placed in the centre of the prism face, and the gemstone lowered onto it. The fluid effectively excludes any air from the gem/prism interface, and because of its high RI does not interfere with gemstone readings (a faint shadow edge, due to the fluid, can usually be seen at 1.79 on the scale and should *not* be mistaken for a gemstone reading!).

The original contact fluid devised by B.W. Anderson and C.J. Payne for the standard refractometer had an RI of 1.81 and consisted of a saturated solution of sulphur in di-iodomethane and tetraiodoethylene. However, as there are now concerns that some of the 'heavy' organic liquids used in gemmological tests may be carcinogenic (see Chapter 7 under 'Safety precautions') this contact fluid has been replaced with a saturated solution of sulphur in di-iodomethane (as mentioned above, this contact fluid has an RI of 1.79).

For extended range refractometers (described under 'Special refractometer versions' later in this chapter) use has to be made of special high-RI contact fluids, and all of these are **highly dangerous** or **toxic** and should only be used by experienced workers in laboratory conditions.

Sources of illumination

When using the refractometer, it is necessary with most models to provide a source of illumination (some instruments have a built-in light source – Figure 9.7). If the illumination used is a 'white' light (i.e. a filament lamp or daylight), then the shadow edge, as viewed on the scale, will not be sharp as it will consist of a narrow band of prismatic

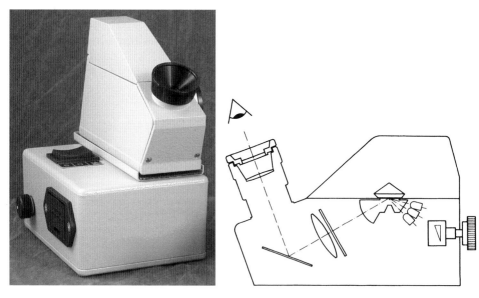

Figure 9.7 (left) A refractometer fitted with a yellow LED light source, which can be mains or battery operated. (Gem-A Instruments Ltd). (right) Sketch of the optical layout of a refractometer with built-in yellow LED illumination, which can be mains or battery operated. The control on the right varies the intensity of illumination. (Eickhorst)

colours (due to the dispersion of the glass prism). Refractive index for gemmological purposes is defined in terms of yellow monochromatic light having a wavelength of 589.3 nm (i.e. sodium light), and if a 'white' light source is used, the RI reading must be taken against the yellow/green boundary in the coloured shadow edge. Alternatively, all but the yellow portion of the shadow edge can be removed by placing a deep yellow filter over the refractometer eyepiece.

For accurate work it is best to use a monochromatic sodium light source, as this gives the sharpest and most easily seen shadow edge. Unfortunately, because of the limited market, a suitable sodium lamp unit has always been expensive, and is now no longer generally available from suppliers of gemmological equipment. For those who are pre-pared to improvise, a usable source of sodium light can be produced by simply placing some common salt (sodium chloride) on a piece of wire gauze and heating it in a weld-ing torch or butane flame. Alternatively, yellow-filter and interference-filter light sources are still commercially available, the latter having a near monochromatic bandwidth as small as 5 nm. Even less expensive, but with a bandwidth of 35 nm not truly monochro-matic, are the refractometer light units employing yellow high-intensity light-emitting diodes (LEDs) with an emission peak of around 590 nm (Figures 9.8 and 9.9).

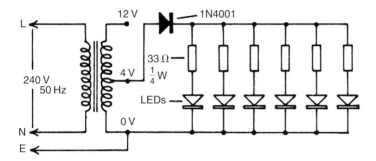

Figure 9.8 Circuit diagram of a typical mains-operated refractometer light source employing six yellow LEDs. The wavelength of the peak emission of yellow LEDs varies, the most suitable being about 590 nm (which is close to the 589.3 of a sodium lamp light source) and can be checked with the aid of a wavelength-calibrated spectroscope

Using the refractometer

As the refractometer prism is relatively soft, care should be taken to avoid scratching the contact surface. On no account should a diamond be knowingly placed on the prism. Even the application of the contact fluid must be made with care, particularly if this is done with a glass-rod type dropper, which should not be brought into actual contact with the prism.

The amount of contact fluid placed on the prism should be limited to a drop 2–3 mm in diameter. Too much fluid with a very small gem will cause it to 'float', while a gem with a large table facet will require a little more to ensure overall contact with the prism. When the test is completed, the fluid should be removed from both the prism and the gemstone. If the fluid is allowed to evaporate on the prism, its surface may be stained and crystals of sulphur will be deposited which may hinder further tests.

Once the refractometer and the source of illumination are set up ready for use (ideally in subdued lighting), first of all clean the stone to be tested. Then place a small drop of contact fluid in the centre of the refractometer prism and carefully place the stone

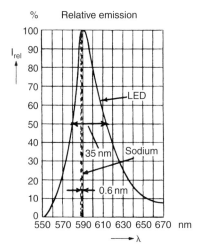

Figure 9.9 (left) Graph comparing the emission bandwidth of a typical LED light source with that of a sodium lamp. (right) A battery-operated Gem-A refractometer light source using a high-brightness LED with a peak emission at 590 nm. An adjusting screw under the front edge allows the light to be aligned with the light aperture of any refractometer

(table facet down) over the fluid (when removing the stone, slide it off the prism first to avoid scratching the glass surface). Check that the gemstone is sitting centrally on the prism to ensure best results. This is particularly relevant with a large stone, which may tilt slightly if overhanging the edge of the prism and produce a spurious reading.

Checking the shadow edge(s)

Close the lid of the refractometer to prevent stray light from entering the rear of the stone, and view the scale through the eyepiece (focus adjustment is usually provided by a pull-out eyepiece). If the RI of the stone is within the range of the refractometer (1.4 to 1.80), a horizontal shadow edge separating the darker top section of the scale from the lower brighter section (see Figure 9.10) should be visible. If the scale remains dark and the only shadow edge visible is that of the contact fluid at 1.79, it is probable that the RI of the gem is above the refractometer's range (this is termed a 'negative reading'). Anyone using the refractometer for the first time should also be aware that

Figure 9.10 (a) The single shadow edge produced by natural spinel (isotropic) at 1.715. (b) The twin shadow edges (in their maximum separation positions) produced by peridot (anisotropic) at 1.653 and 1.690. In both cases the shadow edge at 1.79 is caused by the contact fluid

the visibility of the shadow edge is dependent upon the size and condition of the facet in contact with the prism. Because of this a small or scratched gem will produce a fainter shadow edge than a large undamaged one.

If one or two valid shadow edges are visible, raise the refractometer lid slightly, and using the index finger of both hands rotate the stone while keeping it in the centre of the prism. If only a single shadow edge is visible and this remains at the same point on the scale during rotation of the gem, then the stone is singly refracting*. If the stone is doubly refracting, two shadow edges will either be immediately visible (Figure 9.10(b)) or a single shadow edge will resolve into two during rotation of the stone. In this case either one or both of the shadow edges will move as the stone is rotated.

Measuring double refraction (DR)

If the stone is doubly refracting (i.e. anisotropic), adjust its angular position in small incremental steps until the lower-RI shadow edge is at its lowest RI reading on the scale (ignoring the other shadow edge) and make a note of this value. Then rotate the stone again until the higher-RI shadow edge is at its highest RI reading (once again ignoring the other shadow edge) and record this value. A polarizing eyepiece filter is sometimes supplied with the refractometer and this can be used to eliminate first one shadow edge and then the other when rotating the stone. The value of the stone's DR is obtained by simply subtracting the lower reading from the higher one.

This method of checking the DR of a stone is recommended initially as the movement of the two shadow edges will vary – either one or the other edge will move, while the remaining one is stationary, or *both* edges will move independently (with experience, the double refraction of many stones can be read directly from the scale). The significance of the shadow edge movements will be explained in detail later under the heading 'Optic axes, optic sign and optical character'.

The twin shadow edges of a doubly refracting gemstone can normally be seen quite easily, but occasionally, where the double refraction is very small (e.g. less than 0.01), the separate edges may be difficult to distinguish, particularly if the refractometer light source is not monochromatic. In these circumstances, the polarizing eyepiece filter mentioned earlier can be used. By rotating the filter over the eyepiece the detection of twin shadow edges can be made easier as first one edge and then the other appears and disappears.

There are also some doubly refractive gemstones whose higher refractive index is above the range of the refractometer. This can give rise to some confusion as the single shadow edge gives the misleading impression that the gem is singly refracting; examples are rhodochroisite (1.58, 1.84), smithsonite (1.62, 1.85) and painite (1.727, 1.816). In these cases, reliance must be placed on the double-refraction evidence obtained with a polariscope, an instrument which will be described in the next chapter. On rare occasions, a gem containing more than one mineral (e.g. lapis lazuli) may produce two or more shadow edges on the refractometer due to the RIs of the constituent minerals, and this can be mistaken for double refraction.

Another anomaly seen in some green tourmalines is the presence of four shadow edges on the refractometer. This effect, discovered by Dr C.J. Kerez (and called the 'Kerez effect') is thought to be due to skin-deep alterations caused by local overheating during polishing. The two extra anomalous shadow edges disappear if the table facet is repolished.

---

* Note that polycrystalline and microcrystalline gems, whose individual crystal fibres or microcrystals are doubly refracting, will also show only a single shadow edge – see under jadeite, nephrite and chalcedony in Appendix C.

While on the subject of gemstone identification by refractive index measurement, it should be mentioned that if a stone is proved to be singly refracting and has an RI between 1.50 and 1.70, then it is almost certainly glass (or plastic) as there are no singly refracting *natural* gemstones within this range except for amber, jet and a few rare collector's stones.

The Dialdex refractometer

So far the description of the use of a refractometer has dealt with instruments having a built-in calibrated scale. One version of the standard refractometer developed by the Rayner Optical Company dispenses with the internal scale and instead is fitted with a calibrated control on the right-hand side of the instrument (Figure 9.11). Once a shadow edge is detected through the eyepiece, this control is rotated to bring a black shutter down one side of the scale until its bottom edge coincides with the shadow edge. The RI reading is then read from the calibrated scale on the control. This version, called the Dialex, was developed to facilitate the testing of cabochons by the 'distant vision' method (which will be described later).

Figure 9.11 The Dialdex refractometer with yellow and polarizing eyepiece attachments. (Gem-A Instruments Ltd)

Optic axes, optic sign and optical character

Optic axes

The optic axes in a crystal are directions of single refraction in an otherwise doubly-refracting material. Gemstones in the tetragonal, hexagonal and trigonal systems have one such axis (parallel to the vertical crystal axis) and are described as *uniaxial*. Gemstones in the orthorhombic, monoclinic and triclinic systems have two optic axes (neither of which is parallel to any crystal axis) and are termed *biaxial*.

When the refractive indices of a uniaxial gemstone are measured on a refractometer, only one of the two shadow edges moves when the gem is rotated, while the other remains stationary. The moving shadow edge is produced by the *extraordinary* ray, and the fixed shadow edge is produced by the *ordinary* ray. With biaxial stones, both shadow edges move in response to the stone's α (alpha-lower RI) and γ (gamma-higher RI) polarized rays. These shadow edge movements can therefore also help in identifying a gemstone by indicating to which group of crystal systems it belongs.

At this point it should be mentioned that there are some rare exceptions to the shadow edge movements with both uniaxial and biaxial stones. If the tested facet of a uniaxial stone happens to be cut exactly at right-angles to its optic axis, the shadow edge due to the extraordinary ray will remain stationary in its *full* double refraction spacing from the ordinary ray.

Similarly, it sometimes happens that one of the optic axes in a biaxial stone is at right-angles to the facet being tested. In this case one of the shadow edges will remain stationary when the stone is rotated.

However, despite these rare exceptions, the difference between the minimum and maximum readings on any facet is always a measure of the full double refraction for that stone.

The importance of refractive index and double refraction values in the identification of gemstones has already been mentioned. Sometimes another characteristic, called *optic sign*, becomes equally important in narrowing down the identification possibilities. The convention of designating crystals as optically negative or positive is based on the relative RI values and shadow edge movements measured on a critical angle refractometer.

Optic sign in a uniaxial stone

A uniaxial gemstone has a *positive* optical sign if the moving shadow edge (due to the extraordinary ray) has a higher RI reading than the fixed shadow edge due to the ordinary ray (see Figure 9.12 (left)). If the moving shadow edge has a lower RI reading than the fixed one, then the stone is optically *negative* (see Figure 9.12 (right)). On the rare occasion when a stone is cut with the tested facet exactly at right angles to an optic axis then both shadow edges will be stationary, and it will not be immediately obvious which one is due to extraordinary ray. In this case, the extraordinary ray can be identified by placing a polarizing filter in the north-south orientation (i.e. vertically polarized) over the eyepiece. This will filter out the ordinary ray leaving only the extraordinary ray visible, and from this the stone's optical sign can be deduced as indicated above. An alternative and perhaps simpler solution is to check the stone's shadow edges on another facet, thus avoiding the optic axis.

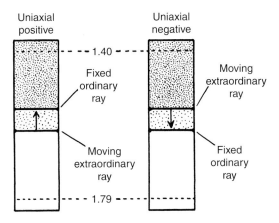

Uniaxial positive

Uniaxial negative

1.40

Fixed ordinary ray

Moving extraordinary ray

Moving extraordinary ray

Fixed ordinary ray

1.79

Figure 9.12 Sketch indicating the refractometer shadow edge movements which occur with a uniaxial gemstone with a positive optic sign (left) and one having a negative optic sign (right)

Optic sign in a biaxial stone

The situation with a biaxial stone is rather more complex as both shadow edges move when the stone is rotated on the refractometer. If the higher RI γ ray shadow edge moves *more* than halfway from its highest reading towards the lowest RI reading position of the other α shadow edge, the stone is optically *positive*. The lowest RI reading of the γ ray is designated β (see Figure 9.13 (left)).

However, if the higher RI γ ray shadow edge moves *less* than halfway from its highest RI towards the lowest RI reading of the α ray, then the stone is optically *negative*. This lowest RI reading of the γ ray is also designated β in Figure 9.13 (right).

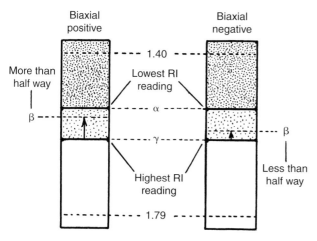

Figure 9.13 With biaxial gemstones both shadow edges move simultaneously when the gem is rotated on the refractometer. If the higher RI γ shadow edge moves more than halfway towards the lowest RI position of the α shadow edge, the optic sign is positive (left). If the γ shadow edge moves less than halfway, the optic sign is negative (right)

To simplify this assessment, it is advisable to use a polarizing filter over the refractometer eyepiece, and to rotate it until one shadow edge disappears. This will make it easier to concentrate on and to record the highest and lowest readings on the other edge. If this is done carefully with both shadow edges, and the highest and lowest readings are recorded in each case, subtraction of the lowest α reading from the highest γ reading will give the full value for double refraction and this can be recorded.

In the rare case of a biaxial stone where the facet being tested is cut at exactly right angles to the vibration directions of the γ or α rays, one or the other of these will appear as a stationary shadow edge, and it will be necessary to repeat the test on another facet in order to avoid these vibration directions.

The optic sign of any anisotropic gemstone may be indicated by placing a plus or minus sign in front of its value of double refraction. Alternatively, a gem's α, β and γ RI values may be quoted, and the optic sign must then be obtained by subtraction as previously indicated. While it may take a few minutes to determine the optic sign of a gem, it can occasionally be the only means of separating two gemstones having similar RIs and SGs. For example, peridot and sinhalite have RIs of 1.654, 1.690 and 1.670, 1.710 respectively (and similar SGs of 3.34 and 3.48). However, peridot has a positive optic sign, while sinhalite's is negative.

Optical character

The terms used to describe the specific optical character of a gemstone are shown in Table 9.1.

Table 9.1

Isotropic	Singly refracting. Non-crystalline, or belonging to the cubic crystal system.
Anisotropic	Doubly refracting. Crystalline, belonging to the tetragonal, trigonal, hexagonal, orthorhombic, monoclinic or triclinic crystal systems.
Uniaxial	One optic axis. Belonging to the tetragonal, trigonal or hexagonal crystal systems.
Biaxial	Two optic axes. Belonging to the orthorhombic, monoclinic or triclinic crystal systems.

Distant vision method

While it is possible to measure the RI of any faceted gemstone within the range of the refractometer (provided a flat facet of reasonable size is accessible), a problem exists if the stone has very small facets, or has been fashioned as a cabochon (although sometimes a reading can be obtained from the base of a cabochon if this is flat). To overcome this problem, L.B. Benson Jr devised a technique which is called the 'distant vision' method in the UK, and the 'spot' method in the USA.

The method varies slightly with the type of refractometer, but for the Gem-A/Rayner instruments consists first of coupling the rounded surface of the cabochon to the refractometer prism with the smallest possible spot of contact fluid. This is best done by placing a drop of the fluid on a flat surface (e.g. the metal plate surrounding the prism), and then lightly touching the drop with the central point on the curved surface of the cabochon, which will then pick up the necessary minimum quantity. In order to pick up the fluid in the centre of the cabochon 'dome', the stone can be positioned more accurately over the spot of fluid if it is temporarily held in tweezers (in particular the 'prong' type). The cabochon is then placed in the centre of the refractometer prism with the spot of fluid acting as an optical coupler.

Next, the refractometer scale is viewed with the eye positioned in line with the eyepiece, but 12–18 inches (30–45 cm) away from it. By carefully adjusting the viewing position it should be possible to see a small 'bubble' superimposed on the limited section of the scale now visible. This bubble is the spot of liquid coupling the surface of the cabochon to the prism. If the eye is now moved slowly up and down in a vertical direction, the bubble will be seen to change from dark to light (Figure 9.14).

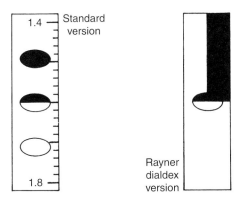

Figure 9.14 When measuring the RI of a cabochon using the distant vision method, the reading is taken against the dark/light 'bubble' of contact fluid (as seen through the refractometer eyepiece at a distance of 12–18 inches (30–45 cm))

When the bubble is dark, this will correspond to a scale reading below the RI of the stone, and when it is light, the scale reading will be higher than the RI of the stone. If the line of vision is adjusted so that the bubble is divided exactly into a light and a dark half, this will correspond to the scale reading for the RI of the cabochon (usually about 0.01 less).

Unfortunately it often happens that when the eye is focused for the best observation of the bubble, the scale is out of focus. The Dialdex refractometer (described earlier in the last paragraph under 'Using the refractometer') solves this problem by dispensing with the scale. For instruments with a built-in calibrated scale, a lens attachment was produced for the refractometer eyepiece which helped to bring both the scale and the bubble into focus. As this is no longer available, an alternative solution is to make a pinhole aperture in a piece of card. If this is placed close to the eye and the bubble and scale viewed through it, this also reduces the problem by increasing the depth of focus; unfortunately, it does not work so well if glasses are worn. Initial attempts at using the distant vision method are best carried out with a shallow cabochon of quartz or chalcedony.

Special refractometer versions

Two variants of the standard glass prism refractometer (which are now no longer pro-duced) were devised by Anderson and Payne. In one of these, the glass prism was replaced by a spinel one which gave a more open scale between 1.30 and 1.68, and had the advantage of an increase in hardness over the glass prism. However, the main rea-son for using spinel was that its dispersion more nearly matched that of the majority of gemstones. This produced a sharper shadow edge when using a white light source than does the standard model. The other variant used blende (with an RI of 2.37) in place of glass to extend the high-reading end of the scale.

A third variant made use of the high refractive index of diamond (2.417) to extend its range and was also proposed by Anderson and Payne. A diamond prism (using glass prisms at each end to transmit the incident and reflected light) was fashioned for the first experimental model from a 6.632 carat 'silver Cape' stone donated by the Diamond Corporation. The weight of the finished prism was 2.505 carats.

Subsequently, a number of diamond-prism refractometers were made to order by the Rayner Optical Company, the weight of the diamond prism being reduced to 1.54 carats to lower the cost. In use, the instrument (now no longer being made) could meas-ure refractive indices from 1.55 to 2.03, the wide range necessitating a sliding eyepiece to cover the length of the scale. In addition to its wide range, the other benefits gained with the diamond refractometer were the optical purity and hardness of the prism, the latter feature enabling it to take a higher polish and to resist abrasion.

Subsequently, other extended range refractometers were developed using strontium titanate (RI of 2.42) and cubic zirconium oxide (2.18) prisms. Unfortunately, to take advantage of the extended range, all of these versions had to use a similarly high-RI contact fluid. One such liquid, formulated for the diamond refractometer, was called West's solution and had an RI of 2.05. It consisted of an 8:1:1 mixture of yellow phos-phorous and sulphur in di-iodomethane. Because of the presence of phosphorous in the mixture, the dried residue was spontaneously combustible, and the liquid had to be handled with care. Equally dangerous are the alternatives – a solution of selenium brom-ide in di-iodomethane, and Cargille refractive index liquids which contain arsenic salts. Because of the high dispersion of these high-RI prisms, it is necessary with all versions to use a sodium light source in order to produce a sharp shadow edge.

Another interesting version of an extended-range refractometer used a strontium titanate prism, and was produced by the Krüss company of Hamburg, Germany (Figure 9.15). This employed a special thermoplastic contact paste (with an RI of 2.22) as a coupling medium. The paste had to be heated to 40°C in a special socket in the instrument's control plinth to melt it. In order to maintain the paste in its liquid state on the prism, this too had to be electrically heated. Because of the very high dispersion of strontium titanate, the control plinth was fitted with a built-in sodium lamp. Sadly, the contact paste, in common with the earlier high-RI contact fluids, also proved to be toxic as it was found to contain arsenic salts.

Figure 9.15 The Krüss high-range refractometer used a strontium titanate prism and a heated thermoplastic paste as a contact fluid. Because of the high dispersion of strontium titanate the refractometer's control plinth contained a sodium light source

Fresnel's reflectivity equation

The lustre, reflectivity or reflectance of a gemstone can be described qualitatively as adamantine, vitreous, resinous, etc. However, it can also be measured in *absolute* terms as the ratio between the intensity of the reflected ray and that of the incident ray:

$$\text{Reflectivity} = \frac{\text{Intensity of reflected ray}}{\text{Intensity of incident ray}}$$

The degree of lustre or reflectivity of a gem (assuming a 'perfect' polish) is due mainly to its refractive index, but is modified by other factors such as its molecular structure and transparency. A simplified equation which relates a transparent *isotropic* mineral's reflectivity in air to its refractive index was formulated by the French physicist Fresnel, and assumes the ideal case where both the incident and reflected rays are normal (i.e. perpendicular) to the reflecting surface:

$$\text{Reflectivity} = \frac{(n - A)^2}{(n + A)^2}$$

where n is the refractive index of the material, and A is the refractive index of the surrounding medium (for air = 1).

Despite the qualifications, Fresnel's equation works well enough for anisotropic materials, and for angles of incidence and reflectance up to 10° to the normal.

If the results of the equation are multiplied by 100, this gives the percentage of incident light which is reflected back from a gemstone's surface. Substituting $n = 2.417$ for diamond (and $A = 1$ for air) produces a reflectivity figure for diamond of 17%. Substituting $n = 1.54$ for quartz indicates a surface reflectivity for this mineral of only 4.5%.

It is interesting to note that if the value for A in the equation is increased above 1, the reflectivity of the gemstone *decreases*. This is the reason why gemstones are often immersed in a high RI liquid when inspecting them under a microscope. The immersion technique *reduces* the amount of light reflected back from the surface of the stone and thus enables the light to enter and illuminate the interior. For the same reason, a colourless stone may virtually disappear during a heavy liquid SG test if its RI is close to that of the liquid!

The reflectance meter

1975 saw the introduction of the first commercial reflectance meters designed for gemstone identification. Perhaps the most successful of these early instruments was the 'Jeweler's Eye' (Figure 9.16). Although subsequently many instruments of this type appeared with analogue-type meters calibrated directly in gemstone names, one of the first models to have its scale calibrated in RI values was the 'Gemeter 75' (see Figure 1.5). The implication that reflectance meters could provide the same precision of reading as the critical angle refractometer was probably the main reason why the performance of these instruments caused some initial disappointment. Further development produced more reliable versions including a reflectance instrument called the 'Jemeter Digital 90' (see Figure 9.17). This, with care, proved capable of measuring refractive indices to an accuracy of 0.008 and displaying them digitally to three decimal places over the range 1.450–2.999. Using a built-in polarizing filter it could also resolve double refraction down to 0.01. With this specification, such an instrument could have been be classed as a 'reflectance refractometer', although some academics would probably have objected to this!

In general, gemstone reflectance meters are designed to indicate *differences* in reflectivity between polished gemstones rather than *absolute* values, and are usually calibrated

Figure 9.16 The Hanneman 'Jeweler's Eye' reflectance meter has two ranges, the top one is calibrated for diamond and its simulants (strontium titanate, diamond and rutile are the highest readings on the scale)

Figure 9.17 The 'Jemeter Digital 90' reflectance instrument measures gemstone RIs to three decimal figures over the range 1.450–2.999. Fitted with a polarizing filter, it can also detect double refraction down to 0.01

with diamond as the standard. With twin-range instruments, synthetic spinel is often chosen to set the calibration of the lower range.

One common denominator of all reflectance meters is the use of miniature solid-state infrared light-emitting diodes (LEDs), which form a conveniently compact and efficient source of incident light. A photodiode is mounted alongside the LED, and this is used to detect the amount of infrared energy reflected back from the flat surface of the gemstone under test, and to display this on a suitably calibrated meter (see the circuit diagram in Figure 9.18).

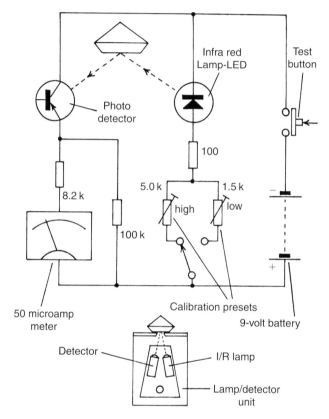

Figure 9.18 Circuit diagram of the 'Jeweler's Eye'. The positions of the infrared LED lamp and the photo detector are shown in the sketch below the circuit diagram

The intensity of the incident beam is assumed to be constant, and the intensity of the reflected beam, although not an absolute measurement of reflectivity, is used to indicate the reflectance of the gemstone relative to the calibration standard.

Because it is physically difficult to have both the incident beam and the reflected beam perpendicular to the gemstone's surface, the angles of incidence and reflection are each offset by 10° to the normal. For this reason, and because factors such as double refraction, absorption and surface finish are not taken into account in Fresnel's simplified formula, the relationship between reflectance and refractive index may not be precisely as indicated in the formula.

In addition, the use of infrared light having a wavelength of around 930 nm for the incident beam can result in misleading readings for highly dispersive stones. This is because RI values are quoted in terms of yellow monochromatic light at 589.3 nm, and the RI (and therefore the reflectance) of a gemstone having a high dispersion is much lower at 930 nm than 589.3 nm. The effect of this 'dispersion error' can be seen in the gap that exists between strontium titanate and diamond on the scale of a reflectance meter (Figure 9.16). Although both stones have almost identical RI values, the dispersion of strontium titanate is more than four times that of diamond. In this case the difference is useful as it enables the two materials to be distinguished from each other.

Because the reflectance meter is, in effect, measuring the lustre of a gemstone, anything that reduces that lustre, such as dirt, grease or surface scratches, will produce a misleadingly low reading. It is therefore important to check that the surface of the gemstone is in good condition and is thoroughly clean before making a test (it is also important to keep the test aperture free from dust). If extraneous light enters the back of the gemstone under test, this can also produce incorrect readings as it may increase the amount of light reaching the photodiode. To prevent this, most reflectance meters are provided with an opaque cap which must be placed over the stone when making a test. Some instruments use a *pulsed* source of infrared incident light to make the instrument impervious to uncoded external light (see Figure 9.19 and the block diagram in Figure 9.20).

Figure 9.19 A single-range reflectance instrument whose meter is calibrated for diamond and its simulants. It uses pulse-coded infrared illumination and detection to minimize errors caused by ambient light

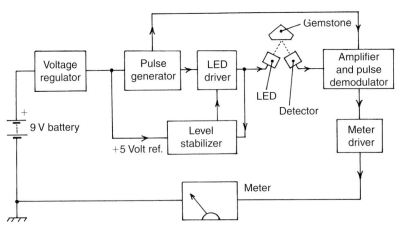

Figure 9.20 Block diagram of a reflectance meter using pulse-coded infrared illumination and detection

Occasionally, the cut of the stone may result in a high reading due to stray internal reflections from the pavilion facets. To avoid being misled by spurious readings of this type, several readings should be taken over the area of the table facet (which must necessarily be large enough to cover the test aperture completely) and any unusually high or low readings disregarded.

Reflectance meters are particularly useful for checking gemstones whose refractive indices are too high to be measured on the standard refractometer. Used in this way they enable diamond to be separated quite easily from its many natural and synthetic simulants. In fact the technique is far better employed for this purpose than in covering the lower RI gemstone range, where overlaps in reflectance values can sometimes cause misidentification.

In this latter context, the critical angle refractometer is superior to the reflectance meter in the accuracy of its readings, and in the extra information which can be extracted from them (i.e. double refraction, optical character, optic sign). However, the gap is beginning to narrow, for example, the sensitivity of some reflectance meters has made it possible to detect the difference in lustre between hand-polished gems (such as sapphires and rubies) and their machine-cut synthetic counterparts.

Other methods of RI measurement

Approximation by immersion

As previously mentioned, when a colourless transparent gemstone is immersed in a liquid having an RI close to that of the gem, it virtually disappears. Even if the gemstone is coloured, its facet outlines (or its shape) will become indistinct. This provides perhaps the simplest method of approximating a stone's RI, and consists of inserting the gem in a series of small containers containing liquids of various known refractive indices. The RI of the stone being tested will be nearest to that of the liquid in which the stone's outline appears most hazy.

As with SG determinations using heavy liquids, this method should not be used with gemstones having a *porous* surface (i.e. opal and turquoise) or with substances which might be soluble in the test liquid! Suitable immersion fluids and their RIs are shown in Table 9.2.

Table 9.2

Water	1.33	Bromoform	1.59
Alcohol	1.36	Iodobenzene	1.62
Petrol	1.45	1-bromonaphthlene	1.66
Benzene	1.50	Iodonaphthalene	1.70
Clove oil	1.54	Di-iodomethane	1.74
		Refractometer contact liquid	1.79

Approximation by Becke line method

Although this technique was originally developed for the measurement of powdered samples of unknown materials (or inclusions), it was subsequently adapted by R.K. Mitchell for use with faceted gemstones. The method requires the use of a microscope having light-field illumination (i.e. transmitted light), an iris aperture adjustment and a magnification factor of between 30$\times$ and 40$\times$.

The gemstone under test is placed, table facet down, in an immersion cell (Figure 9.21) containing a liquid of known RI. The immersion cell is placed on the microscope stage so that the pavilion edges are visible, and the iris control closed so that the light is restricted to the area of the gemstone.

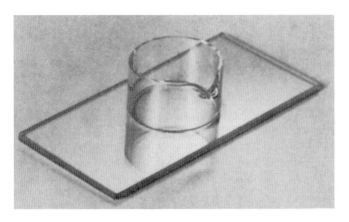

Figure 9.21 An immersion cell used for Becke line estimation of RI

The microscope is then focused down from the liquid into the body of the stone. If the facet edges change in appearance from *light* to *dark* as the microscope is focused into the stone, then the RI of the gemstone is *greater* than that of the liquid (see Figure 9.22). However, if the opposite occurs, and the facet edges change from *dark* to *light* then the RI of the stone is *less* than that of the liquid. By progressively changing the liquid in the immersion cell for one of a higher or lower RI, a close approximation to the gemstone's RI can be obtained. In some cases the SG of the stone will be less than that of the test liquid, and it will be necessary to weight the stone down while making the test.

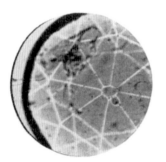 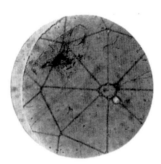

Figure 9.22 Photographs of a ruby (RI 1.77) immersed in di-iodomethane (RI 1.74). (left) With the microscope focus raised just above the stone, the facet edges appear light. (right) With the focus lowered into the stone, the facet edges appear dark. (Courtesy of R.K. Mitchell)

Direct method of measurement

This method can only be used on a transparent polished gemstone, and depends on the availability of a microscope having a calibrated focus adjustment or a vernier height scale (alternatively a dial gauge or an electronic gauge can be fitted to the microscope – Figure 9.23).

The microscope is first used to measure the *apparent depth* of the gemstone under test, and is then used to measure its *real depth*. The refractive index of the stone can then be calculated by dividing the real depth by the apparent depth (in a doubly

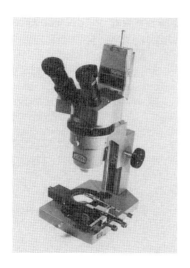

Figure 9.23 A Krüss microscope fitted (at top rear) with an electronic gauge for RI measurement by the 'direct' method

refracting gemstone this will be the RI due to the *ordinary ray* – see earlier in this chapter under 'Optic axes, optic sign and optical character').

Measurement of a gem's RI using the direct method is limited to an accuracy of plus or minus 1%, but has the advantage that it can be used to determine the refractive index of high RI stones such as diamond and zircon. Unlike the use of the refractometer and the reflectance meter, it is also independent of the quality and flatness of the stone's surface finish.

When using the method, the gemstone is positioned on the microscope with its culet (i.e. the tip of its pavilion) in contact with the stage, and with its table facet parallel with the stage. A small piece of plasticine or 'blue tac' can be used to secure the stone in this position. Using maximum magnification (to obtain a shallow depth of focus), the microscope is carefully focused on the surface of the table facet, and the position of the focus setting read from the scale ((A) in Figure 9.24). The microscope is then focused down through the stone until the culet is sharply defined, and a second reading is taken (B). If the second reading is subtracted from the first, the result will be the apparent depth of the stone.

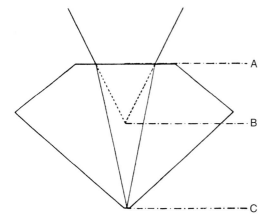

Figure 9.24 Sketch showing the three focus positions A, B and C used for determining the RI of a gemstone by the 'direct' method

The real depth can be arrived at by moving the stone to one side and focusing the microscope on the surface of the stage. If this third reading (C) is subtracted from the first, the result will be the real depth of the stone (alternatively, this can be measured separately with a Leveridge type gauge). The RI of the stone is then obtained by dividing the real depth by the apparent depth.

$$RI = \frac{A - C}{A - B}$$

Measurement using a table spectrometer

Also called a goniometer, the table spectrometer (Figure 9.25) consists basically of a central specimen table, a radially pivotable telescope and a fixed collimator/light source (used for producing a narrow beam of parallel light rays). The telescope is fitted with an eyepiece having a cross wire graticule, and its angular position relative to the collimator can be read from a scale on the specimen table.

Figure 9.25 A student-type table spectrometer made by the Krüss Company

Provided that the specimen to be measured has suitable surfaces or facets which can serve as the two faces of a prism, its RI can be determined to better than three places of decimals. This is done by first measuring the angle between the two chosen facets or faces, and then measuring the *angle of minimum deviation* of the prism formed by these faces. As long as the angle between the faces is not greater than twice the critical angle of the gemstone, there is no upper limit to the refractive index that can be measured by this method.

The angle between the two selected prism faces is measured as follows.

First, adjust the collimator for as fine a slit as possible by focusing it on the telescope cross wires. Then position the gemstone specimen in the centre of the table so that its prism facet edges are exactly vertical, and so that the light from the collimator falls across the adjacent faces of the prism whose angle is to be measured. Pivot the telescope round until the image of the collimator slit, reflected from one of the prism faces, is centred on the telescope cross wires, and make a note of the angular reading on the table scale (V). Then rotate the telescope to view the image of the collimator slit reflected from the other prism face, again noting the scale reading (W). The prism angle (A) is equal to half the difference between the two scale readings.

$$\text{Prism angle } A \; = \; \frac{V - W}{2}$$

To measure the prism's angle of minimum deviation, first remove the gemstone specimen from the table, rotate the telescope until the collimator slit is centered in its cross wires, and take its angular position reading from the scale (X). Replace the gemstone on the table and position it so that it receives the light from the collimator on only one of the selected prism faces. This light will be refracted by the prism. Readjust the position of the telescope to receive the refracted image of the collimator slit. If a white rather than a monochromatic light source is used, the image of the slit will be seen to be dispersed into a spectrum of colour, and for the following steps the cross wires of the telescope must be aligned on the red part of this spectrum (this being more easily visible).

The next operation is to find an angular position for the prism that produces the smallest angle of deviation between the incident light from the collimator and the refracted light seen through the telescope. To do this, look through the telescope, and rotate the gem about its vertical axis so that the refracted slit image or spectrum moves towards the line of incident light emerging from the collimator; follow this image round by rotating the telescope. A point will be reached where the image will appear to stop and then to reverse its direction of travel. In the position where the image just stops, adjust the telescope position so that the cross wires again coincide with the red section of the spectrum. Read the telescope's angular position on the scale (Y).

The angle of minimum deviation is obtained by subtracting the in-line reading (X) from the refracted image reading (Y).

$$\text{The angle of minimum deviation} = Y - X = B$$

$$\text{The gemstone's RI} \; = \; \frac{\sin(A + B)/2}{\sin A/2}$$

Because there is usually only one suitable orientation of the gemstone on the spectrometer table, with uniaxial specimens it is normally only possible to obtain an RI for the ordinary ray. With biaxial stones, the RI reading may be anywhere between that of the α and γ rays, as these both vary with orientation. For isotropic minerals, of course, no such problem exists.

Although the measurement of a mineral's RI by this method is time consuming and is only suitable for specimens of a reasonable size, its inherent accuracy and the fact that its range is unlimited makes it the first choice for newly discovered species. It has the added advantage of being able to measure the dispersion of a specimen by using a monochromator (or suitable interference filters) to evaluate the RIs at the B and G Fraunhofer wavelengths of 686.7 nm and 430.8 nm.

The student-type spectrometer shown in Figure 9.25 is generally adequate for gemmological purposes. The more elaborate modern version of the instrument shown in Figure 9.26 is designed for analytical spectroscopic work as well as the measurement of refractive index and dispersion.

Measurement by the Brewster angle of polarization

Brewster's law states that when monochromatic light meets the flat surface of an optically denser medium, the reflected ray becomes polarized in the horizontal plane of that

Figure 9.26 A more modern version of the goniometer. The addition of a wavelength scale and a gemstone holder (far right) enables the instrument to be used as a spectroscope. The light source in the base of the unit is channelled to the gemstone holder via a glassfibre light guide. (Krüss)

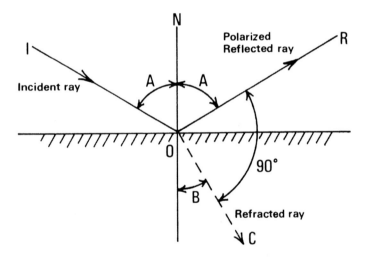

Figure 9.27 At the Brewster angle A, only horizontally polarized rays are reflected from the surface of a denser medium. This occurs when the reflected ray is at right-angles to the refracted ray

surface when its angle is normal (i.e. perpendicular) to the associated refracted ray in that medium (Figure 9.27). This is why vertically polarized sunglasses are used to reduce the glare from horizontal surfaces.

If the Brewster angle of polarization is A, then (as RI = sine A/sine B, and $A + B = 90°$) the RI of the reflecting medium is equal to tan A, and this provides an interesting additional method of arriving at a gemstone's RI.

A battery-operated portable Brewster-angle meter (developed by the author and marketed by GAGTL) is illustrated in Figures 9.28, 9.29 and 9.30. The instrument uses a 5 milliwatt 670 nm vertically polarized laser as a light source. To measure its Brewster angle a gemstone is placed over a test aperture, and a control knob calibrated in

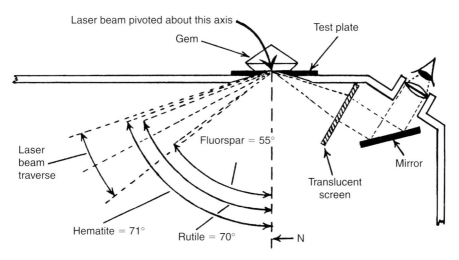

Figure 9.28 Diagram showing the use of the Brewster-angle of polarization to measure a gemstone's RI. The laser beam is vertically polarized, and at the Brewster angle the reflected light imaged on the translucent screen drops to a minimum

Figure 9.29 A Brewster-angle meter (developed by the author for Gem-A Instruments Ltd). The control is used to adjust the angle of the laser beam for minimum reflected light from the gemstone under test. The stone's Brewster angle is read from the control's scale and the gemstone identified from a set of tables.

degrees is used to rotate the vertically polarized laser beam around the test aperture (Figure 9.30). As the beam approaches the gem's Brewster angle, the intensity of the reflected ray as viewed on a translucent screen reduces to a null (the red laser light is vertically polarized and only horizontally polarized light is reflected from the gem's

Figure 9.30 View of a gemstone on the test platform. The laser beam can only be activated when the circular test platform cap is in position and the knurled test button on the right of the cap is pressed

surface at the Brewster angle). At the precise Brewster angle this null is seen as a dark horizontal bar across the translucent screen. This angle is read from the control scale and the gemstone is identified from a set of tables which take into account the laser wavelength and the dispersion of the gem.

Although only capable of an accuracy of around ±0.01, the Brewster-angle meter has the advantage over the critical angle refractometer of a wide RI range (1.40 to 3.2) without the need for a contact fluid. This wide range encompasses diamond and all its simulants including synthetic moissanite at 2.65, 2.69. Unlike the reflectance meter it is not oversensitive to the surface condition of the stone (which causes a diminution of the null reading rather than an error in RI). By rotating a gemstone on the test aperture, double refraction can be measured providing this is greater than 0.01 in value.

Chapter 10

Polarization and pleochroism

In Chapter 8, light was described as either a form of energy radiated as a wave motion (according to Maxwell's theory), or as a large number of travelling wavelets or photons (according to Einstein's quantum theory). Leaving aside the finer academic points of propagation, we now must look at the way in which light waves *vibrate* during propagation.

The polarization of light

Polarized light has already been mentioned briefly when discussing double refraction in Chapter 9. Ordinary unpolarized light waves vibrate in *all* directions at right angles to their line of travel (Figure 10.1). However, if unpolarized light passes through a doubly refracting material (such as a gemstone) it emerges as two separate polarized rays. These rays now vibrate only in a *single* plane at right angles to each other and to their direction of travel. If we can devise a method which allows us to separate out one of these rays we will have produced a source of *plane polarized light*.

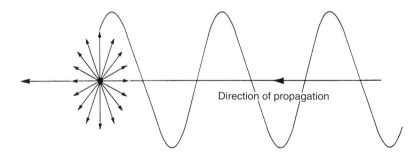

Figure 10.1 An unpolarized ray vibrates in all directions at right angles to its direction of propagation

Polarizing filters

Before the development of present-day polarizing filters, one of the ways of producing plane polarized light was to pass unpolarized light through a suitably cut section of a doubly refracting mineral. In 1813, T.J. Seebeck made a polarizing filter by cutting a section from a brown tourmaline crystal parallel to its length (i.e. along its *c* axis).

Light entering this section produced two plane polarized rays, one of which (the ordinary ray) was absorbed by the gem, while the other one (the extraordinary ray) emerged as plane polarized light. The disadvantage of this method was that the polarized light was much reduced in intensity due to the colour of the tourmaline.

Later, in 1828, W. Nicol discovered that if a rhomb of optically clear calcite (known as Iceland spar) was cut diagonally and the two sections cemented together again with 'Canada balsam' (a tree resin), this could be used to produce polarized light. The modified rhomb was subsequently known as a *Nicol prism* and for many years provided the principal means of producing polarized light.

The way in which the Nicol prism works is quite ingenious, and, like the refractometer, is based on the critical angle of total reflection. Calcite, with RIs of 1.66 and 1.49, has a very large double refraction (0.17), and light passing into the rhomb splits into two widely-divergent plane polarized rays. The RI of the balsam layer (1.54) is such that it produces total internal *reflection* of the more divergent ordinary ray (responsible for calcite's RI of 1.66) because this ray meets the layer outside the critical angle resulting from these two RIs. The extraordinary ray is responsible for calcite's RI of 1.49, which is *less* than the RI of the balsam, and this ray is therefore *refracted* into the layer and passes on through the rhomb (Figure 10.2). The unwanted ordinary ray is normally absorbed by the black coating around the sides of the calcite rhomb.

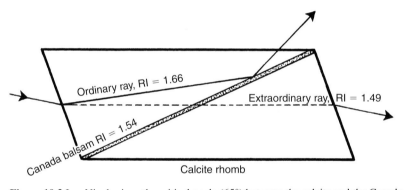

Figure 10.2 In a Nicol prism, the critical angle (65°) between the calcite and the Canada balsam layer causes the ordinary ray to be reflected back from the layer. As calcite's RI for the extraordinary ray is less than that of the Canada balsam, this ray is refracted through the layer into the right hand half of the rhomb, from which it emerges as a source of plane polarized light

Yet another early method of producing polarized light used the phenomenon of the *Brewster angle of polarization* mentioned at the end of the previous chapter. At the Brewster angle for a given material, the reflected light becomes polarized in the plane of the material's flat surface. This is why unpolarized daylight becomes partly polarized in the horizontal plane when it is reflected from a relatively flat horizontal surface such as the sea. Sunglasses, fitted with vertically polarized filters, are used to reduce the glare from horizontal surfaces, and this brings us to the use of such filters for gemmological purposes.

The majority of polarizing filters today consist of a plastic sheet containing either microscopic crystals of quinine idosulphate or, more recently, 'long' molecules. These crystals and molecules are orientated so that they transmit light with minimum attenuation only when it is vibrating in one plane, and progressively absorb those rays which are polarized at increasing angles to this plane. Rays at 90° to this plane experience

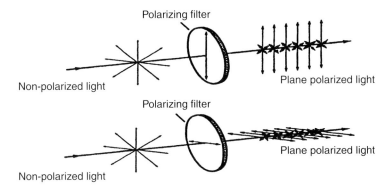

Polarizing filter

Non-polarized light

Plane polarized light

Polarizing filter

Plane polarized light

Non-polarized light

Figure 10.3 Sketch showing how a polarizing filter transmits with minimum attenuation only those rays in unpolarized light that are aligned with the filters plane of polarization

maximum attenuation (Figure 10.3). Plastic filters of this type are often protected by being sealed between plates of glass. In use they imitate the function of the Nicol prism, but are more compact and less expensive.

The polariscope

Perhaps one of the most important applications of the polarizing filter occurs in gemmology when two such filters are combined in an instrument called a polariscope (Figure 10.4). The way in which the polariscope works can best be seen in the diagram in Figure 10.5. The light from an unpolarized light source is first passed through a filter (A) which produces polarized light vibrating mainly in one specific plane. If a second filter (B) is introduced and rotated so that its plane of polarization coincides exactly with that of the first filter, then the polarized light from the first filter will be able to pass through the second filter. However, if the second filter (C) is now rotated by exactly 90° very little light will emerge, and the filters are said to be in the *crossed* or *extinction* position. At any other angle the light transmitted from the second filter will be somewhere between these two extremes.

Figure 10.4 A typical table polariscope (the small light port on the lower front can be used as a refractometer light source). (Gem-A Instruments Ltd)

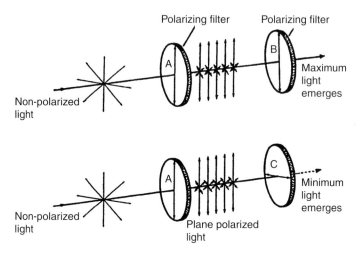

Figure 10.5 (top): A light ray diagram showing two polarizing filters aligned to pass light. (bottom): Filter C has been rotated by 90° into the 'crossed' or 'extinction' position which is the normal arrangement in a polariscope

This crossed position of the polarizing filters is of practical importance to the gemmologist because it provides a very sensitive test for double refraction in a gemstone. There are several other useful tests which can be performed with the polariscope, and these will be described in detail later in this chapter.

As can be seen in Figure 10.6, for all its usefulness the polariscope is a relatively uncomplicated instrument. Of the two polarizing filters, the lower one is called the *polarizer*, and the top one the *analyser*. Although the top one is often rotatable, it is usually locked in the crossed or extinction position. The lower filter is fixed in position, but for convenience and protection it is usually covered by a rotatable glass platform to support the specimen under test and allow it to be easily turned between the two filters.

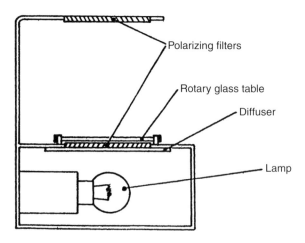

Figure 10.6 Sketch showing the basic components of the polariscope

Using the polariscope

There are three basic tests which can be carried out with the aid of the polariscope. One of these is used to check the optical character (isotropic/anisotropic) of a gemstone; the second indicates the microcrystalline or polycrystalline nature of the stone, while the third reveals strain in the specimen (which may be indicative of a 'paste', i.e. glass, gem).

Optical character

The sample under test (which can be polished or unpolished, but must be transparent or translucent in some degree) is placed on the polariscope's table and rotated through 360° while viewing it through the top filter. If the stone is isotropic (singly-refracting) then there will be very little light visible because the polarized light from the bottom filter passes through the sample without being changed by it and is then blocked by the top filter. If this happens, the sample is either non-crystalline or belongs to the cubic crystal system.

However, if the whole stone appears to alternate between transmitting light and then blocking it (i.e. appearing alternately light and dark four times during a complete 360° rotation), then the stone is anisotropic or doubly-refracting (and belongs to the tetragonal, hexagonal, trigonal, orthorhombic, monoclinic or triclinic crystal systems).

The reason for the variation in light appearing through the top filter is that in one position of the stone its vibration direction is in the same direction as that of the lower filter, and it therefore has no affect on the polarized light. At 45° to this position, however, the stone splits the polarized light into two polarized rays, and a component of both of these rays is now able to pass through the top filter. After another 45° of rotation, the stone appears dark again, and this sequence is repeated for the full 360°.

Note: When making this test it is important to check the sample in at least two positions on the polariscope table if the first position indicates that the specimen is singly-refracting. This is because a doubly-refracting stone viewed in the direction of an optic axis will appear to be singly-refracting (remember, there is one such direction in uniaxial minerals and two in biaxial minerals). With faceted gems it is advisable first to check the stone in the table facet down position. If the results indicate a singly-refractive material, the test should then be repeated with the stone lying on one of its pavilion facets (this eliminates the possibility that the gem might be doubly-refracting with an optic axis lying perpendicular to the table facet).

Microcrystalline and polycrystalline gemstones

If the sample being tested is a transparent or translucent microcrystalline gemstone (such as the chalcedony varieties, agate, chrysoprase, etc.) or a polycrystalline material (such as jadeite and nephrite), it will appear to transmit light constantly when rotated through 360° between the polariscope filters. This is because the small crystals or crystal fibres in these materials are randomly orientated, and some of them will always be in a position where they can modify the angle of the polarized light sufficiently for it to pass through the top filter.

The same effect can sometimes also be produced in a material which contains a series of thin plates formed by repeated twinning (i.e. lamellar twinning). Some specimens of corundum will pass light through the top polariscope filter at any angle of rotation because alternate layers of the material have differing directions of vibration.

Strain

Some materials produce what is termed 'anomalous' double refraction when rotated on the polariscope table. This is most marked in 'paste' (i.e. glass) gemstones as these usually contain very strong internal stresses which show up as dark curved or angular bands through the top filter. These stress bands rotate individually as the sample is turned between the filters.

Many isotropic materials (including synthetics) also show some signs of anomalous double refraction. This usually appears as a vague dark patch or band moving across the sample as it is rotated. Synthetic spinel shows a similar effect which has been described as 'tabby extinction'. With experience, there should be no difficulty in distinguishing between the very positive light/dark effect produced by a doubly-refracting material and one which exhibits one of these anomalous effects. When in doubt, alternative tests should be made (e.g. using a refractometer as described in the previous chapter, or using a dichroscope as covered later in this chapter).

Mention has already been made of the sensitivity of the polariscope in detecting double refraction in minerals. This makes the instrument a very useful screening device for assessing the optical character of a gemstone before attempting to measure its RI on a refractometer. Some stones have a very small double refraction (e.g. DR of taaffeite and apatite) and this may be difficult to detect on the refractometer, particularly if non-monochromatic illumination is in use. However, if double or single refraction is first confirmed on the polariscope, this will avoid any error in identification when using the refractometer (and will also give advance warning of paste or micro/polycrystalline materials).

The main limitations in the use of the polariscope is the need to have some area of the sample which is at least translucent (even the thinner edges of an otherwise opaque cabochon may transmit some light), and the problem of spurious reflection-polarization. The latter can sometimes occur with light reflected from the surface of a gemstone, and may give the impression of double refraction in a singly-refractive stone. This impression can be checked by turning the specimen over onto another facet, or reducing the reflection by viewing it immersed in a liquid having a similar RI. The use of an immersion dish is also mentioned in the following section.

The conoscope (interference figures)

The use of the polariscope to determine the optical character of a gemstone has already been described, but this was limited to identifying the sample as isotropic or anisotropic. To decide whether the stone is uniaxial or biaxial (thus narrowing down the identification of its crystal system), or to identify the position of an optic axis, we need a polariscope fitted with a strongly converging lens. Such an instrument is called a *conoscope* (sometimes spelt konoscope). The lens (or strain-free glass sphere) is placed between the specimen being tested and the top analyser filter on the polariscope (Figure 10.7).

When checking a stone on the conoscope, it should be rotated in the fingers just below the lens and manipulated until coloured bands intersected by dark cross arms or 'brushes' appear. This is called an *interference figure* and is caused by the interaction between the optic axis of the stone (which is in line with the direction of viewing) and the strongly convergent polarized light. Two 'idealized' interference figures are shown in Figure 10.8, and indicate the difference between figures produced by uniaxial and biaxial stones. One gemstone which can be positively identified by this test is quartz,

Figure 10.7 A polariscope fitted with a conoscope lens and supplied with an immersion dish. With the lens pivoted between the two filters the instrument can be used to view interference figures. (Gem-A Instruments Ltd)

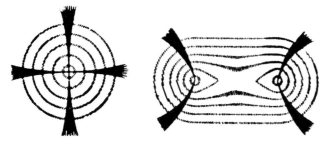

Figure 10.8 A representation of the interference figures seen with the aid of a conoscope. The figure on the left is produced by a uniaxial mineral; the figure on the right is produced by a biaxial mineral. The dark 'brushes' are called isogyres

as its uniaxial interference figure is unique – unlike other uniaxial gemstones, the dark brushes (called *isogyres*) do not meet at the centre.

To gain experience in the use of the conoscope it is best to practice first with a uniaxial mineral such as quartz, as it is generally easier to obtain a recognizable interference figure with this material. An immersion dish may also be provided with the conoscope. This is intended for use in the identification of synthetic gem material as described in Chapter 16 under 'Use of the polariscope (identifying synthetic quartz and Vernueil corundum)'

Pleochroism (differential selective absorption)

In Chapter 9 we saw that a doubly-refracting mineral is able to split the light entering it into two separate rays which are polarized at right angles to each other. In some coloured doubly refracting stones, these two rays (which are travelling at different speeds) may emerge differing in shade or colour. When this happens, the rays are said to have experienced *differential selective absorption* in the gemstone (i.e. a different portion of the visible spectrum has been absorbed from each ray).

The effect is called *pleochroism* ('many coloured'), and when the light passing through the stone is split into two colours or shades the stone is said to be *dichroic*; if three colours or shades are produced, the gem is *trichroic*. Dichroism is associated with coloured *uniaxial* stones, and trichroism with coloured *biaxial* stones.

Because pleochroism only occurs with coloured doubly refracting stones, if detected it can form a useful means of identifying an anisotropic mineral from an isotropic one (e.g. a ruby from a red garnet). With some gemstones, such as andalusite and zoisite, pleochroism is an attractive quality, and the stones are cut so as to bring out all the colours to best advantage. In ruby and blue sapphire, however, one of the dichroic colours is less attractive than the other, and the stone is normally cut with the crown facet at right angles to the c-axis so that this ray is not visible through the crown facets (see also reference to dichroism under 'The Verneuil flame-fusion process' in Chapter 15).

The dichroscope

With the exception of those stones which exhibit strong dichroism (e.g. andalusite, iolite, ruby, blue sapphire, sphene, tourmaline and zoisite) it may not be easy to detect the presence of dichroic or trichroic colours with the unaided eye. As pleochroism can be a useful identifying feature in a gemstone, an instrument called a *dichroscope* was designed to separate the polarized rays and to enable them to be compared side-by-side for signs of colour or shade difference (Figure 10.9).

The dichroscope consists of a cleavage rhomb of optical quality calcite (Iceland spar), which is mounted in a tube having an eyepiece at one end and a square aperture at the other (Figure 10.10). A glass prism is cemented to each end of the calcite rhomb

Figure 10.9 Two calcite dichroscopes. The one at the rear is fitted with a rotatable gemstone mount. (Gem-A Instruments Ltd)

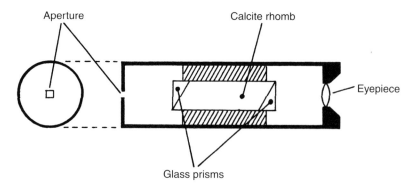

Figure 10.10 Diagram showing the construction of a calcite dichroscope

to allow the light to enter and leave the calcite rhomb in a straight line (alternatively, the ends of the rhomb are ground at right angles to the sides).

The gemstone under test is positioned so that white light passes through the stone and enters the dichroscope aperture. If the gemstone is doubly refracting, the two polarized rays of light emerging from it are separated by the strong double refraction of the calcite rhomb and presented to the eyepiece as side-by-side images of the aperture.

If the gemstone is coloured doubly refracting and pleochroic, and is being viewed in a direction *other* than that of an optic axis, the two images which appear side-by-side will differ in shade or colour. If, however, the images are *exactly* the same colour and shade, then the stone has no pleochroism, and is singly refracting (Figure.10.11). It should be noted that only two colours can be seen at any one time on the dichroscope. With a trichroic gemstone, the third colour (which identifies the stone as being biaxial) will replace one of the other two when the stone is reorientated.

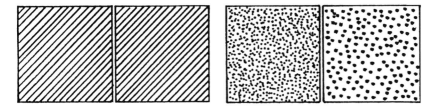

Figure 10.11 A representation of the twin images as seen through a calcite dichroscope. (left): With a singly refracting coloured gem both images show equal colour in any orientation. (right): With a coloured doubly refracting pleochroic gem, two different colours or shades are seen in particular orientations and the gemstone is probably uniaxial. If the stone is biaxial and trichroic, further rotation will cause a third colour to replace one of the other two colours. However, some pleochroic biaxial stones may only show a total of two different colours

Note: If only two shades or colours are visible, the gem under test is probably uniaxial, although it may be biaxial as some coloured biaxial gems do not show three colours.

When checking a stone for pleochroism, it is important that the gem is viewed in several positions (in the same way as when checking for single refraction or DR on the polariscope). This covers the possible situation where the stone is viewed in the direction of an optic axis (which will prevent the detection of pleochroism) or where the direction of polarization of the gemstone's rays and the calcite rhomb are at 45° to each other.

Dichroism can also be detected by using a polarizing filter. If the gem is rotated and viewed through such a filter, first one and then the other polarized ray will become visible. However, it is not easy to detect differences in shade by this means as only one shade is visible at a time. Commercial dichroscopes which use a polarizing filter in place of a calcite rhomb do, however, provide for simultaneous viewing of the colours or shades. This is achieved by cutting the filter into two pieces, and rotating one of these pieces through 90° into the crossed position. These are then mounted side-by-side in the holder. An example of a filter dichroscope is shown in Figure 10.12.

One of the disadvantages of the polarizing filter type dichroscope is that it splits the image of the gemstone between the two sections of filter. The calcite rhomb version, on the other hand, produces two images of the same field of view. This means that when viewing a small stone using the filter dichroscope the gem must be positioned

Figure 10.12 A dichroscope using a polarizing filter. The filter consists of two sections joined side-by-side, with the polarizing plane of one section rotated by 90° to the other. (Gem-A Instruments Ltd)

carefully about the centre line of the two filters to ensure that dichroism, if it is present, can be detected. In an Australian filter dichroscope (developed by J. Snow) this disadvantage was overcome by splitting the image in two with a small glass prism and placing the two filter sections, suitably orientated, on the exit faces of the prism.

Where relevant, details of gemstone pleochroic colours are given in the list of inorganic gem constants and characteristics in Appendix C.

Chapter 11

Spectroscopy

As mentioned briefly in Chapter 8, the perceived colour of most objects is the result of their ability to absorb certain wavelengths or colours in the light passing through them or reflected off their surface. This suppression of parts of the spectrum in the illuminating light is known as *selective absorption* and plays an important part in the identification of some gemstones.

Absorption and emission spectra

In the majority of gemstones, colour is due to the presence in the stone of one or more of the eight transition elements (see Chapter 8). These elements are either present as trace impurities (e.g. chromium in emerald and ruby, nickel in chrysoprase) or as an integral part of the gem's chemical composition (e.g. copper in malachite, manganese in rhodonite). In either case, the resulting colour is due to the selective absorption of wavelengths in the light illuminating the gemstone.

In order to discover which wavelengths have been absorbed by a gem it is necessary to inspect it using an instrument called a *spectroscope* which spreads out the light from the gemstone into its spectral colours. By this means it is possible to see the absorbed wavelengths as dark lines or bands across the range of spectral colours. This is called an *absorption spectrum*, and if distinctive enough can sometimes positively identify the gemstone even if it is an unpolished specimen (see Figure 11.1). The reason for this is that most of the transition elements which are responsible for a gemstone's colour produce characteristic absorption *bands*, *lines*, or *doublets* (two closely spaced lines). A transition element can also produce its absorption features in significantly different positions of the spectrum with different gem species (e.g. the principal iron bands in sapphire and almandine garnet in Figure 11.1), and this provides another useful identification aid.

In some cases, it is even possible for the light which is illuminating the gemstone to stimulate the colouring elements in the stone so that instead of absorbing characteristic wavelengths in the incident light, they emit light at these same wavelengths to produce *fluorescent* lines. When this happens the result as viewed through a spectroscope is called an *emission* spectrum. The most important of the gemstones which show fluorescent emission lines in their spectrum are ruby and red spinel. With both of these stones, the emission lines appear at the red end of the spectrum and are caused by chromium. In the case of red spinel they are striking enough to be described as 'organ pipes' (see Figure 11.1). Emission lines are best seen with the specimen illuminated with blue-filtered light.

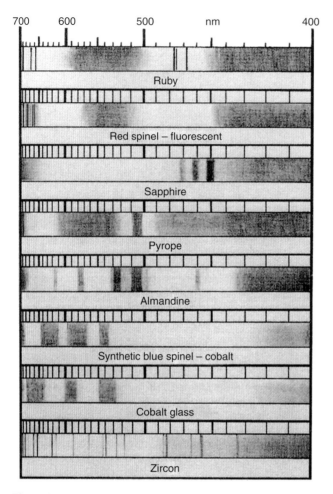

Figure 11.1 A selection of principal gemstone spectra as seen using a prism spectroscope

The prism spectroscope

In 1666 Sir Isaac Newton was the first to demonstrate that white light was composed of a continuous spectrum of colours ranging from red at one end to violet at the other. His equipment was very simple and consisted of a circular aperture to admit a beam of daylight into a darkened room and a glass prism to disperse this light into its component colours (see Figure 8.6). It was not until 1814 that a Bavarian optician and scientist named Fraunhofer improved on Newton's demonstration by using a smaller aperture and viewing the prism's dispersed light through a telescope. In this way he obtained a more detailed spectrum and was able to document in great detail the many spectral absorption lines in sunlight which had earlier been discovered by Dr Wallaston (who had also used a fine slit aperture). These 'Fraunhofer' lines will be described in more detail later in this chapter.

Following Fraunhofer's early experiments with spectroscopy, other scientists including Zantedeschi, Kirchoff and Bunsen devised forerunners of the present-day spectroscope*. Their instruments all used a small aperture or slit to admit the light, a converging lens, a prism and a viewing telescope. The most important refinement in these early spectroscopes was the *collimation* of the incident light. This was effected by placing the aperture in the focal plane of a converging lens which ensured that only *parallel* rays entered the prism. The result of this modification was the production of a 'pure' spectrum (i.e. one in which there was no overlap or contamination between the component colours of the incident light).

The main limitation in these early spectroscopes was that the angle of dispersion in the single prism was too small to allow for a detailed examination of the spectrum. One way of increasing the angle of dispersion was to use several prisms. However, the angle of deviation between the incoming incident light and the emergent spectrum becomes greater as each prism is added, and sets a practical limitation to this method of increasing the instrument's angle of dispersion.

In order to achieve a compact spectroscope which had a reasonable dispersion as well as an 'in-line' relationship between the incident light and the centre-line of the emerging spectrum, it was first necessary to develop a prism which had dispersion *without* deviation. In 1860, Amici achieved this objective with a compound prism using crown and flint glass components of differing refractive indices (Figure 11.2).

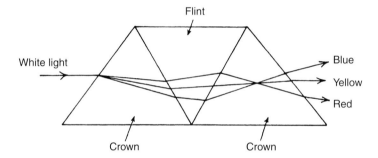

Figure 11.2 An Amici compound prism designed to give zero deviation at yellow wavelengths

The construction of a modern 'direct vision' spectroscope using a triple-element *Amici* prism can be seen in Figure 11.3. This type of instrument disperses the visible spectrum over an angle of 7°. More expensive instruments are available containing a five-element prism producing around 10° of dispersion (greater dispersions than this can be achieved but as explained later this can result in reducing the visibility of faint absorption lines or bands).

One of the important parts of the spectroscope is the aperture or slit through which the light to be analysed passes. The spectrum produced by the instrument is in fact a series of images of this slit. If the slit is too wide, these images will overlap (as they did in Newton's experiment) and the resulting spectrum will be blurred and the colours not pure. However, if the slit is too narrow, not enough light will enter the prism to produce a visible spectrum. Because of these constraints, most spectroscopes are fitted with an adjustable slit which can be set for a resolution appropriate to the spectrum being

---

* It is interesting to note that Fox Talbot, one of the pioneers of photography, was the first to express the concept of chemical spectrum analysis in terms that anticipated the discoveries of his contemporaries Kirchoff and Bunsen.

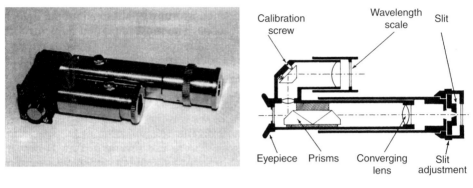

Figure 11.3 (left) A wavelength-type prism spectroscope. (right) A sketch showing its construction

viewed. The setting of the slit width is usually a compromise between obtaining the best resolution of fine absorption lines and letting enough light in to make the spectrum visible. The method of adjusting the slit and of focusing the instrument will be covered later in this chapter.

As mentioned earlier, another requirement for the production of a pure spectrum is that the light rays entering the first prism section are parallel (i.e. collimated). This condition is met by placing a converging lens between the slit and the prism so that its focal plane is coincident with the slit.

An added refinement often fitted to the prism spectroscope is a calibrated wavelength scale. This scale, which is sometimes illuminated by a separate light source, is superimposed on the image of the spectrum by means of additional optics (see Figure 11.3). A calibrating screw adjustment enables the image of the scale to be moved relative to the spectrum. The scale is usually provided with a vertical line at the 589.3 nm point, and the calibrating screw is adjusted by the manufacturers to coincide this with the emission line from an industrial sodium lamp.

While the prism type instrument is recommended for initial practical work on spectroscopy because of the brightness of its spectra, it has the disadvantage that the spectrum it produces is not evenly spaced out across the range (Figure 11.4a). This is due to the dispersion characteristic of the prism which compresses the spectrum at the red end and increasingly spreads it out towards the violet end. Very faint absorption lines and bands become increasingly difficult to detect if they are spread out, and as the eye

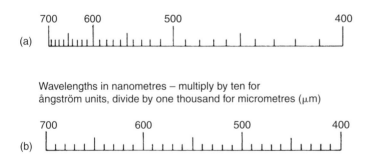

Figure 11.4 The wavelength scale of a prism spectroscope shown at (a) is cramped at the red end and spread out at the violet end. The scale of a diffraction grating spectroscope at (b) is linear

is less sensitive at the blue/violet end of the spectrum, this can sometimes be a limitation. Because refraction in the prism is dependent on the wavelength of the light, the focus of the spectroscope also needs resetting slightly when viewing different areas of the spectrum. For these reasons some gemmologists prefer the alternative type of spectroscope which is described next.

The diffraction grating spectroscope

Instead of a prism, this type of 'direct vision' spectroscope uses a *diffraction grating* to disperse light into its spectral components. The grating usually takes the form of a plate of glass on which are printed a series of very fine equidistant parallel lines. The pitch of the lines is in the region of 15 000 to 30 000 to the inch, and these are printed on the glass photographically by optical reduction from a much larger master negative.

Although the diffraction grating spectroscope (Figure 11.5) may appear to be a simpler instrument that its prism counterpart, the optical theory behind its operation is quite complex. The following explanation, although a simplification, is sufficient to give a general understanding of the production of spectra by this method.

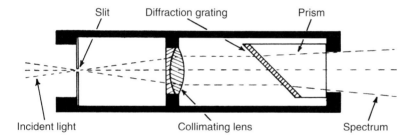

Figure 11.5 Sketch showing the construction of a diffraction grating spectroscope

The diffraction grating spectroscope uses a combination of *diffraction* and *interference* effects to disperse light into its spectral colours. Diffraction is the bending or fanning out of light rays when they pass through a narrow aperture, and interference is caused by the differing path lengths between parallel rays (see Figure 8.5). When light waves are restricted by an aperture (such as a slit), some of the light spreads out into the region not directly in line with the source. If a series of slits are used (as in a diffraction grating) the diffracted rays interfere with each other and produce a spectrum.

Figure 11.6 indicates the progressive increase in path length difference from A,B to C,D in a set of parallel rays bent from their original path by a diffraction grating. Light is propagated as a wave motion, and because of path length differences, rays emerging from the grating will undergo mutual interference. Some rays will have a wavelength (or colour) where they are in phase (i.e. in step) with each other and will reinforce that colour, and some will be 100% out of phase and will cancel each other. If the light falling on one side of the grating is collimated, a spectrum of colours of increasing wavelength will be produced on the other side as the path length differences progressively increase across the grating.

As can be seen in Figure 11.5, the diffraction grating spectroscope consists of a slit, a converging collimator lens, a diffraction grating and a prism/eyepiece. The slit lies in the focal plane of the converging lens which produces parallel rays. Corresponding

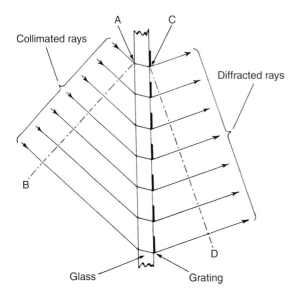

Figure 11.6 Sketch showing the diffraction of rays as they emerge from a grating. The resulting difference in path length (AB to CD) produces a spectrum by means of interference effects between the rays

to each wavelength in the incident beam emerging from the grating a series of diffracted beams is produced, the main one being displayed through the eyepiece.

The resulting spectrum is not so pure or bright as that obtained with a prism spectroscope, as multiple spectra are also produced on each side of the main one and tend to dilute it. The spectrum produced by the diffraction grating spectroscope does however have advantages over that of the prism type in that the wavelengths are both evenly distributed (Figure 11.4b) and in focus across the spectrum. Because of its constructional simplicity, this type of spectroscope is also less costly than the prism version and is often produced as a small portable unit with a fixed slit and focus.

Using the spectroscope

Before putting the spectroscope to use, the slit and focus adjustments (if fitted) should be set. With the slit adjustment partially open, focus the instrument on either the Fraunhofer lines by pointing it at daylight, or on the emission lines in a sodium or fluorescent light source. (On most instruments, focusing is effected by sliding the eyepiece tube in or out of the body of the spectroscope).

One of the most important pieces of ancillary equipment for use with the spectroscope is a high-intensity source of white light. With some spectroscopes (Figure 11.7), this is built in as part of the instrument together with a cooling fan and an infrared heat filter to prevent the specimen under test from being overheated. Even with these refinements it is not advisable to leave a heat-sensitive stone under the light source for too long. Flexible fibre-optic light guides are often used in these combination units to channel the light to the specimen.

When a separate light source and fibre-optic light guide is used, it is prudent to check that neither of these has an absorption spectrum of its own which would superimpose spurious bands on the spectrum of the specimen. With unmounted 'hand-held' spectroscopes such as the multi-slit prism model (Figure 11.8), a bench mount was also available which was fitted with a rotatable specimen table covered in black non-reflecting material.

Figure 11.7 A prism spectroscope unit with built-in fibre-optic illumination. The disc at the end of the light guide is fitted with colour filters to improve the contrast of spectra. The light port at the rear right-hand side is for refractometer illumination

Figure 11.8 A hand-held prism spectroscope mounted on a stand

If a microscope is available, it is also possible to use this as a convenient mounting and source of illumination for a hand-held spectroscope. The specimen is placed on the microscope stage, the microscope set for its lowest magnification factor and the eyepiece removed. The source of illumination, position of the specimen and the microscope focus are adjusted so that the microscope viewing tube is filled evenly with the body colour of the gem (it is advisable to place a piece of ground or opal glass over the viewing tube to prevent being dazzled by the light when making these adjustments). The spectroscope is then inserted in the viewing tube and, depending on its dimensions, secured with a paper wedge or a mounting ring.

If the microscope is fitted with a built-in light source, set this for light-field illumination (i.e. transmitted light). If an iris-type diaphragm is fitted to the stage, adjust this so that the minimum of light escapes round the edges of the specimen (failing this, a metal washer can be used as a diaphragm aperture). For opaque gems, it is necessary to use an incident (reflected) top light source, and this method of illumination can also be used to advantage with translucent and transparent stones. Faceted transparent stones should be placed table facet down; the light can then be directed in through one side of the stone's pavilion, reflected off the inside face of the table facet and out through the opposite side of the pavilion (Figure 11.9). An alternative method is to inject light from a 'pin-point' source vertically over the rear pavilion facets. Either method increases the path length of the light in the stone and the depth of coloration of the emerging light.

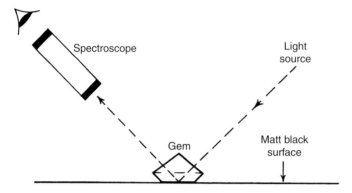

Figure 11.9 Sketch showing the use of reflection from the interior of a transparent polished gem to increase the path length and hence the depth of coloration of the transmitted light

With strongly dichroic stones there will be a difference between the spectrum produced by the ordinary ray and that produced by the extraordinary ray. Because of this, such a stone should be rotated so that any change in the spectrum can be noted before making an identification.

Many students of gemmology have problems initially in seeing absorption spectra, particularly when using a hand-held spectroscope and a separate light source (these problems are often due to poor illumination and incorrect settings of the slit and focus adjustments). While it is easier to achieve the correct illumination and viewing angles with a built-in spectroscope unit as illustrated in Figure 11.7, with patience equally good results can be obtained using a hand-held instrument provided that the spectroscope and light source can be securely positioned and their angles easily adjusted. For portable use, quite good results can be obtained with a small diffraction grating spectroscope and a pen torch. The gemstone can be held on the end of the torch using 'blue tack', which also serves to prevent light from escaping around the edges of the stone.

Successful use of the spectroscope also depends on the sensitivity of the eye, and this can be optimized by taking note of the following suggestions:

1. Before using the spectroscope, allow the eyes to become partially adapted to the dark, and keep the surrounding level of illumination as low as practical. When adjusting the slit control, avoid looking at a brightly lit spectrum for too long.
2. Spectral lines and bands can sometimes be seen more clearly by moving the eye an inch or two back from the eyepiece. The amount of spectrum that is visible will be restricted, but this can be used to advantage as moving the head from side to side will allow sections of the spectrum to be inspected separately.
3. For very faint bands, try looking at them using the more sensitive peripheral area of vision. Cupping both hands round the eyepiece will also help to cut out spurious light and increase the contrast of the spectrum.
4. To reduce the amount of spurious light entering the eye, place the spectroscope/light source on a large sheet of black material (e.g. black cloth or cartridge paper).

5. As the eye is often more capable of distinguishing horizontal patterns, the unorthodox trick of rotating the spectroscope through 90° and viewing the spectrum in a vertical orientation can sometimes help to make faint lines more visible.
6. When checking the emission lines in ruby or red spinel, they can usually be seen more clearly by placing a blue filter in front of the source of illumination so that the red/orange end of the spectrum is blocked out. This technique will also make absorption lines and bands more visible in the blue end of a gem's spectrum.

One further spectroscope technique, used mainly by the larger gem testing laboratories, is the enhancement of faint absorption lines and bands by cooling the specimen. This is most effective with very low temperatures in the region of −160°C (i.e. towards the temperature of liquid nitrogen). The method is used to help detect signs of irradiation treatment in fancy coloured diamond, and is described in Chapter 14.

Figure 11.1, and Plates 10 and 11, show absorption spectra (with red on the left as is the standard in Europe) for a selection of gemstones. The spectra in Plates 10 and 11 are shown as they appear when using a diffraction grating spectroscope and represent classic specimens having well defined spectra. The spectra in Figure 11.1 are as seen using a prism spectroscope. Wavelengths which were formerly quoted in ångström units are now normally given in nanometres (1 nm = 10 Å). It should also be noted that the absorption spectrum of several stones including green aquamarine, emerald, iolite, peridot, ruby and sapphire vary with the direction of viewing (see under 'Inorganic gemstones' in Appendix C).

In practice very few specimens will produce such a strongly defined spectrum as illustrated in Plates 10, 11, and Figure 11.1. Students of gemmology who are attempting for the first time to see absorption bands in coloured gemstones are advised to make use initially of reasonable size specimens of synthetic ruby, synthetic blue spinel, green and brown Sri Lankan zircons, and almandine garnet. They should also remember that among coloured gemstones, only about 20% possess a diagnostic spectrum which can be identified with a hand-held spectroscope.

Many other coloured stones certainly possess selective absorption because this is the mechanism by which they appear coloured, but the amount of absorption is much less dramatic and exists as only a small reduction in brightness across a section of the visible spectrum. As the eye is not capable of detecting these partial absorptions it becomes necessary to use a more sophisticated instrument such as the spectrophotometer which is described later in this chapter, and in Chapter 16.

The overall 'picture' of an absorption spectrum is generally of more significance for identification (and exam) purposes than the wavelengths of the individual absorption bands, lines and doublets, and it is only necessary to be aware of their relative strength, width and colour position in the spectrum. However, there are a few stones such as Cape series diamond, emerald, ruby, sapphire, red spinel and zircon where it is often useful to know the wavelengths of the principle absorption features (these are listed in Appendix C).

Fine line spectra

Some gemstones such as apatite, the man-made products YAG (yttrium aluminium garnet), coloured strontium titanate and some coloured CZ (cubic zirconium oxide) exhibit what is termed a 'fine line' spectrum. This consists of a series of thin absorption lines distributed across the spectrum. These lines are caused by the presence of one or more of the 'rare earth' elements (e.g. neodymium and praseodymium – known collectively as didymium – in apatite). Although zircons may also exhibit a similar spectrum, due in this case to a minute trace of uranium, the absorption lines in the spectrum of these stones are thicker and are more accurately described as bands. In colourless specimens which show this type of spectrum (in particular zircon and YAG), the bands or lines are so evenly distributed across the spectrum that they have little overall effect on the stone's colour.

Although apatite is one of the few natural gemstones that exhibit a fine line absorption spectrum, the strength of this spectrum is much weaker than in those man-made gems which owe their colour to the rare earths (see the 14 rare earth elements marked with an asterisk in the Table of elements in Appendix H). For this reason, any strong fine line spectrum can be taken as a positive indication of a man-made 'synthetic' gem (see Figure 11.10). Unlike the transition elements, the rare earth elements used in man-made gems each produce their own distinctive pattern of fine lines irrespective of the host crystal (i.e. a pink CZ doped with erbium has the same absorption spectrum as a similarly doped pink YAG).

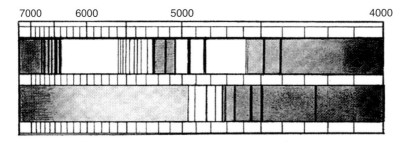

Figure 11.10 Fine line spectra. The upper spectrum is produced by the rare earth erbium, and can be seen in pink CZ and in pink rare earth garnets (YAG). The lower spectrum is caused by dysprosium in yellow/green rare earth garnets

Fraunhofer lines

The name Fraunhofer has featured several times in this chapter. When the Bavarian scientist Fraunhofer refined Newton's demonstration that proved light was made up of a continuous spectrum of colours, he was able to observe in detail the solar spectrum that had been discovered by Dr Wallaston. What he saw was a spectrum consisting of a mass of fine lines stretching from the far red across to the violet end of the spectrum. As there were no photographic methods of documentation available in the early nineteenth century he recorded the solar spectrum by making a drawing showing some 576 of these lines (Figure 11.11).

Since those days, the solar spectrum has been well explored and documented (see Appendix I 'Table of Principal Fraunhofer Lines'), and its absorption lines have been

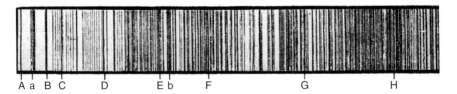

Figure 11.11 Reproduction of one of Fraunhofer's drawings of the solar spectrum

named Fraunhofer lines. The majority of the lines are due to the absorption of characteristic wavelengths by vaporized elements in the chromosphere surrounding the Sun. The lines designated A and B are produced by oxygen in the Earth's atmosphere.

In gemmology, the Fraunhofer B line (686.7 nm in the red) and G line (430.8 nm in the blue) are normally used as the standard wavelengths for measuring the dispersion of a gem. The standard wavelength at which refractive indices are measured is the sodium vapour emission line at 589.3 nm (this is the mean value of the Fraunhofer D_1 and D_2 sodium vapour lines at 589.0 and 589.6 nm).

Other spectroscope versions

Other versions of the standard prism and diffraction grating spectroscopes have been developed over the last few decades. These include a reversion model developed by Professor H. Hartridge to improve the ease and accuracy of measuring the wavelength of absorption lines or bands. This was achieved by providing two spectra, one above the other but with the direction of wavelength reversed in one of them. A calibrated control enables the two spectra to be moved simultaneously in opposite directions. The wavelength reading for a particular line or band is made by setting the appropriate line or band in one spectrum against the same feature in the other spectrum, and this can be done with great accuracy.

Several commercial scanning spectroscopes have also been developed in which the viewed spectrum can be moved relative to a hairline cursor by means of a control knob. The knob is coupled to an electrical transducer whose output is displayed as a digital wavelength reading of the position of the hairline cursor on the spectrum.

The Nelson Comparison Spectroscope, produced by McCrone Research Associates Ltd., was designed to overcome the parallax error normally experienced between a superimposed wavelength scale and the spectrum on a conventional wavelength spectroscope. Instead of a wavelength scale, a comparison spectrum from an appropriate gem material is introduced. This allows a direct comparison to be made between the gemstone under test and a plane-polished section of any transparent gem material. Such sections, mounted in special slides, serve as permanent spectrum references. By a suitable choice of materials having known absorption wavelengths, accurate measurements can be made of the line and band wavelengths in the specimen under test.

Yet another version (produced by Gemlab and by Hanneman Gemological Instruments) combined the spectroscope with a video camera to display gemstone spectra on a television monitor. This produced a large magnification of the spectral image over which a wavelength scale could be superimposed. Its other advantage was that a video camera is more sensitive at the blue/violet end of the spectrum than the eye, and as a result it is easier to detect and display absorption lines and bands in this area.

Raman spectroscopy

When a high intensity light (such as a laser beam) impinges on a surface, a small amount of the light undergoes *Raman scattering*. This scattered light has components which are longer or shorter in wavelength than the incident beam (described as *Stokes* and *anti-Stokes* wavelengths – see Stokes' law in Chapter 12). The effect is caused by molecular vibrations in the surface of the material under test. The degree of wavelength difference, or *Raman shift*, between the incident beam and the scattered light depends on the nature of the surface atoms and their structural bonding. This spectral shift enables diagnostic absorption spectra in the infrared to be seen in the visible region of the spectrum by means of a standard spectrophotometer (see next section). The use of Raman spectroscopy in gemmological research is increasing as the data-bank of Raman spectra produced for a range of materials is built up.

Spectrophotometers

One of the limitations in the use of spectroscopes is the eye's inability to detect minor reductions in the intensity of wavelengths in the viewed spectrum. Many gemstones owe their colour to relatively modest absorptions of parts of the spectrum rather than the dramatic black absorption bands or lines illustrated in Figure 11.1. Because of the eye's insensitivity to small differences in intensity, some of these much weaker spectral features cannot be detected at all with the spectroscope. For this reason, only around 20% of all coloured gemstones have an easily visible absorption spectrum, a fact that often causes students of gemmology to despair!

When it becomes important to detect weak absorption spectra, particularly in research work, use must be made of the spectrophotometer. This instrument employs a monochromator which can be tuned manually or automatically scanned over the visible spectrum (some instruments also cover the near infrared (IR) and ultraviolet (UV) sections). The resulting transmitted or reflected absorption spectrum is detected with a photocell, and either displayed on a video monitor or plotted as a graph.

Because of the instrument's sensitivity it is able to display relatively minor attenuations in a spectrum. For example, although it is difficult to see even the 450 nm iron absorption band in many Sri Lankan sapphires (because of their low iron oxide content), this band, together with the other two iron bands at 460 and 470 nm, is easily detectable with a spectrophotometer. In addition, IR/visible/UV versions of this instrument have the advantage of being able to reveal differences between some natural and synthetic gemstones in the near IR and UV sections of the spectrum. A few of the uses of the spectrophotometer in the detection of synthetic stones are described under 'Laboratory equipment and methods' in Chapter 16.

Chapter 12

Luminescent, electrical and thermal properties of gemstones

While not often capable of producing precise numerical data, the luminescent, electrical and thermal properties of gemstones can still provide useful qualitative information to help identify a gemstone.

Luminescence (fluorescence and phosphorescence)

When certain materials acquire energy in one form or another, below the level which would cause incandescence or burning (i.e. not sufficient to produce a flame, or a glow due to heat), they convert this energy into a 'cold' radiation whose wavelength usually lies in the visible section of the spectrum.

The mechanism producing this cold radiation, or *luminescence*, is associated with the excitation of atoms within the material. The surplus energy acquired by luminescing substances is used up in moving electrons out of their normal orbital shells (known as the 'ground' state) into orbits of a higher energy level (known as the 'excited' state). When these electrons eventually return to their more stable orbits, they give up the surplus energy in the form of electromagnetic radiation (i.e. visible, or very occasionally ultraviolet, light). In terms of the quantum theory, the electrons discharge their energy in the form of *photons* of light (i.e. discrete 'packets' of energy).

In all of the forms of luminescence which will be described in this chapter, the light emitted is either due to some intrinsic property of the material (e.g. the lattice defects in diamond) or to the presence of luminescent impurities called *activators* (e.g. chromic oxide in ruby).

In the case of luminescence due to crystal defects, the exciting energy is sufficient to cause an electron to leave its orbit and become a free electron. As the electron has a negative charge, it leaves a gap in the crystal structure which has a positive charge. The free electron travels through the structure until it is captured by another positive gap and releases its acquired energy as radiation. This radiated energy usually falls within the visible spectrum and on occasions is partially or wholly absorbed by the presence of iron oxides.

Luminescence activators are impurities which have unpaired electrons in the outer orbit of their nuclei. If these electrons absorb energy from a source of stimulation they are raised to a higher energy level. This excited state is not stable, however, and the electrons immediately return to a lower level and release their surplus energy, which usually falls within the visible spectrum but may be partially or wholly absorbed if the host material contains iron oxides.

In a luminescing material, the movement of electrons between their normal and higher energy level orbits takes place in a random fashion. If there is virtually no delay between the electrons acquiring energy and then releasing it as visible light, the phenomenon is called *fluorescence*, and ceases immediately the source of excitation is switched off. If, however, there is a discernible delay before the electrons give up their surplus energy, the phenomenon is called *phosphorescence*, and can be seen as an 'afterglow' when the excitation ceases. With fluorite, the mineral that gave its name to fluorescent phenomena, phosphorescence, may sometimes occur in the ultraviolet (UV) range rather than at visible wavelengths. When this happens it can be detected electronically or photographically.

It is also possible for a given material to both fluoresce and phosphoresce. When this happens there is sometimes a change in luminescent colour when the radiation is switched off and the phosphorescent effect is seen on its own. Such a change can form a useful identification feature with certain minerals. Other materials have the ability to retain the energy they acquire from electromagnetic radiation, and then release it again when heated. This effect is called *thermophosphorescence* or *thermoluminescence*.

Photoluminescence and Stokes' law

Of the many varieties of luminescence that can occur, *photoluminescence* is probably the most useful for gemmological purposes. With this type of luminescence, the excitation energy is in the form of *electromagnetic radiation* and can be visible light, UV radiation or γ/X-rays. However, the important feature of photoluminescence is that whatever the wavelength of the excitation energy, the resulting luminescence is in the *visible* range of the spectrum. The relationship between the wavelength of the excitation energy and that of the resulting luminescence is specified in *Stokes' law*. This states that a material's luminescence is always of a longer wavelength than that of the original excitation.

The various forms of electromagnetic radiation available to the gemmologist include visible light, long-wave (LW) UV, short-wave (SW) UV and X-rays. Each of these forms of energy has its particular application in the identification of gemstones by means of their photoluminescent properties which may vary in colour depending on the type of radiation. This is particularly useful for identification purposes when the luminescing colour and intensity varies between LW and SW UV. Table 12.1 indicates some of the colours and intensities seen when certain gemstones are exposed to SW UV, LW UV or X-rays. The techniques and equipment associated with each of these four forms of radiation will now be discussed in more detail.

Crossed filters

With some gemstones, even the shorter wavelengths at the blue end of the visible spectrum can be made to produce a longer wavelength fluorescence at the red end of the spectrum. This particular phenomenon was first put to practical use by G.G. Stokes in 1852 in a test devised to detect the presence of chromium. The test, known as the *crossed filter* method, was later adapted by B.W. Anderson in the Gem Testing Laboratory of the London Chamber of Commerce and used for the identification of gemstones. To avoid any confusion it must be pointed out that the term 'crossed filters' has nothing to do with polarizing filters but involves the use of colour filters.

The crossed filter technique is one of the simplest of all the methods used to test for photoluminescence. All that is required is a strong source of white light, a blue filter and a red filter. These filters can be of the simple gelatine variety sandwiched between

Table 12.1

Gemstone	LW UV	SW UV	X-rays
Apatite (yellow)	Lilac	Lilac/pink	Pinkish-white/yellow
Apatite (blue)	Dark blue to light blue		Faint pinkish straw
Apatite (green)	Mustard yellow	Weak mustard	Yellowish-white
Apatite (violet)	Greenish yellow	Pale mauve	Bright greenish-yellow with phosphorescence
Alexandrite (natural, synthetic and corundum simulant)	Red	Weak red	Inert
Benitoite	Inert or dull red	Bright blue	Blue
Chrysobetyl	Red	Weak red	Inert
CZ (cubic zirconium oxide)	Apricot orange or yellowish	Apricot orange and yellowish	White
Danburite	Blue	Weak blue	
Diamond	Blue (with yellow phosphorescence), green, pink, red, yellow	Similar weaker colours	Chalky blue, green and yellow
Emerald (natural)	Red	Red	Red
Emerald (synthetic)	Strong red	Red	Strong red
Fluorite (Blue John variety is inert)	Blue, violet	Weak blue and violet	Blue, violet (some specimens show phosphorescence)
GGG (gadolinium gallium garnet)	Faint yellow	Peach	Lilac
Glass (colourless)	Inert	Chalky white	Inert
Hydrogrossular garnet	Inert	Inert	Orange
Kunzite	Orange	Inert	Orange
Lapis lazuli	Orange spots	Greenish, whitish	Inert
Moissanite (synthetic)	Inert to weak orange	Inert	Inert
Opal	White, green (some), often with phosphorescence	White, green (some)	Green (some)
Paste (glass)	Faint blue and green	Blue (some), green	Blue (some), green
Ruby (natural and synthetic)	Red	Red	Red (synthetic shows phosphorescence)
Sapphire (pink, natural and synthetic)	Red	Red	Red
Sapphire (green synthetic)	Red	Inert	Inert
Sapphire (orange synthetic)	Red	Red	Red
Sapphire (white)	Orange	Inert	Orange
Sapphire (yellow Sri Lankan)	Apricot	Apricot	Apricot
Sapphire (blue synthetic)	Inert	Green/blue (some)	Green/blue (some)
Sapphire (blue Sri Lankan)	Inert	Pink	Pink
Scapolite	Yellow/orange	Pink/orange	White/violet/orange
Scheelite	Inert	Blue	Blue
Spinel (white synthetic)	Inert	Blue/white	White (some)
Spinel (red, natural and synthetic)	Red	Red	Red
Spinel (pale blue natural)	Green (some)	Inert	Green (some)
Spinel (blue, green, synthetic)	Pink	Pink	Pink
Spinel (yellow, and some green synthetic)	Bright green	Green	Green
Strontium titanate	Inert	Inert	Inert
YAG (yttrium aluminium garnet)	Yellow	Inert	Yellow/mauve
Zircon	Brownish-yellow	Inert	Inert

sheets of glass to protect them from the heat of the lamp. Alternatively, a flask of copper sulphate solution can be used in place of the blue filter, and this will serve additionally as a heat filter and a lens to concentrate the light.

The blue filter (or flask) is positioned between the light source and the gemstone. The stone should ideally be placed on a black non-reflecting surface and shielded so that only the blue-filtered light reaches it. The gemstone is then viewed through the red filter. If it is seen to glow pink or red when viewed through the red filter then the stone must be fluorescing, as the incident light illuminating it contains no red component (see Figure 12.1).

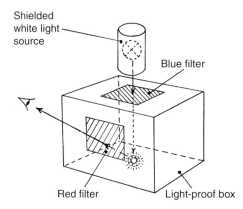

Figure 12.1 Sketch illustrating the use of the crossed filter technique to verify luminescence

Gemstones which demonstrate their fluorescence most readily under blue/red crossed filters are those owing their coloration to chromium impurities (i.e. activators). The principal stones in this category are ruby, red spinel, the rare alexandrite variety of chrysoberyl, emerald and pink topaz. However, it should be remembered that the presence of iron oxides in a gemstone may diminish or completely extinguish any fluorescence. The synthetic versions of ruby, red spinel, alexandrite and emerald also owe their colour to chromium and will also glow red under crossed filters. As many of these synthetic products are completely free from iron oxides they often fluoresce more strongly than the natural stone. Pyrope garnet and jadeite, which contain traces of iron as well as chromium, do not fluoresce at all.

Some stones have a yellow or orange fluorescence (e.g. yellow Sri Lankan sapphires) and this can be made more visible by using a yellow or orange filter in place of the red one.

Once luminescence has been confirmed with the crossed filter test, more specific diagnostic information can be obtained by checking the fluorescent spectrum of the stone using a spectroscope and blue-filtered illumination. In this way, for example, red spinel can be distinguished from ruby (see Figure 12.2).

LW and SW UV radiation

Mercury vapour lamps are used as a source of ultraviolet radiation as they have strong emission lines ranging from yellow to the far ultraviolet section of the spectrum. The dominant mercury line which is used for long-wave ultraviolet (LW UV) work has a wavelength of 365 nm, while the dominant line used for short-wave ultraviolet radiation (SW UV) has a wavelength of 254 nm.

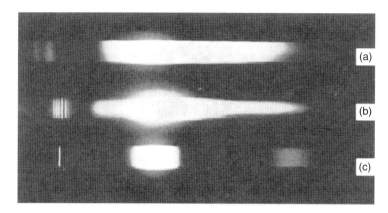

Figure 12.2 Fluorescence spectra of (a) emerald; (b) red spinel and (c) ruby, as seen on a spectroscope using a blue light source

The mercury lamp used to produce LW UV is a high-pressure discharge type, and this is fitted with a filter to block out most of the lamp's visible radiation. The filter is made from Wood's glass which contains cobalt and a trace of nickel (a commercially available version is a Chance filter type OX1). The mercury lamp used for SW UV is a low-pressure type using a quartz tube. With this lamp, visible and other unwanted wavelengths are attenuated by a Chance type OX7 filter (*Note*: some SW UV filters and lamps may begin to deteriorate after 100 hours of use).

While the higher energy level of SW UV may be necessary to stimulate the orbital oscillation of electrons in some materials (e.g. benitoite, colourless synthetic spinel and paste), in general it produces a weaker fluorescence than LW UV in most gemstones. This applies equally to diamonds, of which only about 15% fluoresce strongly under LW UV. However, if a diamond fluoresces blue under LW UV, and when the energy is switched off it phosphoresces yellow, this provides a positive test for diamond, as no other blue-fluorescing mineral exhibits a yellow afterglow.

The majority of UV test units combine both LW and SW lamps which can be operated either separately or together (Figures 12.3 and 12.4). The combination of both types of lamp in the one unit allows for the rapid comparison of a material's fluorescence at these two wavelengths. To enable luminescent effects to be detected in normal ambient lighting conditions, darkroom viewing cabinets are also available (Figure 12.5).

Although UV lamps supplied for gemmological work have a relatively low UV emission, it is not advisable to look directly at them unless special protective glasses are worn. As well as filtering out the harmful UV rays, these glasses also improve the

Figure 12.3 A combined LW and SW UV lamp. The sliding chrome cover is used to bring the appropriate filter into operation. (Gem-A Instruments Ltd)

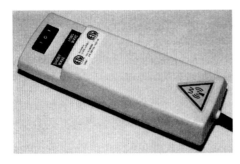

Figure 12.4 (above) A combined LW and SW UV lamp using two 4-watt tubes

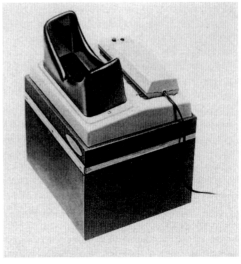

Figure 12.5 (right) A darkroom viewing cabinet suitable for viewing luminescence of minerals as well as gemstones

contrast of a luminescing stone by removing the violet haze passed by the lamp's filter (even ordinary glasses will give some protection against SW UV which is the more potentially harmful of the two forms of radiation).

Some gemstones can also suffer damage in the form of fading when exposed long term to UV radiation. Blue heat-treated zircon is susceptible to fading after long exposure to the UV in sunlight (its colour can be fully restored by heating the stone to a dull red temperature). Gadolinium gallium garnet (GGG, a diamond simulant) fluoresces a peach colour under SW UV radiation but can turn brown if left too long under the SW UV lamp (it reverts to colourless if left for several hours in the dark).

X-ray radiation

Owing to the size and cost of most X-ray units, and the precautions necessary to protect users against radiation dangers, these sources are generally only found in the larger gem testing laboratories. For identification purposes, X-rays are used either as a source of energy to stimulate luminescence in gemstones, or (because of their high penetrating powers) to provide information on the internal structure of gem materials.

The wavelength range of X-rays extends from the vicinity of the shortest UV wavelength of around 20 nm down to a millionth of a nanometre. γ-rays are indistinguishable from X-rays, the only difference being their method of production. X-rays are generated in a vacuum tube when electrons, accelerated in a high-potential electric field, bombard a tungsten target (the intensity of the radiation increases, and the wavelengths become progressively shorter, as the electric field potential is increased). γ-rays are produced by radioactive materials and cover the same range of wavelengths as X-rays.

While diamonds show a variable degree of luminescence under UV radiation, almost all diamonds exhibit a chalky blue fluorescence under X-rays (with no phosphorescence). This forms the operating basis of the X-ray separators used in many diamond mines to extract the gem from the gravels and crushed rock.

Because of their short wavelength, X-rays are able to penetrate with ease those materials whose constituent elements have a low atomic weight. Diamond, for example, which is composed of carbon atoms, is virtually transparent to X-rays, while all other gemstones (including diamond simulants, both natural and man-made) show varying degrees of opacity depending on the atomic weights of their constituents elements. An X-ray unit marketed by the Gemmological Association of All Japan used the X-ray transparency of diamond to distinguishing it from its simulants (Figure 12.6). Test specimens are placed on a reflecting fluorescent plate, and when the operate button is pressed, the internal low-energy X-ray tube is energized and the X-ray opacity of the specimen is seen as a shadow profile through the viewing window. (X-ray spectroscopy and X-ray powder diffraction analysis are subjects discussed in Chapter 16).

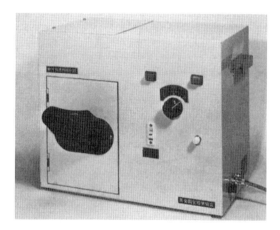

Figure 12.6 A low-energy X-ray unit for distinguishing diamond from its simulants by means of its transparency to X-rays

Other forms of luminescence

While photoluminescence is produced by electromagnetic radiation in the visible, UV and X-ray parts of the spectrum, energy sources which cause *triboluminescence*, *cathodoluminescence* and *electroluminescence* are quite different.

A triboluminescing material is one which glows when it is rubbed or abraded. This sometimes occurs when a diamond is being faceted or sawn, and results in an intense blue or red fluorescence. If the stone is taken out of contact with the polishing or sawing surface, the fluorescent effect immediately vanishes, proving that it was due to friction and not to heat.

Cathodoluminescence is a fluorescent effect displayed by some materials when they are bombarded with a beam of electrons in a vacuum. The phosphors on the inside surface of a television tube screen are everyday examples of cathodoluminescing materials. However, the application of this form of luminescence for analysis purposes was discovered when viewing specimens in an electron beam microscope. It has since been developed into a useful research tool for detecting the presence of rare earths and specific minerals in agglomerates. It is also used for investigating crystallographic features and nitrogen 'platelet' aggregates in diamond crystals. Its use in the detection of synthetic gemstones is described at the end of Chapter 16.

Yet another form of luminescence can be produced in some materials by passing an electric current through them, and when this occurs the substance is said to be *electroluminescent*. Natural Type IIb blue diamonds are *semiconductors* (i.e. their electrical conductivity lies somewhere between that of a conductor and an insulator). One of the tests used to distinguish them from artificially coloured diamonds (which do not conduct electricity) is to apply a direct or alternating voltage to them via a current indicator such as a meter (see circuit diagram in Figure 12.8). During this test, natural blue diamonds often exhibit electroluminescence due to the current flowing through the crystal lattice.

SW UV transmission test

Mention has been made of the inhibiting effect that the presence of iron oxides has on luminescence in a material. The presence of these oxides also determines how transparent the material is to SW UV. As most natural gemstones always contain some degree of impurity in the form of iron, this not only diminishes the intensity of luminescence in minerals containing activators such as chromium, but in most natural stones it greatly reduces their transparency to SW UV.

Until the early 1980s, manufacturers of synthetic gemstones such as spinel, ruby, sapphire, alexandrite and emerald experienced difficulty in introducing iron oxides into their products (see Chapters 15 and 16). As a result, these synthetics were much more transparent to SW UV than their natural counterparts. This fact was exploited in a technique known as immersion contact photography. The technique had originally been developed by B.W. Anderson to reveal differences between the RIs of gemstones, but was extended by N. Day to distinguish between natural and synthetic emeralds and rubies.

Using this method, the specimen under test is placed (together with a natural stone as a reference) table facet down on a piece of photographic paper in a darkened room. The stones and the paper are then placed in the bottom of a shallow dish containing water, and exposed to SW UV light for a few seconds (the water enables the UV to enter the stone rather than be reflected from its facets).

When the photographic paper is developed, the reference stone will appear white (i.e. it will have absorbed the SW UV) while the test specimen, if it is a synthetic, will appear black with a white rim. Although later productions of synthetic stones containing iron oxides (e.g. Kashan rubies) failed this test and absorbed UV like the natural comparison stone, on those occasions when the test stone appears highly transparent to SW UV the result is still valid.

A more convenient test of the SW UV transparency of a stone was later provided by the Culti company of Japan in their 'Color Stone Checker' (Figure 12.7). This is a

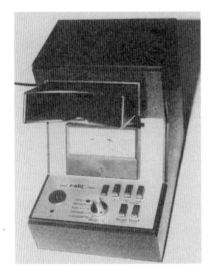

Figure 12.7 The Culti 'Color Stone Checker' contained both a LW and a SW UV lamp as well as a reflectance meter. The SW UV lamp was also used to check the transparency of gemstones at this wavelength

combination instrument comprising a reflectivity meter, a LW and SW source of UV, and a UV transparency detector. It was designed to enable the less experienced jeweller to distinguish between natural alexandrites, emeralds, rubies and sapphires and their many simulants and synthetics.

The relative transparency of natural and synthetic stones is measured by placing the gemstone over a test aperture in the instrument's UV compartment. A piece of scheelite beneath the aperture fluoresces as the SW UV passing through the stone reaches it, the intensity of this fluorescence depending on the UV transparency of the stone. A photodiode alongside the scheelite converts the fluorescence into an electrical signal which is displayed on a meter. Suitable calibration is provided which enables the gem under test to be identified as natural (absorbing SW UV) or synthetic (transparent to SW UV). As with immersion contact photography, synthetic stones containing iron oxides fail this test as they produce readings similar to those of natural stones.

Near-UV transparency tester

The main manufacturer (C3 Inc.) of synthetic moissanite also produced a transparency tester to separate this simulant from diamond. The C3 tester uses the transparency/opacity of suspect stones to near-UV as generated by a small high-intensity halogen lamp. Diamond transmits this near-UV radiation but synthetic moissanite absorbs it. However, an unknown near-colourless stone must first be tested with a standard thermal conductivity probe, as only those samples which have high thermal conductivity (i.e. diamond and synthetic moissanite) need to be tested on the C3 unit, which will then indicate whether the stone is synthetic moissanite or diamond.

Electrical properties

Although most gemstones are electrical insulators, there are a few, such as hematite, synthetic rutile and natural Type IIb blue diamond, which will pass a current if a voltage is applied across them (see Figure 12.8). The semiconductor property of natural blue diamond has already been mentioned, and is due to the presence of impurity atoms of boron in the crystal lattice. In the case of blue diamond, the use of either the simple

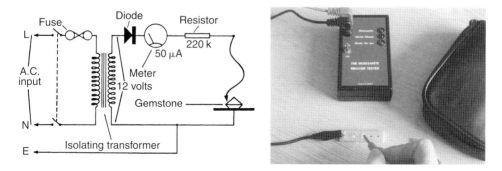

Figure 12.8 (left) A test circuit for checking the electrical conductivity of gemstones. (right) The battery-operated Synthetic Moissanite Megger Tester Mk II detects the semiconductor property of this diamond simulant (see text). The tester also verifies the semiconductor property of natural blue diamonds and synthetic boron-doped blue diamonds (diamonds whose blue colour has been artificially induced by atomic irradiation are non-conductors of electricity). (Gemmological Instruments)

circuit or the tester illustrated in Figure 12.8 will identify the naturally coloured stone (which passes an electrical current) from an electrically non-conducting blue diamond whose colour has been artificially induced by atomic particle radiation. (*Note*: synthetic blue diamonds which have been boron-doped are also semiconductors).

The diamond simulant synthetic moissanite is also a semiconductor but conducts electricity much less readily than natural Type IIb diamond and requires a more sensitive tester to identify it (see the Megger Tester in Figure 12.8). Moissanite's semiconductor property also appears to be due to impurities. However, some moissanites contain fewer surface impurity atoms and test as non-conductors. If these stones are irradiated by LW UV while being tested with the Megger Tester this will stimulate conduction and still give a moissanite reading on the tester. Because of the simulant's variable semiconductor property it is recommended that all suspect synthetic moissanites are tested while being irradiated by LW UV.

Another electrical property exhibited by some minerals, and in particular by quartz and tourmaline, is the *pyroelectric* effect. When pyroelectric minerals are heated, they develop an electric charge across the opposite ends of their crystal axis. Tourmaline is particularly sensitive in this respect, and if warmed by the sun or the lights in a jeweller's display window, will attract dust particles. Tourmaline and quartz also possess *piezoelectric* properties (an effect which was discovered by the Curies in 1880), and become electrically charged when stressed mechanically in certain directions. When a piece of quartz which has been cut along a certain axis is compressed, positive and negative electric charges appear on its opposite faces. The polarity of these charges reverses when the crystal is placed under tension (i.e. when it is 'stretched').

The piezoelectric properties of quartz make it particularly suitable for use as an electrical resonator in oscillator circuits. For this application a plate of the crystal is cut in either of the directions indicated in Figure 12.9. When an alternating voltage is applied to metal electrodes secured to the two flat sides, the crystal will be set in mechanical vibration. When the frequency of the alternating current coincides with the natural mechanical oscillation of the plate (which is controlled by its dimensions) the amplitude of vibration greatly increases. In a suitable electrical circuit, this can be used to control precisely the frequency of oscillation of that circuit.

This is also the principle of operation used in the quartz watch and is the reason why the synthesis of quartz has been pursued with such vigour. Most natural quartz is twinned

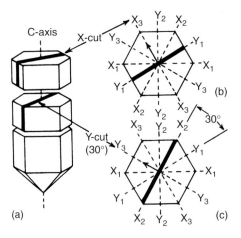

(a)   (b)   (c)

Figure 12.9 The piezoelectric effect in a quartz crystal (a) produces mechanical stress along the Y axes when an electrical charge is applied across the X axes (b) Plates are cut from the crystal as shown for use in electrical oscillators. The 30° Y-cut plate (c) oscillates more readily, particularly at lower frequencies, than the X-cut plate

and is unsuitable for piezoelectric applications so it is necessary to obtain untwined material by artificial means (the detection of twinned and untwined quartz is discussed in Chapter 16 in connection with synthetic amethyst). The other main application for quartz's piezoelectric property is as a mechanical/electrical generator of sound waves in mobile telephones, buzzers and hi-fi tweeters (the 'crystal pickup' application has now been replace by the laser in CD equipment).

Another electrical property possessed by some minerals is *photoconductivity*. When this is present, the mineral's normally high electrical resistance falls when it is exposed to electromagnetic radiation such as UV and γ-rays. Semiconducting diamonds containing boron are photoconductive to γ-rays and are used as radiation detectors in environments which require a tough corrosion-resistant sensor material (e.g. radiation monitors, similar to geiger counters, for use in the core of an atomic pile).

Amber is a *triboelectric* material which develops a negative electrostatic charge on its surface when rubbed. When this happens it is capable of picking up small fragments of tissue paper. However, as several of amber's plastic imitators possess the same property, this should not be relied upon as an identifying feature without other confirmatory tests.

Thermal conductivity and thermal inertia

Metals are normally very good conductors of both heat and electricity, while most gemstones tend to be poor heat conductors and equally poor conductors of electricity. The most dramatic exception to this rule is diamond, which is able to conduct heat many times better than copper or even silver and, with the exception of natural blue diamond, remains a non-conductor of electricity.

This ability of diamond to conduct heat has long been exploited in the electronics industry as an efficient means of channelling heat from high-power semiconductor devices into copper heat sinks. However, it was not until 1979 that the first gem test instrument to use this phenomenon for the identification of diamond became commercially available (Figure 1.6; also see Figure 12.10 for later version).

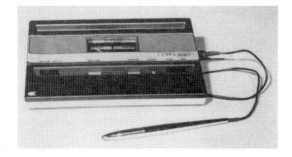

Figure 12.10 The Diamond Probe II thermal conductance tester. Two test surfaces beneath the meter enable calibration to be tested. The test probe uses miniature thermistors

In order to compare the relative thermal properties of gemstones it is first necessary to define these properties. Thermal conductivity is a measure of the ability of a material to conduct heat, while the rather more complex concept of thermal inertia indicates a material's response to a periodic (i.e. cyclically varying) source of heat. In order to explain the occasional anomalies that occur with some instruments designed to compare the thermal conductivity of gemstones, Dr B. Hoover has suggested that these instruments are in fact measuring thermal inertia rather than thermal conductivity. Because of this it has been agreed that the term 'thermal conductance meter' more accurately describes these instruments. However, it should be remembered that several of the simpler units use a constant rather than a varying source of heat, and the stones under test conduct the heat from the probe into either the metal mount or, in the case of unmounted stones, into a metal heat sink provided with the instrument.

Thermal conductivity is measured in units of watts per metre per degree celsius (W m^{-1} $°C^{-1}$). The thermal conductivity of diamond at room temperature varies from 1000 W m^{-1} $°C^{-1}$ for Type I material to 2600 W m^{-1} $°C^{-1}$ for Type IIa material. As the next highest gemstone to diamond in thermal conductivity is synthetic moissanite at around 200–500 W m^{-1} $°C^{-1}$, followed by corundum at 40 W m^{-1} $°C^{-1}$, this property forms a very effective test for diamond and a means of separating it from its various simulants (with the exception of synthetic moissanite) as indicated in Table 12.2. The table also indicates that some crystalline materials exhibit directional thermal conductance.

Table 12.2

Mineral	Thermal conductivity (W m^{-1} $°C^{-1}$)
Diamond	1000–2600
Synthetic moissanite	200–500
Silver	430
Copper	390
Gold	320
Platinum	70
Corundum	40*
Zircon (high)	30*
YAG	15
GGG	8
Rutile	8*
Quartz	8*
CZ	5
Glass	1

* Mean value of thermal conductivity between c axis and a axis directions.

Thermal conductance testers

A typical thermal diamond tester (see Figure 12.11) consists of a test probe whose metal tip is electronically heated, and a control box which contains circuits to detect the large fall in the temperature of the test tip when it is placed in contact with a diamond's surface. No similar temperature drop is produced by any diamond simulant (natural or man-made) as their ability to conduct or absorb heat is less than that of diamond (although synthetic moissanite, with a thermal conductivity closer to diamond than any other simulant, may give a false diamond reading on most thermal testers). The loss of heat from the probe tip due to diamond is indicated either by a meter reading, a digital display or an indicator lamp. An audible signal is sometimes provided to reinforce the visual indicator.

The method of heating the probe tip and sensing its temperature ranges from the use of one or two thermistors (an electrical component whose resistance changes in proportion to its temperature) to the use of a resistive heater and a thermocouple (a welded junction made of two dissimilar metals which develops a voltage proportional to its temperature). A metal test plate is usually provided to act as a heat sink for unmounted gems of less than half a carat. With mounted stones, the heat passing through the stone (if it is a diamond) is dissipated by the mount.

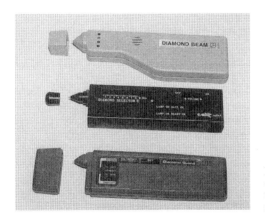

Figure 12.11 Three hand-held battery-operated thermal conductance testers designed to discriminate between diamond and its simulants

The metal used in gemstone mounts has a thermal conductivity approaching that of diamond (see Table 12.2) and if the test probe tip inadvertently touches the mount this may result in a false diamond indication. To avoid this type of error, some testers are fitted with a circuit which sounds a warning buzzer should the probe tip touch the metal mount. This circuit applies a small voltage between the test tip of the probe and the metal body of the probe. If the stone mount is held in one hand and the probe in the other hand, a minute current passes through the body of the person testing the stone when the tip touches the metal mount. This current is detected and activates the warning buzzer. Because the advantages and disadvantages inherent in thermal conductance meters nicely complement those found in reflectance meters, several manufacturers have combined both techniques in the same instrument (Figure 12.12).

Although the majority of thermal testers are designed to distinguish between diamond and its simulants, a few instruments such as the Alpha-test (Figure 12.13) have been marketed which are able to identify both these and other gems by means of their

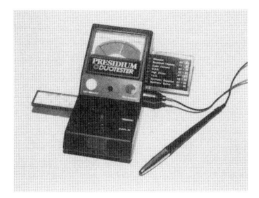

Figure 12.12 The 'Duotester' is a combined reflectance meter and thermal conductance tester. The slide drawer on the left contains a range of diamond simulants for use when checking calibration. The pull-out panel on the right lists typical reflectance readings which appear on an LED display in the front edge of the instrument

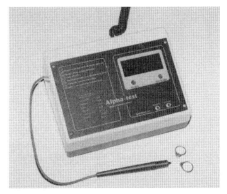

Figure 12.13 The Alpha-test instrument is a thermal tester designed to check the thermal conductivities of a range of gemstones (including diamond and its simulants). (H.S. Walsh & Sons Ltd)

thermal conductivity. With the exception of diamond, gemstones occupy a very limited range of thermal conductivities which extend from glass at 1.0 to corundum at 40. Because of this, such instruments have to be very sensitive in order to discriminate between the much smaller thermal differences, and their use involves several operating constraints.

The test environment must be free from draughts and within a specified temperature range. The gemstone must also be stabilized at a fixed temperature before testing (usually body temperature), and the probe tip applied at right angles to the surface at a preset pressure.

The method used to sense the thermal conductivity of gemstones in this narrow range is quite ingenious. When the temperature of the probe tip starts to fall on contact with a gemstone, it first reaches a preset level at which a timer circuit is turned on. At a second lower preset temperature the timer is turned off again, and the comparative thermal conductivity of the gemstone is displayed digitally as a time interval between the two temperature levels. The greater the thermal conductivity of the stone, the shorter is this time interval and the smaller the readout number. However, in the same way that many gemstones have similar or overlapping SG and reflectivity values, there are many stones with similar or overlapping thermal conductivity values. The use of this type of instrument is therefore limited to the identification of only a relatively small selection of stones, and to providing guidance in the case of very small mounted stones that cannot be tested on more conventional equipment.

Although in general there is little difference between the thermal conductivity of natural gemstones and their synthetic versions, this more sensitive type of instrument can, however, be used to distinguish between natural emerald and the Chatham, Crescent Vert, Gilson and Lennix flux-melt synthetic products as the latter have around twice the thermal conductivity of the natural stone. Hydrothermally produced synthetic emeralds (such as the Biron and solid Lechleitner/Linde) also have better thermal conductance than the natural stone, but the difference is not as great as with the flux-melt synthetics.

The Alpha-test can also be used to separate greenish-brown peridot from sinhalite whose SG, RI and double refraction constants are close enough to each other to make identification difficult (sinhalite's thermal conductivity is twice that of peridot and gives a lower reading on the Alpha-test instrument).

Chapter 13

The hand lens, microscope and Chelsea filter

At the end of Chapter 1, mention was made of the wide range of gem testing equipment now available to the gemmologist/jeweller, and some of this has already been described and illustrated in earlier chapters. Despite the sophistication of much of this equipment, perhaps the most useful and frequently used of all these items is the hand lens, or *loupe* as it is often called.

Types of hand lens

The ideal magnification factor for a hand lens is $10\times$ (i.e. the length of an object magnified by ten) as this is sufficiently powerful to reveal the majority of a gemstone's internal and external features (it is also the standard magnification used for diamond clarity grading). Greater magnifications (up to $25\times$) are available, but these lenses have a very critical focus and a limited field of view and working distance (the distance between the lens and the gem), all of which makes them more difficult to use.

Image distortion and colour 'fringing' are problems associated with single element lenses having magnifications greater than $3\times$. Because of this, all high-quality loupes, particularly those intended for use in diamond grading, are fitted with compound lenses consisting usually of three elements. These loupes are called 'triplets', and are corrected for both *chromatic* and *spherical* aberration (Figure 13.1).

Chromatic aberration, or distortion, causes colour fringes to be produced round an image as the result of the optical dispersion of the glass used in the lens (Figure 13.2). In the quality loupe this is corrected by making the lens in two sections, each having a different dispersion. One lens is a bi-convex and the other is a bi-concave. Because of the different dispersion of the two lenses, all the colour components in white light can be brought to a common focus. This type of compound lens is termed *achromatic*.

Spherical aberration, which occurs in strongly curved lenses of high magnification, is caused by the focus of rays passing through the edges of the lens lying in a slightly different plane to those which pass through the central area of the lens (Figure 13.3). This results in a hazy and sometimes distorted circumference to the image as seen through the loupe. Spherical aberration is corrected by using more than one lens element (with surfaces of different curvatures) to achieve the desired magnification. Such a lens is termed *aplanatic*. Alternatively, the outer circumference of the lens can be blanked off. The term 'apochromatic' is used for a lens system which has been corrected for both chromatic and spherical aberration.

Even between high-quality triplet hand lenses of the same magnification there can be differences in the field of view, and where the instrument is in constant use, the final choice may be for a loupe fitted with a wide-field lens system.

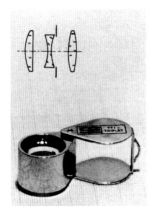

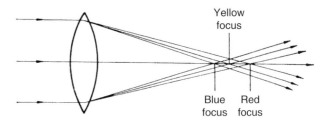

Figure 13.2 The dispersion of a lens produces chromatic aberration

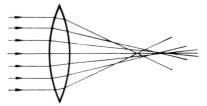

Figure 13.1 The components of a triplet hand lens

Figure 13.3 The outer periphery of a lens produces spherical aberration

The gemmology student may initially experience some problems in the use of the 10× hand lens. This arises from the need to keep the gemstone, the hand lens and the head all steady and in the correct alignment in order to maintain focus. The best way out of these difficulties is first to hold the lens close to the eye (the hand holding the lens can be steadied against the cheek or nose), and then, using the other hand, to move the gemstone (held in tweezers) or the piece of jewellery towards the lens until it comes into focus. Wavering of the hand holding the gem can be prevented by bringing it into contact with the hand holding the lens (Figure 13.4). Movement and fatigue can be further prevented by resting both elbows on a convenient surface. If spectacles are used, the lens should be held as close to these as possible.

Figure 13.4 The author demonstrating the method of holding the hand lens and tweezers for maximum steadiness

One piece of physiological advice is relevant here; if the hand lens is going to be used for several hours a day on a continuing basis, it is advisable to learn the technique of keeping the unused eye open (and to ignore the image from this eye). This will prevent any long-term changes that can occur to an eye whose functions are static for long periods in relation to the working eye.

With all loupes, the correct illumination of the stone under inspection plays a vital role. The lamp used to illuminate the stone should be adjusted so that its light is angled into the side of the stone and not directly into the eyes. Any internal features of the gem will then appear brightly visible against the darker background of the stone. This type of lighting is called 'dark-field' illumination and will be referred to again later in this chapter when discussing the microscope.

Some hand lenses have been combined with battery-powered light sources to provide built-in dark-field and incident (top-reflected) illumination (Figure 13.5). Most hand lenses are triplets with a magnification of 10× (Figure 13.6). Where long periods are spent on close inspection, there are lens combinations that can be clipped to one side of the spectacles, and a head loupe which although of a lower magnification

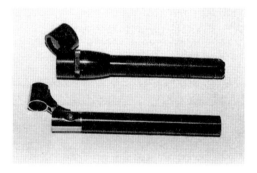

Figure 13.5 Two illuminated 10× lenses which can be pivoted at the appropriate angle to the battery-operated light source to give dark-field and incident illumination

Figure 13.6 A range of triplet hand lenses. The second one from the top is fitted with a rubber grip

than 10× has the advantage of binocular vision (Figure 13.7). Both of these versions have the added bonus of leaving the hands free to manipulate the object under inspection.

Types of microscope

Where magnification, mechanical stability or illumination become limiting factors in the use of the loupe, the microscope is the next choice. For many gemmologists this is the preferred instrument, although for initial inspection the loupe has the great advantage of flexibility and portability. It has been claimed, with some justification, that gemmological analysis depends largely on the trio of instruments formed by the microscope, the refractometer and the spectroscope.

Figure 13.7 The Optivisor head loupe. (R. Rubin and Son)

With the trend in modern gemmological microscope design favouring the binocular stereo model, the length of time that can be spent without strain in exploring the interior of a gemstone has made this instrument a valuable diagnostic tool, particularly when dealing with the more sophisticated synthetics. The diagram in Figure 13.8 indicates the principal components of the standard stereo microscope. As with the hand lens,

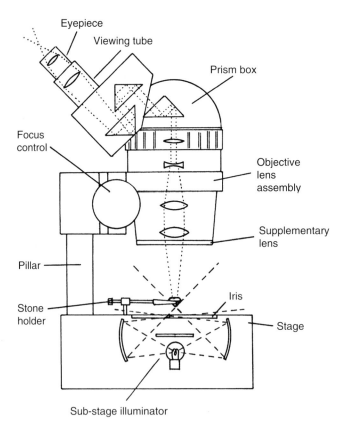

Figure 13.8 Sketch showing the components and one of the two ray paths in a typical stereo microscope. The illuminator in the sub-stage section is set for dark-field work

there are several important features to be considered when choosing a microscope (excluding the price!). These are the magnification, field of view, working distance and illumination.

The greater the magnification of the microscope, the smaller the field of view and the working distance (i.e. the distance between the bottom of the objective lens and the specimen). In addition, the depth of focus (i.e. the distance over which the image is within focus) becomes smaller as the magnification is increased. For these reasons, the maximum magnification for gemmological work is in the region of $60\times$ to $80\times$, with the majority of work being done in the region of $15\times$ to $30\times$. Microscopes (like hand lenses) which are used for the clarity grading of polished diamonds must also cover the standard magnification of $10\times$.

For gemstone inspection, a small but important microscope accessory is the stone holder. This should be capable of holding the gem under inspection securely in all positions and of allowing it to be rotated and moved laterally in the field of view. The stone holder should ideally be detachable so that the stone can be more easily inserted in the jaws of its tweezers away from the confines of the microscope.

Although the majority of modern microscopes are of the binocular type, these are divided between those which have a single objective lens system whose image is shared between the two eyepieces, and the more expensive stereo microscopes which have two objective lens systems each coupled to its respective eyepiece. There are also two basic types of stereo microscope, one using the Greenough system with two converging sets of optics, and the more expensive combined-objective system with two paralleled optical systems. Both produce an erect (correctly orientated) image.

Many microscopes are available with a zoom objective lens system which gives a stepless range of magnification. However, for comparative work where a knowledge of dimensions is important, the known fixed magnifications of the multi-objective turret lens may be preferable to the zoom system. As a compromise, the magnification controls on some zoom microscopes are provided with click stops at specific magnifications.

The overall magnification of a microscope is arrived at by multiplying the magnification factor of the eyepiece with that of the objective. If a supplementary or adaptor lens is fitted to the objective, the overall magnification factor must then be multiplied by the factor of this additional lens. The working distance and the diameter of the field of view are both reduced in direct proportion to the increase in the magnification factor of the objective lens. Changing the power of the eyepiece similarly affects the field of view, but not the working distance.

While some detail has been given here on the construction of a typical gemmological microscope, students taking the British Gemmological Association's courses are advised that all that is asked for in the examinations is a description of the microscope and its many gemmological applications – a detailed diagram of the instrument is *not* called for.

Light-field, dark-field and incident illumination

Apart from the simplest of monocular models, most microscopes are provided with a built-in means of illuminating the specimen on the stage. This can be as basic as a lamp under the stage and a diffuser plate. An added refinement may take the form of an iris control to vary the area of illumination. In the more expensive models, a choice of *incident*, *light-field* and *dark-field* illumination is provided, the latter two being contained beneath the stage (see Figure 13.8, 13.9 and 13.10).

Figure 13.9 A low-cost stereo microscope using a turret lens system for 10× and 30× magnification using 10× eyepieces, a range which can be doubled with 20× eyepieces. (Gem-A Instruments Ltd)

Figure 13.10 A typical stereo zoom microscope with overhead and light/dark-field illumination and a magnification of 0.7–45× using 10× eyepieces, a range which can be doubled with 20× eyepieces. (Gem-A Instruments Ltd)

   With light-field illumination, light is transmitted up through the specimen and into the microscope's objective. When dark-field illumination is chosen, a baffle plate is pivoted over the lamp to block the direct light path to the objective, and light from the lamp is directed into the sides of the specimen by means of a radial sub-stage mirror. Alternatively, dark-field illumination can be achieved by means of one or two horizontally positioned spotlights or fibre-optic light guides. Dark-field illumination is a major requirement for serious gemmological work as it makes a gemstone's internal features (inclusions, stress flaws, etc.) more clearly visible against the darker back-ground of the stone.

Incident illumination is often provided by means of a separate lamp (or lamps) mounted above the stage and fitted with a reflector or a lens to provide a focusable spot. Alternatively, incident light can be channelled to the specimen from the sub-stage lamp (or from an external source) by means of flexible fibre-optic light guides. For diamond colour grading purposes, the incident illuminator should be a colour-corrected fluorescent lamp giving simulated daylight illumination (described as 'north daylight' in the northern hemisphere).

Pin-point illumination and shadowing techniques

Pin-point illumination is yet another technique which enables light to be injected into a stone to reveal inclusions. With normal large area incident illumination, much of the light is reflected back from the stone's surface. With pinpoint illumination, a fibre-optic light guide probe is used to inject a small diameter high-intensity beam into the stone with minimum backscatter of light.

The *shadowing* technique employs an opaque black light shield that is introduced between the light source and the specimen. The improvement in image contrast obtained by shadowing can be seen in its simplest form in the use of a microscope's iris control to restrict the area of illumination. However, J.I. Koivula (AGTA Testing Center, Carlsbad, USA) has developed the technique to the point where the diffraction and scattering of light rays from the edge of a light shield can be used to enhance the contrast of any inclusion which has a refractive index different from the host crystal.

The scattered light rays produced at the edge of the shadowing light shield travel through the host crystal until they reach an inclusion having a different optical density. Some of these rays are reflected away from the microscope's objective and cause a darkening of that area of the image, while those areas of the inclusion which refract the rays into the microscope appear light, the overall effect enhancing the contrast of the inclusion. Shadowing is achieved by using transmitted light and inserting an opaque shield of appropriate shape and dimensions between the light source and the specimen. The shield is then slowly moved across the field of view until shadowing begins to take place and the inclusion under inspection appears to stand out in sharp contrast to the background of the stone.

Immersion techniques

In Chapter 7 mention was made of the 'disappearance' of a gemstone when it is immersed in a heavy liquid having a refractive index close to that of the gem. This phenomenon is caused by the relationship between a material's reflectivity and its refractive index as revealed by the Fresnel equation:

$$\text{Reflectivity} = \frac{(n - A)^2}{(n + A)^2}$$

where n is the RI of the specimen and A is the RI of the surrounding medium.

When the RI value A for an immersion liquid approaches the value of n, the reflection of light from the gemstone's surface falls towards zero, and unless the stone has a deep body colour it becomes difficult to see. For the purpose of inspecting features such as inclusions inside the gem, however, this situation can be turned to advantage. One of the problems encountered when illuminating a gemstone is that a significant proportion of the light is reflected back from its surface. This can make it particularly difficult to light and inspect the interior of a faceted stone. If, however, the gem is immersed in a liquid having a similar RI, the light rays will no longer be reflected back from the facets but will enter the stone and illuminate its interior.

This reduction of surface reflection is achieved by placing the gemstone in an immersion cell (Figure 13.11) containing a liquid having the appropriate RI (even

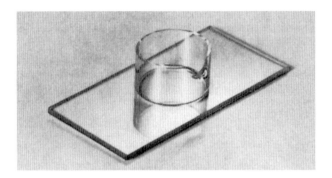

Figure 13.11 A glass immersion cell for use with upright microscopes

water will produce a worthwhile improvement). The following list includes a number of suitable immersion liquids together with their RIs:

Table 13.1

Immersion liquid	RI	Immersion liquid	RI
Water	1.33	Bromoform	1.59
Alcohol	1.36	Iodobenzene	1.62
Turpentine	1.47	Monobromonaphthalene	1.66
Benzyl benzoate	1.56	Di-iodomethane (methylene iodide)	1.74

Note: Porous stones such as opal and turquoise, or those with surface-reaching flaws, should not be immersed in these liquids as discoloration may result. Liquids such as benzyl benzoate are strong solvents and may dissolve the cement used in some composite stones. To prevent floating particles from confusing the image of an immersed stone it is advisable to filter the liquid before using it.

With some microscopes intended specifically for immersion work, the lens assembly has been designed so that the light path into the objective is horizontal (Figure 13.12). This is done to allow the stone holder to be inserted vertically into a deep immersion cell. With the more conventional upright microscope the height of the cell has to be limited to allow for the introduction of tweezers to hold the stone.

Additional uses for the microscope

Many microscope manufacturers design their instruments on a modular basis, the various optical heads, stands, sources of illumination and accessories being interchangeable. This enables microscopes to be assembled to meet the needs of a variety of

Plate 1 Pink tourmaline and diamond suite on quartz crystal (with tourmaline needle inclusions). (Courtesy of P.J. Watson)

Plate 2 Pink topaz and cultured pearls. (Courtesy of P.J. Watson)

Plate 3 Zoisite (tanzanite) crystals, gemstones and jewellery. (Courtesy of P.J. Watson)

Plate 4 Brazilian topaz and diamond ring. (Courtesy of P.J. Watson)

Plate 5 Sri Lankan yellow sapphire and diamond ring. (Courtesy of P.J. Watson)

Plate 6 Malachite rough and polished

Plate 7 Polished rhodochrosite specimens

Plate 8 Andalusite ring stone showing pleochroism

Plate 9 A group of CZ crystals and polished specimens

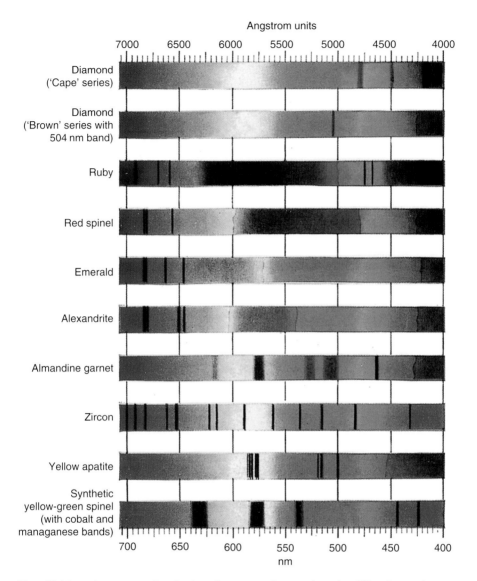

Plate 10 Absorption spectra of a selection of gemstones (as seen through a diffraction grating spectroscope)

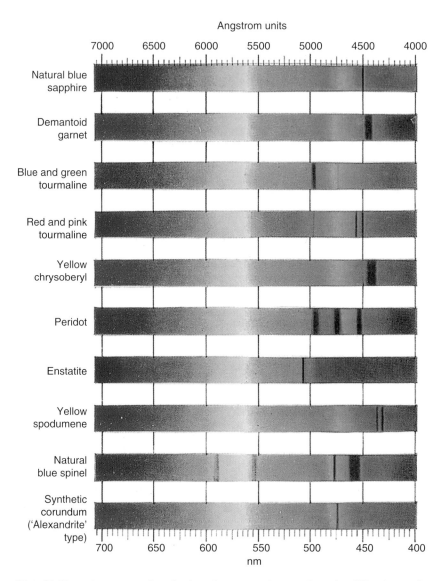

Plate 11 Absorption spectra of a selection of gemstones (as seen through a diffraction grating spectroscope)

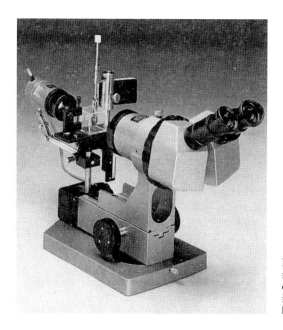

Figure 13.12 The Krüss KA11 stereo zoom microscope has a basic magnification range of 10–40×. It is designed for horizontal immersion work, but can easily be converted for upright operation

Figure 13.13 A twin polarizer attachment that can be fitted over a microscope's immersion cell to act as a polariscope. (Eickhorst)

applications. Microscope accessories are equally numerous, and include polarizing filters (Figure 13.13), colour and neutral density filters, camera and video attachments (Figure 13.14), tracing attachments and projection screens.

Of these items, the most useful for gemmological work is the polarizing filter. The microscope can be converted for polariscope analysis by inserting one filter (as a polarizer) between the light source and the specimen, and attaching a second one (as an analyser) to the objective. The two filters can then be set to the extinction position to reveal internal stresses (referred to as anomalous birefringence), single/double refraction and the microcrystalline or polycrystalline structure of a gemstone. If the polarizer filter is removed, the remaining one on the objective can be used to check for dichroism, or to enhance the visibility of inclusions.

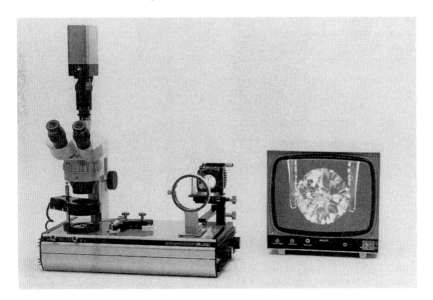

Figure 13.14 A video camera has been attached to the trinocular port of this microscope to give a greatly enlarged image on a monitor. (Eickhorst)

Using the various attachments, a microscope can become a very versatile instrument with the following wide range of applications:

1. Interior examination of gemstones for inclusions or growth features which identify the gemstone as natural or synthetic.
2. Surface inspection of gemstones to assess the quality of cut.
3. The detection of dichroism (using a polarizing filter on the objective).
4. The detection of double refraction using two polarizing filters as polarizer and analyser.
5. Spectrum analysis using a spectroscope in place of an eyepiece.
6. Colour grading of diamonds using a daylight-type incident light source.
7. Clarity grading of diamonds using 10× magnification.
8. Approximation of RI using 'Direct Method'(*real* divided by *apparent* depth).
9. Measurement of a gemstone's proportions and interfacet angles using an eyepiece graticule.
10. Photomicroscopy with camera attachment, and videomicroscopy with video camera attachment.

Care in the use of the microscope

1. To avoid the possibility (when focusing) of lowering the objective onto the specimen and thus scratching the lens, it is good practice always to start with the objective just clear of the specimen. The initial focus setting can then be found by racking the objective away from the specimen. Focusing is more easily carried out by using the lowest magnification power of the microscope and then increasing this as required.
2. Care should be taken when using immersion liquids. Some of these are corrosive and any spillages should be cleaned up immediately.

3. Keep the instrument clean and free from dust. The microscope lenses should not be touched directly with the fingers – if necessary use a lens tissue to remove dust from the external lenses on the eyepieces and objective.
4. When not in use protect the microscope with a suitable cover.

The Chelsea filter

A simple test instrument called the Chelsea filter (Figure 13.15) was developed in 1934 by B.W. Anderson and C.J. Payne in collaboration with the gemmology class at the Chelsea College of Science and Technology, London, to provide a means of separating emerald from its many simulants.

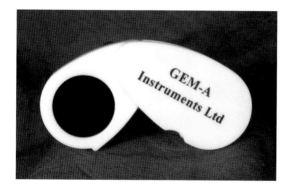

Figure 13.15 The Chelsea filter transmits light in the red and green wavelengths to match emerald's red transmission and yellow-green absorption. (Gem-A Instruments Ltd)

The original filter consisted of a combination of two carefully selected gelatine filters whose combined transmission response matches the deep red transmission and yellow-green absorption of emerald. When green stones are strongly illuminated and viewed through the filter, emeralds appear distinctly red or pinkish (depending on the depth of colour of the stone), while emerald simulants (i.e. imitations) appear green. This red/pink appearance of emerald is due to the stone's colouring element, *chromium*. Subsequently, a single gelatine filter by Ilford was used which had both a red and a green transmission 'window'.

When synthetic emeralds appeared on the market, the Chelsea filter became less reliable as a test instrument as the synthetic product also contains chromium and appears red through the filter (although the red in this case is usually more distinct – there are also a few natural emeralds, in particular those from South Africa, which *fail* to show pink through the filter).

The Chelsea filter also fails with demantoid garnet and green zircon which like emerald appear pink through the filter. However, both of these stones produce a 'negative' reading on the refractometer (i.e. they are above its range) and unlike emeralds will sink when placed in bromoform. Green tourmaline, especially when heat-treated, may occasionally be used as an emerald simulant. Although some green tourmalines contain chromium, and like emerald also appear pink through the Chelsea filter, they can easily be identified by taking a refractometer reading. Emerald will also float in bromoform, while tourmaline sinks.

Despite the present-day emerald testing shortcomings of the Chelsea filter, it does have a useful secondary role. Cobalt-coloured synthetic and man-made materials (such as blue synthetic spinel and blue cobalt glass imitating aquamarine, sapphire and blue zircon) appear red through the filter, unlike the gems they simulate. With the exception of pink smithsonite and a rare blue spinel, there are no natural gemstones that contain cobalt, so this test can provide an early warning that the stone under examination could be a synthetic. Even with this test there is the occasional exception – some Sri Lankan blue sapphires contain enough chromium to appear pink through the filter.

When using the Chelsea filter it is important to illuminate the gem under test with a strong tungsten light source. For best effect, the filter should be held close to the eye (not at arm's length!). Test results obtained with the filter should never be relied upon as a positive identification of a stone, but must always be backed up with other confirmatory tests.

Handling gemstones

Tweezers, or *stone tongs* as they are also called, are one of the basic handling tools of the gemmologist. As can happen when attempting to use chopsticks for the first time, the student may find it difficult initially to handle tweezers with confidence. The problem areas with both are similar. First, it is necessary to be able to pick up a gem so that it is secure in the tweezers; then it is necessary to apply just the right amount of pressure to the arms of the tweezers to retain the stone securely. Too much pressure often causes a poorly positioned stone to fly out of the tweezers (a situation best avoided in a practical exam!).

While the beginner may feel safer with the self-closing prong type stone holder (Figure 13.16), this does tend to mask parts of the stone, and in the case of diamond grading may affect the assessment of a gem's clarity. Continued reliance on this type of holder will also prevent the user from developing the skill to use the more versatile tweezers.

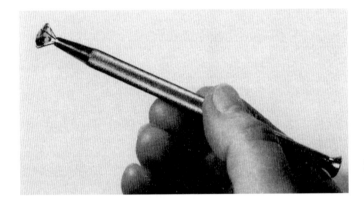

Figure 13.16 The self-closing four-prong stone holder

When choosing tweezers, the student may be surprised at the variety which are available (Figure 13.17). For gemmological use it is best to start with a pair which are about 150 mm long and have slightly rounded (rather than pointed) tips. The inside surface of these tips should have a roughened or milled surface to provide a better grip on the stone.

The best method for picking up a stone is first to lay it, table facet down, on a flat firm surface, and to position the tweezers parallel to this surface. The stone should then be lightly gripped with the tips of the tweezers just over the centre line of the girdle. As an aid to gripping the girdle more firmly, a groove can be filed on the inside faces of the tips, in line with the length of the tweezers.

A gemstone scoop is another very useful handling tool, particularly when dealing with a number of stones, as this allows them to be rapidly gathered up in a single operation and

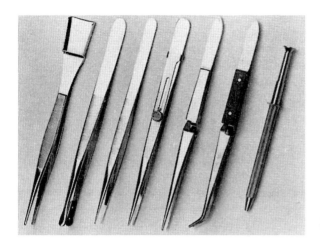

Figure 13.17 A selection of stone tweezers. The middle one has a locking slide to maintain the tweezer grip without the need for finger pressure. The two on the right of it have press-to-open reverse action

transferred back to their packet. A range of small chrome-plated brass scoops is available from suppliers of gemmological accessories (Figure 13.18).

Anyone associated with the gemstone trade must also become familiar with handling gems in stone papers. These stone or diamond papers are manufactured in several sizes,

Figure 13.18 A selection of stone scoops. (R. Rubin and Son)

the most popular being the No. 2 paper which measures 4×2 inch (102×51 mm) when folded, and can hold up to 50 one-carat size stones. Stone papers are usually provided with an inner tissue liner to provide further protection for the contents, and these are sometimes tinted to provide the appropriate background for coloured gems.

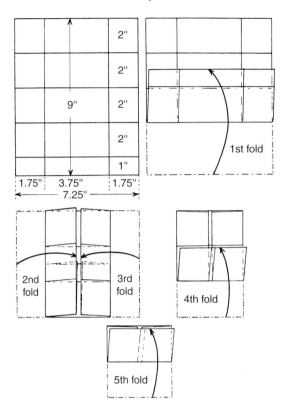

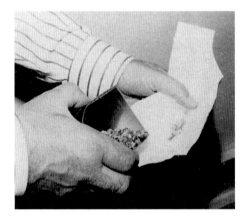

Figure 13.19 Dimensions and folding sequence for making a No. 2 size stone paper

Figure 13.20 Holding a stone paper open to receive stones from a scoop (see text)

Stone papers are of simple construction, and this makes it possible for the gemmologist/jeweller to fabricate them easily should a fresh supply be needed in a hurry (see Figure 13.19). When emptying gems from a stone paper, this can be unfolded until only the left and bottom flaps remain in position to retain the stones. They can then be tipped out directly into a scoop or onto the bench. Similarly, stones can easily be poured from the scoop into the half-opened packet (as indicated in Figure 13.20) before refolding it.

Chapter 14

Gemstone enhancement

Any newcomer to the study of gemmology might reasonably assume that man-made improvement of a gemstone's appearance is rooted in our present-day hi-tech society. Although many of the more sophisticated treatments used to enhance the colour and value of gemstones have been developed comparatively recently, the artificial improvement of gem materials began a very long time ago.

In the first century A.D., Pliny the Second published his 37-volume *Natural History* in which were documented many gemstone treatments, including the use of foils and the oiling and dyeing of stones. Many centuries later, in 1502 Camillus Leonardus published his *Speculum Lapidum* in which he expanded on the earlier writings of philosophers such as Aristotle and Pliny. Like Pliny before him, Leonardus was also well aware of bubbles in glass imitation gems and even knew of composite stones!

Foils, colour backing and dyeing

The use of coloured foil or paper behind a poorly coloured or colourless stone, usually in a closed setting, was commonplace in antique jewellery. Mirror backs were also employed to lighten a dark stone. The foil or mirror back was sometimes embossed or scribed to produce a star effect (alternatively, the base of a cabochon was engraved to produce the same effect). The fact that these stones are mounted in a closed setting to conceal the backing material gives warning of possible deception, and this can often be confirmed with the aid of a hand lens or microscope.

Another relatively simple way of improving or changing a gem's appearance is to dye or stain it. Microcrystalline and polycrystalline gem materials have a slightly porous surface which makes them suitable for this type of treatment. Originally, organic dyes were used, but these deteriorated with age. Today, more stable inorganic dyes are commonly employed. Because of their porous surface, agates can be stained to change their colour and increase the contrast in their banding. A black onyx simulant has been produced by boiling chalcedony in a sugar solution and then treating the stone with sulphuric acid.

Jasper can be stained to imitate lapis lazuli, when it is called *Swiss* or *German* lapis. White or poorly coloured jade is stained to simulate the more valuable green variety, and the colour of turquoise can be similarly improved. With all of these techniques it is usually possible to detect the presence of staining under magnification, particularly where there are surface cracks which can reveal a concentration of the dye. In the case of stained green jadeite, a spectroscope test will usually show a tell-tale broad absorption band in the red due to the dye, and an absence of typical chromium lines in the same part of the spectrum. Dyed mauve jadeite may exhibit a bright orange fluorescence under LW UV. Most dyes can be removed with the aid of a cotton swab moistened with a solvent such as acetone. Poor-quality pearls are dyed black using a silver nitrate solution, which on exposure to UV deposits metallic silver in the gem's nacreous outer layers. Under crossed filters (see Chapter 12) these treated pearls appear inert, unlike the natural black pearl which appears pinkish. Painting the pavilion of a gem has also been used to enhance its colour. With yellow cape series diamonds, a thin translucent coat of blue or violet paint on the pavilion will make the stone appear less yellow (violet is the complementary colour to yellow). While most dyes can be removed by washing the gem in a suitable solvent, the bluish *fluoride* coating (as used on camera lenses) is more tenacious. When it is found on a diamond's pavilion it can only be removed by using an abrasive or by boiling in water.

Impregnation

Some gem materials such as turquoise are legitimately impregnated with colourless paraffin wax (or plastic, which gives a more permanent result) to stabilize them and prevent attack from acidic perspiration. Less legitimate is the use of coloured impregnants to increase the value of, for instance, colourless or pale turquoise. Veins of turquoise that are too irregular or thin to be made into cabochons are often backed with a metal-loaded epoxy resin. The turquoise is then set in a closed mount which hides the resin backing (the value of the backed turquoise is much lower than that of unbacked material of equal colour quality).

Colourless oils are also used to hide surface cracks, and coloured oil serves the double purpose of hiding surface flaws and improving the colour appearance of emeralds, rubies and sapphires (both polished and rough). With emeralds, the oil is sometimes drawn into the cracks under a vacuum by capillary action. In one method of emerald treatment, the flaws and cracks are first thoroughly cleaned out by soaking the stones in hydrochloric acid in a vacuum. The stones are then washed in an ultrasonic cleaner (a practice not normally to be recommended as it can damage flawed emeralds) and immersed in a pot of cedar oil which has an RI close to emerald's 1.57. The oil is heated to lower its viscosity and the pot is placed in a vacuum chamber to draw the air out of the cracks and force the oil in. With some treatments, the emeralds are then heated for several hours at 83°C to 'lock in' the oil.

Several commercial techniques are also commonly used for impregnating surface-reaching fractures in emerald. These techniques include the use of Canada balsam (a natural resin), or a colourless or green epoxy resin such as Opticon. In a more permanent process, similar to the Yehuda method used to fill diamond flaws (described later in this chapter under 'Glass Filling'), a colourless glass has been used to fill the fracture. With the Opticon method, cleaned faceted stones are immersed in the epoxy resin at a temperature of about 95°C for 24 hours. A hardening agent is then wiped across the surface of the stone and left for 10 minutes before the excess hardener is cleaned off.

In another version, the 'LubriGem' kit for colour treating emeralds uses an oil-based dye together with heat (but without vacuum) to induce penetration of the filler into the surface reaching flaws. The kit includes a thick-walled cylinder that is half filled with the choice of oil. Cedarwood oil is said to work best with Colombian emeralds, paraffin oil with Zambian (Opticon can also be used). The filler is heated to 95–100°C, and the emeralds are placed in a perforated glass cup and immersed in the filler. A spring-driven metal piston is inserted into the top of the cylinder and tightened by hand to pressurize the contents. After treatment, which ranges from 30 minutes to 12 hours depending on the emeralds, the heat is turned off. The emeralds are extracted and the surface oil removed with a cloth.

Careful use of low-power microscopy, with diffused transmitted light, should reveal the presence of unnatural fillers in surface-reaching fractures. *Warning* – the ultrasonic cleaning of emerald jewellery should be avoided as this can be deleterious to the stones whether they are oiled or not! Opals also suffer from surface cracks, and lapidaries often subject them to an oiling process to improve their appearance. The oil used here has an RI around 1.41, but since its viscosity is lower than the oil used for emeralds, it tends to evaporate more quickly and the cracks soon become noticeable again. Amber is also enhanced by being treated with oil (see Chapter 18).

With most treatments of this type the oil tends to dry out eventually and reveal the true condition of the gem, and this may occur sooner than intended if the gemstone is cleaned! The presence of oiling can often be detected by immersing the gem for some hours in warm water, after which the tell-tale sign of a thin interference layer of oil can be detected on the surface of the water. The greasy feel of such stones, both polished and rough, can also be detected with the fingers when handling a parcel of oiled gems.

The oiling of stones, especially rubies, is frequently carried out by dealers in the mining areas, and it not unusual to see a bottle of 'red ruby oil' on the desk of traders in Thailand (Figures 14.1 and 14.2)!

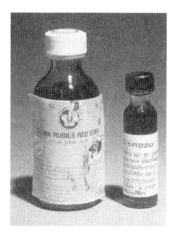

Figure 14.1 Coloured oil is used to improve the appearance of rough and polished rubies in Thailand. (Courtesy GIA)

Figure 14.2 A dealer rubbing red oil over ruby rough in the Bo Rai mining area, Thailand. (R.C. Kammerling)

Heat treatments

General

An increasing number of gemstones are today subjected to various forms of heat treatment to improve or change their colour. Some stones even experience several heat treatments as they are purchased and resold, each owner attempting to improve the stone's appearance and value, with occasionally disastrous results! The various treatments described here and under 'Irradiation methods' cover the most commonly used processes. However, there are many other variations which have been used (and other gemstone species and varieties which have been treated) both commercially and experimentally. For more information on these treatments, reference should be made to the appropriate books listed in the bibliography in Appendix A.

Among the centuries-old heat treatments are those carried out on quartz, tourmaline, topaz, zircon and corundum. Many citrines, for example, are the result of the heat treatment of poor colour amethyst to around 450°C, and are known as 'burnt amethyst'. Sometimes this heat treatment results in a green quartz (called *prasiolite*).

Reddish-brown zircon from Cambodia can be transformed to the more popular blue colour by heating it to around 1000°C in a reducing (i.e. oxygen-free) atmosphere. Although this is regarded as a permanent change, the intensity of the resulting colour can fade slowly over a period of years if exposed to strong sunlight. The depth of colour can be restored by heating the stone to a dull red temperature in air – a treatment it sometimes gets as a bonus during repairs to the metalwork of a mounted stone!

If reddish-brown zircons are heated to around 900°C in air, their colour can be changed to colourless, golden brown or red. Some low (metamict) and intermediate zircons, whose crystal structures have been damaged by alpha particles emitted from isomorphous traces of uranium and/or thorium, can be lightened in colour (and their SG and RI values raised towards that of high zircon) by heating them to 1450°C for around 6 hours. This causes the randomly ordered silica and zirconia produced as the result of the alpha particle irradiation to recombine as crystalline zircon.

Green and blue-green beryl (aquamarine) is heated to around 450°C to produce the more popular shades of blue aquamarine. Orange and apricot coloured beryl is heated to 400°C to produce the pink morganite variety.

Pink topaz can be produced by heat treating yellow and yellow-brown topaz, containing a trace of chromium, to around 550°C (at this temperature the topaz turns colourless, but changes to pink on cooling). If heated in the region of 1000°C, most topaz remains colourless on cooling.

Dark green tourmalines from Namibia can be improved in colour to an emerald green by heating, and most blue zoisite is produced by heat treating brownish-green material to around 370°C. The purple pleochroic colour of natural blue zoisite can be reduced by heating to produce a sapphire blue stone.

Corundum

The most commercially important heat treatment process is that applied to corundum. The blue colour in a sapphire containing titanium can be developed or deepened by heating it to the region of 1600°C in a reducing (i.e. oxygen-free) atmosphere. This converts *ferric* oxide to *ferrous* oxide:

$$Fe_2O_3 \rightarrow FeO$$

Heating the stone to a similar temperature in air lightens the colour by changing *ferrous* oxide to *ferric* oxide:

$$FeO \rightarrow Fe_2O_3$$

This deepening or lightening in colour is due to the development or reduction of intervalence transfer absorption between ferrous/ferric oxides and titanium oxide.

Some dark Australian sapphires can be marginally lightened by much lower temperatures in the region of 1200°C, while a similar process is used to improve the colour of brownish-red Thai rubies (or to remove the blue tint from purplish rubies and pink sapphires) by heating them in air to around 1000°C.

Corundum, rendered translucent to opaque by rutile needles, can be clarified by heating to around 1600–1900°C for an extended period. This forces the rutile to dissolve in the alumina, and rapid cooling prevents it from recrystallizing ('geuda' sapphire is a particular example of a translucent corundum containing undissolved rutile and is dealt with in more detail later in this section). If this (or other corundum material) is heated to 1300°C for an extended period (i.e. annealed), rutile will come out of solution (ex-solve) and produce asterism.

Pale-yellow or near colourless sapphires (containing some ferrous iron oxide) can be deepened in colour to an attractive golden yellow/brown by extended heating at 1000–1450°C in air. This converts the ferrous iron oxide to ferric oxide, but unlike the heating of titanium-rich sapphire mentioned earlier, the result is a deepening or development of the colour. Progressively shorter heating periods are possible at temperatures approaching 1900°C. Unlike the irradiation treatment of pale-yellow sapphires described later in this chapter, this colour is permanent.

Geuda sapphires are a milky-coloured translucent corundum which was once thought to be worthless. Then, in Thailand, a heat treatment technique was developed in the early 1980s which turned this unattractive geuda stone into a top colour transparent blue sapphire. The cloudiness of geuda material is due to undissolved rutile (TiO_2 – containing the transition element titanium which, with ferrous iron FeO, produces blue in sapphire). When the stone is heated in timed steps through 1200°C to around 1700°C in an oxygen-free atmosphere, a process which may take up to 30 hours, the titanium dissolves in the alumina and produces a transparent blue stone.

High temperature treatment furnaces range from electrically powered temperature-controlled kilns (Figure 14.3) to converted oil drum arrangements (Figure 14.4). Temperatures in the 1600–1700°C range are achieved in the cruder furnaces by using firebrick-lined coke-filled oil drums heated by kerosene which is pumped under pressure

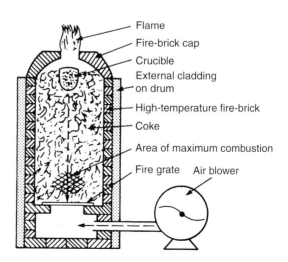

Figure 14.3 An electric muffle furnace with thermocouple temperature measurement and control

Figure 14.4 Sketch of an oil-drum furnace as used in Thailand

to a blower/burner (this originally gave rise to the description 'pressure heat treatment' for the geuda process). An oxygen-free reducing atmosphere is achieved by placing the rough or preformed stones in a sealed porcelain or alumina crucible containing charcoal. The absence of any temperature control in the simpler furnaces sometimes resulted in the melting of the corundum material. Details concerning precise temperatures, heating periods and the addition of chemicals to the crucible are often carefully guarded trade secrets!

Identification

Stones subjected to high-temperature treatments of 1700°C and above can often be identified by the presence of pitting on the girdle (Figure 14.5), and by the effect of these high temperatures on any inclusions. These may have expanded (often turning white) and produced a tell-tale circular stress fracture (Figure 14.6). With these heat-treated sapphires, the 450 nm iron absorption band is also completely absent, although with untreated Sri Lankan sapphires this is often too weak to be detected with a spectroscope, and its presence or absence can only be confirmed by the use of a spectrophotometer. Other tell-tale signs are the virtual absence of any silk (rutile needles, which are dissolved in the stone by the process), plus a dispersion of any hexagonal colour banding, and a chalky-white fluorescence under SW UV. Sri Lankan sapphires heat treated to a golden yellow/brown can be detected by heating them for 15 minutes under a 150 W lamp. This causes them to darken and then return to the pre-test colour on cooling. Because of the effects of heat treatment on inclusions, the presence of two and three-phase inclusions in the host corundum is considered to be conclusive proof that the gem is not only natural but that it has not been heat treated.

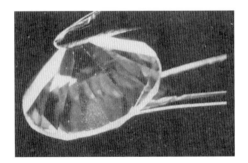

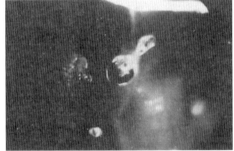

Figure 14.5 Pitted facet (only partly polished) on a corundum subjected to high-temperature heat treatment. (Courtesy GIA)

Figure 14.6 Circular stress flaw surrounding an inclusion which has expanded and turned white in a heat-treated geuda sapphire. (A. Good)

Surface diffusion of corundum

Pale or colourless corundum can be transformed into the rich colours of ruby or sapphire by first packing the pre-formed stones (faceted except for the final polish) in a clay mixture containing the appropriate transition element additives. For ruby this is chromic oxide (Cr_2O_3), and for blue sapphire both ferrous oxide (FeO) and titanium dioxide (TiO_2) are necessary. The stones are then heated to the region of 1750°C for a period of several days to induce the colour-producing elements into their surfaces. With earlier treatments, the resulting surface-diffused colour was only a few thousands of an inch (less than a tenth of a millimetre) deep, but thicker layers of diffusion were subsequently produced. Surface *asterism* can be induced by the same process, using a paste containing titanium dioxide.

Because of surface pitting caused by the high temperature, final polishing of the faceted stone or the cabochon is carried out after the diffusion process. The thinness of the coating, and the possibility of it being removed if the stone ever needs repolishing, makes surface induced enhancement an unacceptable treatment. It can be detected by the absence of normal colour zoning in the stone (the colour distribution appears to be too even, or may be completely missing over a section where it has been polished away). Other indications are the presence of a pitted facet (which may have been overlooked or only partly polished during the final polishing – Figure 14.5).

If immersed in water, the diffused stone usually shows signs of a stronger colour concentration along the stone's facet edges due to the thinning of the facet layer during polishing (Figure 14.7), but this may be less obvious with thicker layers of diffusion. Because of the high temperatures involved in the process, some inclusions may have expanded (often turning white) to produce tell-tale circular stress fractures similar to those seen in geuda sapphires (see Figure 14.6). With sapphires the 450 nm iron absorption band is also absent, and other indications are as described for heat-treated geudas.

Figure 14.7 Spine effect at facet edges seen on immersing a surface-diffused sapphire in water. (A. Good)

Lattice diffusion of corundum

Also known as 'bulk diffusion' and 'beryllium lattice diffusion', this heat treatment process was first introduced around 2001 in Thailand where it was reported that light pink sapphires from Madagascar and other sources were being heated in an oxygen atmosphere to produce the lovely 'padparadscha' orange colour. While details of the new process were not immediately forthcoming, examination of facetted samples revealed that the enhancement was different from the skin-deep surface diffusion process typically applied to pale corundums. The temperatures involved were clearly higher as zircon inclusions (which survive the normal treatment temperatures) were destroyed in the orange samples. In some of the treated stones the colour also permeated from 10%–80% of the stone. The use of secondary ion mass spectrometry (SIMS) analysis proved that the enhanced stones contained significant amounts of *beryllium* which was not present in the untreated material.

It is thought that this beryllium diffusion process was first discovered accidentally by the chance presence of pebbles of chrysoberyl in the corundum rough, or by the presence of beryllium inherent in the crucible itself. Deep penetration of the colour was then produced by experimentally extending the treatment period. Other colour samples of corundum rough (when diffused with beryllium) produced the following colour changes; blue corundum became light to dark yellow; yellow-green material turned yellow, and dark red stones turned orangy-red.

Identification of lattice diffused corundum is by inspection. The high temperatures used may cause surface pitting and inclusions may have expanded and have circular stress flaws (see also surface diffused and geuda corundums). For partly diffused stones, colour zoning will be evident, e.g., orange stones may have a pink core; with yellow stones this core may be nearly colourless.

HPHT enhancement of diamond

In the late 1970s, researchers at the General Electric Laboratory in the USA discovered that yellow Cape series diamonds (and synthetic diamonds) could be lightened in colour by heating them under high pressure for an extended period (now called HPHT treatment).

The yellowness of Cape series diamond is due to nitrogen atoms which are dispersed through the crystal lattice and replace carbon atoms. Diamond of this nature is referred to as *Type Ib*. In *Type Ia* diamond, the nitrogen atoms are grouped in clusters (referred to as platelets or aggregates) and do not affect the diamond's colour. By heating yellow Cape series diamond (which contains Type Ia and Ib material) under the high temperature and pressure available in a diamond synthesis plant, it was found possible to make dispersed nitrogen atoms migrate together into clusters. This converted some of the Type Ib material to Type Ia, thus reducing the yellowness of the stone. (In this context it should be mentioned that *Type IIa* diamonds contain no nitrogen or other impurities, and *Type IIb* diamonds contain only boron atoms as an impurity).

In 1999, General Electric and Pegasus Overseas Ltd (GE POL) announced a new HPHT treatment that improved the colour and brightness of some brown Type II diamonds (stones suitable for this treatment amounted to less than 1% of all natural diamonds). Of these diamonds, treated brown Type IIa material became colourless or pink; treated brown Type IIb material became blue. The girdles of the treated diamonds were then laser inscribed 'GE POL' (in 2000, these were later rebranded and inscribed 'Bellataire').

Since the late 1980s DTC had been planning strategies for the identification of HPHT treated diamonds, and of HPHT synthetic diamonds (including the more recent chemical vapour deposition (CVD) synthetic stones described in Chapter 15). This eventually resulted in DTC's introduction of the DiamondSure and DiamondView instruments and, in 2003, of a third instrument, the 'DiamondPlus' which was primarily designed for the identification of GE POL HPHT treated brown Type II stones. These instruments are described in Chapter 16.

Irradiation methods

General

The mechanism by which irradiation increases or modifies colour in gemstones is tied up with the production of *colour centres* (discussed in Chapter 8), which can then be altered by subsequent heat treatment. Blue topaz is one of the principal gems which has its colour enhanced by means of irradiation. *Gamma (γ) rays*, generated by a radio-active source such as *cobalt-60*, *high-energy electrons* developed in a linear acceler-ator, or *neutrons* produced in a nuclear reactor are the types of irradiation used in this treatment, which turns certain types of colourless material brown. The stones are then heat treated at around 250°C to produce a stable blue colour. Unless some residual radioactivity is detected (which has happened when high energy radiation has been used to produce deeper colours and the stones released on the market too soon) there is no simple way of distinguishing between natural blue topaz and the irradiated stone. Irradiation is also used to modify colourless synthetic quartz to a citrine colour, and to produce an amethyst colour in synthetic quartz which has been doped with iron oxide.

Pale yellow sapphires have been deepened in colour by means of γ-ray irradiation, but this improvement in colour is not stable and can fade on exposure to sunlight and UV (and if the stone is heated). As with heat-treated sapphires, the diagnostic iron absorption band at 450 nm disappears after irradiation. With untreated Sri Lankan sap-phires this is often too weak to be seen with a spectroscope, and its presence or absence can only be confirmed with the aid of a spectrophotometer.

Irradiation of diamond

Because of the large price differential between a Cape series yellowish diamond of poor colour, and a naturally coloured 'fancy' diamond having an attractive pink, golden yel-low, cinnamon brown, green, purple or blue tint, there has been much research into the artificial coloration of diamond by irradiation. Sir William Crookes was the first to use a *radium* source to artificially colour diamonds in the early part of the twentieth century. After exposure to radium, the diamonds had turned a dark green. This was only a skin-deep colour, but surviving specimens have remained radioactive up to the present day!

Diamonds have subsequently been artificially coloured by irradiation with high-energy *neutrons* in an atomic reactor. This produces a homogeneous green body colour which can be changed to yellow or golden yellow/brown (or with rarer types to pink, red or purple) by means of heat treatment at around 800°C. The stones are intensely radioactive immediately after treatment, but this dies away rapidly.

A colour change can also be produced by means of *electron* bombardment in an electron accelerator (Figure 14.8). With this process, some diamonds turn a pale blue or a bluish-green. The colour is only skin-deep and can be polished off if not suitable, and the process repeated. Because of this option, there has been a trend towards the use of low-energy neutrons which, like electron bombardment, produce only a skin-deep colour.

The *protons, deuterons* and *alpha particles* generated by a cyclotron can also be used to change a diamond's colour, the resulting hues after heat treatment to around 800°C being only skin-deep, and the diamonds rapidly losing their initial radioactivity. Although any of these methods can be used to modify the colour of a diamond, it can-not *lighten* the colour (methods to lighten the yellowness and brownness of diamonds have been covered earlier in this chapter under 'HPHT enhancement of diamond').

Identification

Diamonds that have been turned blue by irradiation can be distinguished from naturally coloured blue diamonds by testing them for electrical conductivity (see Figure 12.8). Natural blue stones (Type IIb) contain *boron* which replaces some carbon atoms in the

Figure 14.8 A voltage-multiplier electron accelerator. (Bell Laboratories)

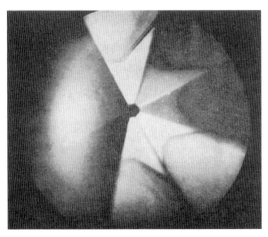

Figure 14.9 The 'umbrella' effect seen round the culet of a cyclotron-treated diamond when viewed through the table facet

crystal lattice and allows an electrical current to pass. Artificially irradiation-coloured blue diamonds are electrically *non-conductive*.

If a diamond has been cyclotron-irradiated through the pavilion (i.e. from the culet end), an 'opened umbrella' effect will be seen around the culet when the stone is viewed through the table facet (Figure 14.9). If the stone has been irradiated through the table, a dark ring will be visible around the girdle. Stones which have been treated through the side of the pavilion will show a concentration of colour around the girdle nearest to the irradiation source.

In the larger gem laboratories, the detection of treated diamonds is mainly done by spectrographic analysis. With irradiated and heat-treated stones, a narrow absorption band is visible at 594 nm in the deep yellow. This band is often very faint, but can be made more easily detectable by cooling the diamond down towards the temperature of liquid nitrogen. This is achieved by passing a stream of low-temperature nitrogen gas (boiling off from the liquid nitrogen) over the diamond. The diamond is contained in a double-walled glass tube to reduce moisture condensation, and the enhanced spectrum is viewed through the glass with a spectroscope (even the use of an aerosol refrigerant will improve the visibility of faint absorption bands).

In 1978, however, Dr A.T. Collins of Kings College, London, showed that if an irradiated diamond is heated to 1000°C, the 594 nm band disappears completely. In 1985, he and Dr G.S. Woods developed a test for treated diamonds which is based on absorption bands in the infrared at 1936 nm and 2924 nm (these are detected using a spectrophotometer, and are often reasonably sharp even with the sample at room temperature). If these bands (and the 595 nm band) are absent, the diamond's colour can safely be identified as natural.

Glass filling

Surface cavities and pits are a common feature on faceted rubies and sapphires, particularly with high value stones. The reason for this is that if these surface defects were ground away it would seriously affect the weight and thus the value of the stone. In 1984, rubies began to appear on the Thai market with the surface pits repaired by an infilling of glass. The glass had been fused into the cavities, rendering them less visible and improving the overall appearance of the stone.

Because of the difference in the reflectivity of glass and corundum, the deception is easily visible as a marked contrast between the two materials (Figure 14.10). This can be made more obvious if the stone is placed in a high RI immersion liquid such as di-iodomethane (methylene iodide) and viewed under diffused light-field illumination. Gas bubbles are often visible within the glass filling. Similar glass repairs also appeared in sapphires.

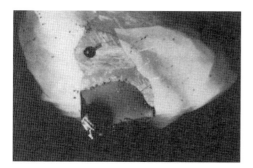

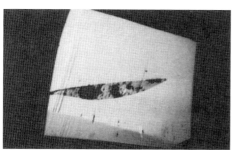

Figure 14.10 The glass-filled pits and cracks of rubies have a lower lustre compared with the surrounding corundum. Gas bubbles in the glass appear as black circular areas. (R.W. Hughes)

A similar glass-filling method is used for diamond, but in this case the aim is to improve the diamond's clarity grade. Surface-reaching fractures, which lower a diamond's clarity and therefore its value, are filled by a process developed by Zvi Yehuda in Ramat Gan, Israel. The low melting point *lead oxychloride glass* used in the process has an RI close to that of diamond. However, this treatment caused controversy in the trade, and some laboratories have refused to grade diamonds whose clarity has been improved by the Yehuda process.

Identification of glass infilling with diamond is mainly visual. In dark-field illumination, there is what has been described as an interference 'flash effect'. The yellowish-orange appearance of the fracture in dark-field illumination suddenly changes to a vivid electric blue as the stone is rotated slightly to a position where the background becomes bright. There may also be flattened gas bubbles trapped in the glass. Where the filling material is thick, this has a light brown to orangy-yellow appearance. If the filled areas are substantial they will also be opaque to X-rays.

Laser drilling of diamond

The 'lasering' of diamond is a cosmetic technique used to improve the stone's clarity. It consists of laser drilling a fine hole (less than 0.005 inch in diameter) into the diamond to reach an inclusion. In the case of dark inclusions, a bleaching agent is leached through the hole and into the inclusion to whiten it. The hole may then be filled with a transparent gel or epoxy resin to make it less visible.

As the wavelength of the laser beam is in the infrared (usually around 1060 nm), and because diamond is transparent to these wavelengths, it is necessary to initiate drilling by covering the surface just above the target area with an energy absorbing coating (such as amorphous carbon). When this layer is vaporized by the laser beam, the released energy converts the diamond layer beneath it to carbon, and the drilling process continues.

Some diamonds which have been lasered contain more than one hole. When inspected under magnification, the laser hole often has a wrinkled appearance and may change direction slightly along its length (Figure 14.11).

Figure 14.11 A diamond showing laser-drilled holes reaching etched inclusions. (Courtesy GIA – J. Koivula)

Disclosure of enhancements

In the USA, the Federal Trade Commission (FTC) advises that the lasering of diamonds weighing 0.20 carats or more should be disclosed to the purchaser. The Commission also advises that full disclosure of all enhancements has now been changed from an ethical obligation to a legal one. In this context, the vendor is now required to disclose any enhancement if this is not permanent, or if it results in special care requirements for the purchaser, or if the treatment has a significant effect on the gem's value. Such treatments include artificial coloration by coating, irradiation, heating and surface or lattice diffusion. At the barest minimum, such disclosure information should appear on the invoice of the gem or jewellery and be explained to the purchaser at the point of sale.

CIBJO, the international confederation of jewellery and silverware trades, requires that gems whose colour has been altered by irradiation, heat or chemical treatment, or which have been coated or diffused (surface/lattice), must be designated as 'treated', and the mineral name of the gem variety used in their description. CIBJO excludes from the need for disclosure those stones whose treatment results in a permanent and irreversible colour change, i.e., chalcedony (banded agate, cornelian, onyx, green agate, blue agate), beryl (aquamarine, morganite), quartz (citrine, prasiolite), pink topaz, all colours of tourmaline, blue zoisite and corundum. Onyx and cornelian treatments are generally colour stable, and provided that the gemstone is sold as a chalcedony, CIBJO rules do not require mention of enhancement. Stained green chalcedony, produced to imitate chrysoprase, and blue jasper stained to imitate lapis lazuli are two cases where CIBJO requires disclosure.

Both sets of regulations are of course aimed at the prevention of deception, but they also legitimize the status quo situation in which stable treatments, together with treatments which (so far) are not technically detectable, need not be disclosed, although where a gemstone is known to have been treated its disclosure is recommended to avoid confusion and liability.

Chapter 15

Synthetic gemstones and gemstone simulants

The manufacture of man-made gems began around 4000 BC. One of the first products were beads of blue glazed steatite (soapstone) made to imitate lapis lazuli. By 3500 BC, steatite was being replaced by a material called *faience*. This was made from powdered quartz which was mixed with various additives, moulded into beads and heated. The appropriate glaze solution was then applied and the beads fired again.

A brief history of early gemstone synthesis

The man-made products described above were not *synthetics* in the gemmological meaning of the word, but rather *simulants* or *imitations* of other gems. As mentioned in Chapter 1, it was not until the nineteenth century that the first true synthesis of a gemstone (ruby) was made by the French chemist Gaudin. This happened during a period in which mineralogists were attempting to reproduce the type of crystallization found in rocks. Gaudin's success was repeated by other chemists whose combined efforts formed the basis of the present-day *flux-melt* method of growing crystals. This technique makes it possible to dissolve gem constituents having a high melting point in a solvent or *flux* having a much lower melting point.

In 1877, another Frenchman, Frémy, succeeded in using a lead flux to grow large numbers of very small ruby crystals. At that time, the problem with the flux-melt growth of crystals was the difficulty of preventing the multiple nucleation of small crystals. This made it virtually impossible to grow large single crystals.

A few years later, an alternative to the flux-melt process was being used successfully to produce much larger crystals of ruby. This resulted in the commercial introduction of the 'Geneva' ruby. At the time these stones were thought to have been produced by fusing together small pieces of natural ruby, and became known as 'reconstructed ruby' (no doubt an early ploy to avoid using the term 'synthetic'!). Attempts to reproduce the process in the late 1960s proved that such a method could not have been used to grow the Geneva ruby. The synthesis or recrystallization process used was more likely to have been a multi-step melting of ruby powder in a flame.

While this was still not an ideal method for the bulk synthesis of ruby, it served to encourage Verneuil, a student of Frémy's, to dispense with the troublesome flux-melt method and to experiment with the crystallization of ruby by the *flame-fusion* process. This eventually led to the large-scale production of synthetic corundum using *Verneuil furnaces*. Before going on to describe the construction and operation of the Verneuil flame-fusion equipment, and of subsequent processes, it is relevant at this point to define the gemmological use of the terms 'synthetic' and 'simulant'.

Definitions

For the purpose of gemmology, the term 'synthetic' can only be applied to an artificially produced material which has essentially the same chemical composition, crystal system and physical/optical characteristics and constants as its natural counterpart (in practice, small variations in composition due to impurities are accepted). There are, however, several artificially produced gem materials which have no counterpart in nature, and these are more correctly described as *artificial products*. Because it is not possible to synthesize gem materials which are the by-product of a biological growth process, e.g. organic gems such as amber, coral and jet, the definition of the term 'synthetic' is only applicable to inorganic gemstones.

A *simulant*, on the other hand, can be any material which has the external appearance of the gem it imitates. Often, a suitably coloured natural gemstone is used to simulate a more expensive one (e.g. blue tourmaline for sapphire). Similarly, a man-made gemstone such as green synthetic spinel may be used to simulate green tourmaline. Even more frequently, coloured glass (called 'paste') is used as a gemstone simulant. The chemical constituents, constants, etc., of the simulant and the natural stone being imitated are usually quite different, and this makes identification relatively easy (as described in Chapter 17).

Sometimes confusion between the terms 'synthetic' and 'simulant' can result in the description 'synthetic diamond' being applied to a diamond simulant, particularly when this happens to be a synthetic corundum (or the more recent synthetic moissanite), or an artificial product such as CZ or YAG.

The Verneuil flame-fusion process

The Verneuil furnace (Figure 15.1) consists basically of an inverted oxy-hydrogen blowpipe burner, a powder dispenser and a ceramic pedestal. It is suitable, with small modifications, for the manufacture of synthetic crystals of corundum, spinel, rutile and strontium titanate. (*Note*: Although strontium titanate was not previously thought to occur in nature, in 1987 grains of $SrTiO_3$ were found in the former USSR and named 'tausonite'.)

When corundum is being synthesized, the dispenser is filled with high-purity alumina powder. This is produced by recrystallizing ammonium alum from solution in water until it is pure, and then calcining it at 1100°C. The calcining process drives off the ammonia and sulphur dioxide gases to leave pure γ-alumina. Around 2–3% of the appropriate colouring impurity (as indicated below) is added to the alum before it is calcined. These additives are as follows:

Chromic oxide for ruby
Iron and titanium oxides for blue sapphire
Nickel oxide for yellow sapphire
Nickel, chromium and iron for orange sapphire
Manganese for pink sapphire
Copper for bluish-green sapphire
Cobalt for dark-blue sapphire
Vanadium oxide and chromic oxide for a colour-change effect simulating alexandrite (mauve/pale blue instead of red/green)

A vibrator (or a small cam-driven trip hammer) attached to the dispenser causes the prepared alumina powder to be dropped at a controlled rate down the central oxygen

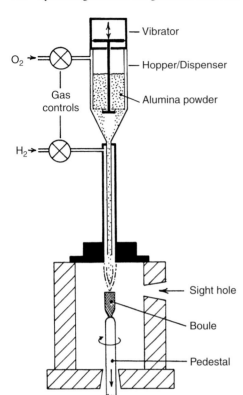

Vibrator

O_2

Hopper/Dispenser

Gas
controls

Alumina powder

H_2

Sight hole

Boule

Pedestal

Figure 15.1 Sketch of a Verneuil furnace for
flame-fusion synthesis

feed tube. As the powder drops through the 2200°C oxyhydrogen flame, it melts and falls
onto the rotating ceramic pedestal which is enclosed in a circular firebrick chamber
(Figure 15.2). When the molten alumina powder starts solidifying, the powder rate
is increased until a corundum *boule* of 15–25 mm diameter begins to form. As the boule
grows, the pedestal is lowered so that the top of the crystal is maintained in the hottest
part of the flame. A typical 40–80 mm long boule weighing between 200 and 500 carats
takes about 4 hours to form (Figure 15.3). The word 'boule' originated because the early
crystals were round, resembling the metal spheres used in the French game of boules.

The rapid growth and subsequent cooling of the boule produces internal stresses
which would cause it to crack if sawn at right angles to its length. Because of this,
the boule (if not already fractured), is split in two lengthwise after its removal from
the furnace so as to relieve these stresses. In order to obtain maximum yield, stones are
usually cut from these two halves with their table facets parallel to the split face of the
boule (i.e. parallel to the c-axis) and this results in making the less attractive dichroic
ray visible through the table facet (natural rubies and sapphires are cut with their table
facets at right angles to the c-axis to avoid this ray and show only the better red or blue
dichroic ray respectively).

The first synthetic rubies produced by the Verneuil process appeared in 1910. World
production of Verneuil corundum is now in excess of 1000 million carats a year, the
bulk of this being used for watch and instrument bearings, watch glasses and as thread
guides for the textile industry. The main producers of synthetic corundum by the
Verneuil process are the Djeva company in Switzerland and Nakazumi of Japan.

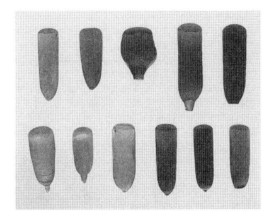

Figure 15.3 A typical selection of corundum and spinel boules grown in a Verneuil furnace. (D. Swarovsky)

Figure 15.2 Interior of a Verneuil furnace showing the finished boule. (D. Swarovsky)

Despite the volume and low cost of gem quality rubies and sapphires produced in this way, for many years it was thought that there would never be a synthetic version of the star ruby and star sapphire. Then, in 1947, the Linde company in the USA began the commercial production of synthetic star corundums (Figure 15.4), and these are now also manufactured in many countries including Germany and Japan.

With the star stone, titanium dioxide is first added to the alumina powder, and the synthetic corundum boule is grown in the normal way in a Verneuil furnace. The finished boule is then heated again to precipitate out the titanium dioxide as needles along the directions of the crystal's three lateral axes This is similar to the technique now used to dissolve and recrystallize randomly oriented rutile needles in natural corundum (see under 'Heat treatments' in preceding chapter). Subsequent analysis of the needles in some synthetic star corundums has shown them to be aluminium titanate (Al_2TiO_5), and not rutile (TiO_2) as originally thought.

Synthetic spinels were first produced by the Verneuil method in 1926 using a mixture of magnesia and alumina (MgO and Al_2O_3). When using the correct 1:1 proportions to reproduce exactly the chemical composition of natural spinel (MgO . Al_2O_3), difficulties were initially experienced in that the boules were very prone to spontaneous fracturing.

Figure 15.4 Synthetic star corundums showing strongly defined asterism. Analysis of the needles in some synthetic star corundums has shown them to be aluminium titanate, and not rutile (titanium dioxide) as originally thought

This was finally overcome by using a ratio of one part of magnesia to between 1.5 and 3.5 parts of alumina. However, this range can vary with the colouring additive (e.g. for stones coloured by chromic oxide, the ratio is 2:1 for greenish brown – $MgO . 2Al_2O_3$; 3:1 for green, $MgO . 3Al_2O_3$; 6:1 for dark green, $MgO . 6Al_2O_3$). The resulting boule contains a mixture of synthetic spinel and γ-alumina, and this causes some strain within the material which shows up between the crossed polarizing filters of a polariscope as *anomalous double refraction*. The composition also makes the refractive index and specific gravity of the synthetic product higher than that of the natural spinel, and this allows the two stones to be easily distinguished from each other. Although the difference in chemical composition and constants between the two stones would, strictly speaking, render the description 'synthetic' spinel invalid, its use has become accepted over the years.

Synthetic spinel boules are not perfectly round like the synthetic corundum boules but usually exhibit slightly flattened sides, evidence of their cubic crystal structure. The main colouring agents used for synthetic spinels are as follows:

Cobalt oxide for blue and green stones
Iron oxide for pink stones
Manganese oxide for pale-green stones
Chromic oxide for green and brown stones
Vanadium and chromium oxides for an alexandrite simulant (green/grey)

None of these colours resembles the tints found in natural spinels, and the stones are mainly manufactured to simulate other natural gemstones (aquamarine, sapphire, tourmaline, emerald, etc.). Red synthetic spinels, omitted from the above list, have proved to be more difficult to make by the Verneuil process (although crystals have been grown successfully by the flux-melt method, which is discussed later in this chapter). However, red synthetic spinels have occasionally been made in a Verneuil furnace, but in order to produce a red rather than a green boule it is necessary to use a 1:1 magnesia/alumina mix. Because the resulting boules fracture easily, only small stones can be cut from them. As a result, Verneuil-grown red synthetic spinels are rarely seen, and, due to the brittleness of the boule, cut stones are usually small. As a result of their 1:1 ratio they have the same SG value as the natural stone, although the RI value is slightly higher as the result of extra chromium.

Strontium titanate* and synthetic rutile, are also manufactured by the Verneuil process. However, as the titanium content of both of these materials tends to lose its oxygen at temperatures close to their melting points, it is necessary to supply the boule with extra oxygen during growth. This is done by replacing the standard Verneuil blowpipe with a tricone burner. The burner has an extra outer tube which is fed with a separate supply of oxygen and enriches the flame with an envelope of the gas. Even with this extra supply of oxygen, the finished boules are black in colour and have to be annealed in oxygen to render them transparent.

The Czochralski method

First developed by J. Czochralski in 1918, this is a 'crystal pulling' process, and today is mainly employed for growing high purity crystals for the optical and laser industries. It uses a seed crystal which is lowered into an iridium crucible containing the

* Once thought to have no counterpart in nature; in 1987 natural grains of the material were found in the former USSR and named 'Tausonite'

molten source material (iridium and platinum are two of the few metals that can withstand the high temperatures and chemical attack associated with this and other synthesis processes). The crucible is usually heated by means of a radio-frequency (RF) induction coil (Figure 15.5).

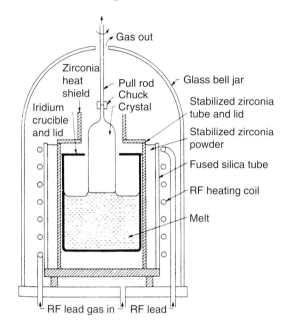

Figure 15.5 Sketch of an apparatus used in the Czochralski 'crystal-pulling' process

When the seed crystal comes into contact with the surface of the melt it is rotated and then slowly raised at a carefully controlled rate. The source material crystallizes on the seed and grows downwards as it is pulled out of the melt. The temperature of the molten source material is critical and is held to a few degrees above the melting point. Too high a temperature results in the seed crystal melting; too low a temperature allows crystals to spontaneously nucleate within the melt.

The technique is used to grow large ruby crystals and rare earth garnets (YAG, GGG, etc.) for use in lasers. It is also used to produce synthetic scheelite, fluorspar, and lithium niobate (once used as a diamond simulant). More recently, the alexandrite variety of chrysoberyl was synthesized by the Kyocera Company of Japan using the Czochralski process. This stone, with the trade name 'Crescent Vert Alexandrite' was marketed as 'Inamori Created Alexandrite' in the USA, and (unlike the synthetic corundum and spinel simulants of alexandrite) has a colour change similar to that of the natural gem. More recently, the same company was the first to market a synthetic cat's-eye alexandrite (the natural cat's-eye alexandrite is among the most costly of all gems). However, value is not the only criterion for the synthesis of a gem material. At a lecture in London where the French synthesizer Pierre Gilson was describing his various synthetic gem products he was asked why he had never synthesized alexandrite. 'Because I do not like the stone' was his answer!

Flux-melt growth

This is a solvent-based process similar to the ones that were being developed by French and German chemists in the latter part of the nineteenth century. Although for some time ruby remained the main objective of the earlier synthesizers, in 1888 Hautefueille and Perry used two fluxes (*lithium molybdate* and *lithium vanadate*) in which they dissolved the appropriate constituents and grew small crystals of emerald.

The techniques in use today are similar to the one originally developed in 1935 by the German dye manufacturing company I.G. Farbenindustrie for the purpose of producing synthetic emerald. This method used a heated platinum crucible (Figure 15.6) in which the constituent gem-forming chemicals (beryllium and aluminium oxides,

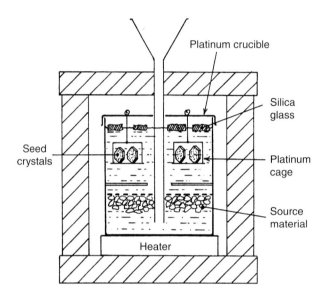

Figure 15.6 The flux-melt apparatus used to grow synthetic emerald crystals in the I.G. Farbenindustrie process

plus chromic oxide as the colouring agent) are dissolved in a solvent or flux of lithium molybdate heated to around 800°C. Slabs of silica glass are floated on top of the melt, and the beryllium and aluminium oxides combine with them to form a beryl solution. Seed crystals of natural or synthetic beryl are then lowered into the solution in a platinum cage, and the temperature of the crucible is slowly lowered to a preset level.

As the beryl solution becomes supersaturated, crystals of synthetic emerald are precipitated out and grow on the seeds. The process is a very slow one and requires accurate temperature control to produce a thermal gradient in the crucible which enables the source material to dissolve at the bottom and to recrystallize at the top. During the growth process the source materials are replenished at regular intervals through an access funnel at the top of the apparatus.

In 1939, the war interrupted the work of the German company, and their 'Igmarald' synthetic emerald was never produced in commercial quantities. Previously, in 1938, the American chemist Carroll F. Chatham also succeeded in synthesizing emerald but this was on a commercial scale. Although the process was kept secret, the emeralds were close enough in character to the German Igmarald to indicate that they were also a product of a flux-melt process. Another synthetic emerald was subsequently manufactured for a brief period by W. Zerfass in Idar-Oberstein, Germany (the process used was also thought to be the same as that developed by I.G. Farbenindustrie).

In 1963, Pierre Gilson in France further improved the flux-melt technique to produce high-quality emerald crystals (he also introduced mechanized faceting of the stones). In the Gilson method, the platinum crucible is divided by a perforated screen

Figure 15.7 (left) Platinum cage containing Gilson synthetic seed plates after 2 months overgrowth. (right) Gilson synthetic emerald crystals after 10 months growth. (Gilson)

into two side-by-side compartments, one of which contains the seed plates (Figure 15.7) and the other the source material. The temperature gradient is arranged so that the seed compartment is cooler than that of the source material so that the flux circulates between the two. Japan is one of the world's major manufacturers of synthetic emeralds and in the 1980s the Nakazumi Earth Crystals Corporation acquired the Gilson process.

The Lennix 2000 emerald process, developed over a period of several years from 1952 onwards by L. Lens of France, was unusual in that it used a series of *sequential* growth stages, each of 5–6 weeks (rather than a single period of 8–10 months). This permitted selection of the best crystals for onward growth. The other unusual feature of the Lennix process was that the crystals were grown in a consumable low-cost *ceramic* container rather than an expensive platinum crucible. This enabled production costs to be cut (although platinum is not chemically attacked during gemstone synthesis, it becomes soft at the high temperatures involved in the process, and suffers from mechanical erosion; this loss of platinum due to erosion is said to account for 80% of the cost of the end product).

Two companies in Japan who also produce synthetic emeralds by the flux-melt process are Kyocera and Seiko. The Kyocera product is called 'Crescent Vert', while that produced by Seiko is mainly marketed in jewellery manufactured by its subsidiary company, Bijoreve. The former USSR has also produced flux-melt emeralds.

The process of crystal growth by the flux-melt method, as already mentioned, is a very slow one, taking between two and ten months (depending on the method) to produce crystals of a size suitable for cutting. In this respect it has a distant connection with the very much longer geological process of gem mineral formation, rapid cooling producing a multitude of small crystals while the production of large crystals requires a very slow and steady precipitation.

Synthetic ruby, spinel, quartz, alexandrite and the rare earth garnets are also produced by the flux-melt process using the appropriate solvent, constituent chemicals and colouring agent. Synthetic alexandrites produced by Creative Crystals in the USA are coloured by iron and chromium to produce a colour change close to that of the natural Russian variety. Kashan flux-melt rubies appeared on the market in 1969, later productions being distinguished by having variable amounts of iron oxide added to them, which made them less easy to separate from natural rubies by the SW UV transmission test (Kashan synthetic rubies have since ceased production).

In 1982 the Knischka flux-melt ruby appeared on the market, the manufacturing process being described as 'Luxury synthesis' to distinguish it from the less costly Verneuil method. A more interesting gemmological feature of the Knischka ruby was the shape of the uncut crystal which took the form of a squat hexagonal bi-pyramid with multiple secondary faces. The Knischka company also offered, for a surcharge, to produce synthetic rubies using crushed natural rubies as the source material (this is similar in concept to the Pool emerald described later in this chapter under 'The hydrothermal process').

In the USA Carroll Chatham has produced ruby and sapphire by the flux-melt method. The Ramaura synthetic ruby, also made by the flux-melt process in the USA, originally had a fluorescent dopant added to the source material to make the product more easily identified. Sadly, this dopant tended to migrate to the surface skin of the crystal and was not always easily visible in cut stones. When compared with a natural ruby under LW UV, the fluorescent colour of the Ramaura ruby (no longer in production) appears to be shifted more towards the yellow-orange. Large synthetic red spinel crystals in the 10–20 carat range have been successfully grown by the flux-melt method in the former USSR. This has resulted in the appearance of much larger cut stones than were possible from red spinels produced by the Verneuil flame-fusion process.

Zone melting

Zone melting can be used as a method of refining (i.e. increasing the purity of a material) or of growing high-purity crystals from powder or partially fused powder. The equipment consists of a radio-frequency induction coil which is traversed along the length of the source material, melting it. As the coil moves on, the material cools and crystallizes or (if it is being refined) recrystallizes. Alternatively, the induction coil can be stationary, and the source material moved through it.

For zone refining (Figure 15.8, left), the crystal is held vertically and rotated as it moves down through the radio-frequency heating coil. The molten zone area is suspended within the diameter of the material by surface tension. As the crystal moves down through the coil, any impurities in the source material are carried along in the molten zone to the end of the crystal. Seiko synthetic rubies, sapphires and alexandrites are produced by a variant of this process called 'floating zone melting' (Figure 15.8, right). Crystals produced in this way are usually free from inclusions and growth features.

The hydrothermal process

Unlike other processes used to synthesize gemstones, the hydrothermal method grows crystals from an aqueous solution of the source material. If water is heated under pressure in a pressure cooker, its temperature can be raised well above its normal boiling point of 100°C. Using a more substantial vessel called an *autoclave* (Figure 15.9), the boiling point of water can be raised even higher. At a pressure of around 21 000 psi (pounds per square inch; 144.8 MPa), the boiling point of water rises to 400°C, and at this temperature water and the associated superheated steam acts as a solvent for many minerals including quartz. Because of the reactive nature of the high-temperature aqueous solution, the inner wall of the autoclave has a liner made from a noble (inert) metal such as silver.

The hydrothermal method of synthesis effectively duplicates the process by which quartz crystals (and others) form in nature. By using the solubility of source materials

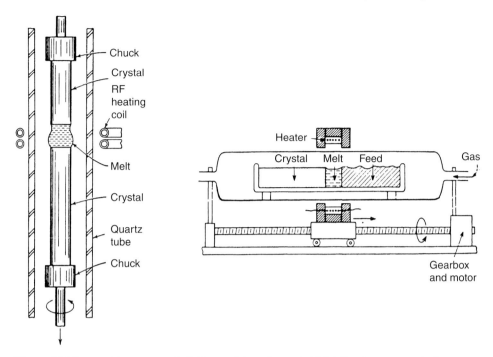

Figure 15.8 Zone melting process. (left) Zone refining of crystal. (right) Floating zone method as used by Seiko. (After K. Nassau)

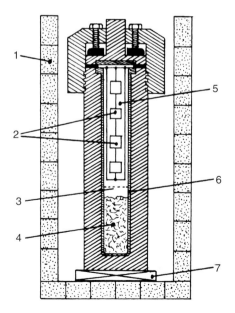

Figure 15.9 Simplified sketch of a silver-lined autoclave for the production of hydrothermally grown quartz: 1, thermal insulation; 2, seed plates; 3, baffle; 4, source material; 5, silica-rich aqueous solution; 6, silver liner; 7, electrical heater

in superheated water and steam it is possible to produce a supersaturated aqueous solution from which gem material can be precipitated and grown on suitable seed crystals.

To synthesize quartz, crushed silica is placed at the bottom of the autoclave as the source material, and prepared slices of quartz are hung in the upper part of the vessel to act as seed plates (see Figure 15.9). The autoclave is then 85% filled with slightly alkaline water containing 1% by volume of sodium hydroxide – a mineralizer which increases water's solubility to quartz.

The autoclave is heated, and by the time its temperature has reached 200°C, most of the water has turned to superheated steam. When the temperature reaches 300°C, the pressure inside the autoclave is around 20 000 psi (137.9 MPa) (more than 1000 times greater than atmospheric pressure) which explains why the autoclave is also known as a bomb! Without the addition of a mineralizer to the water it would be necessary to raise the temperature still higher to achieve the necessary solubility for quartz. The crushed quartz at the bottom of the autoclave now dissolves in the mixture of super-heated steam and water and begins to rise. The temperature in the region of the seed plates is about 40°C lower than at the base of the vessel, and when the silica solution reaches this cooler area it becomes supersaturated and crystallizes out onto the seed plates.

Colourless quartz crystals up to 50 mm wide by 150 mm long (mainly for use in the electronics industry) can be grown in 3–4 weeks by this method (Figure 15.10). Synthetic coloured quartz for use in jewellery is produced by adding cobalt (for blue) or iron (green or yellow) to the solution. Synthetic amethyst is grown by adding iron and irradiating the resulting crystals with a radioactive source.

Synthetic emerald and ruby are also grown using the hydrothermal process. In 1960, J. Lechleitner of Innsbruck, Austria, was the first to use the method to deposit a thin coat of synthetic emerald onto an already faceted and polished natural beryl gemstone of poor colour. The crown facets of the coated stone were then lightly polished, but the pavilion facets were left in a matt condition to retain as much of the colour as possible. The emerald-coated beryl gemstones were first marketed under the name 'Emerita' (which was later changed to 'Symerald'). They were subsequently produced by the Linde Division of the Union Carbide Corporation of America as 'Linde synthetic emerald' (Figure 15.11). In 1964, Lechleitner produced hydrothermally grown synthetic emeralds from seed plates (this was followed in 1965 by a similar product from Linde) and in 1985 he produced synthetic rubies and sapphires using Verneuil seeds with a hydrothermal overgrowth.

The main differences between the hydrothermal production of synthetic emerald and quartz is that the source material for emerald is placed both at the bottom and at the top of the autoclave, and the seed plates of beryl are suspended in the centre (oxides of aluminium, beryllium and chromium are contained in the lower part of the vessel, while crushed quartz is held in a separate perforated container at the top). Another difference between the hydrothermal production of quartz and emerald is that it is necessary to use acidized water as the mineralizer for emerald in order to keep the chromium oxide in solution. The emerald constituents are dissolved in the superheated water/steam at around 600°C, and react together in the centre of the autoclave to form the emerald solution which then crystallizes on the seed plates.

Unlike the flux-melt process, it is not possible to replenish the source material on a continuous basis because the autoclave is sealed, and this limits the size of the crystals. Larger crystals are achieved by repeating the process several times over, using the same crystals each time as seeds.

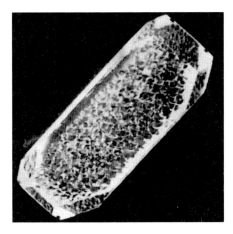

Figure 15.10 A synthetic hydrothermally grown quartz crystal showing the characteristic pebbly surface which is never seen in natural quartz

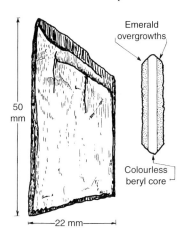

Emerald overgrowths

50 mm

22 mm

Colourless beryl core

Figure 15.11 Sketches of Linde 'Lechleitner' type synthetic emeralds (Sinkankas). (left) A tabular crystal as removed from autoclave. (right) A cross-section showing synthetic emerald overgrowth on a thin seed of colourless beryl

Some synthetic emeralds, such as the Regency product of Vacuum Ventures in the USA, contain sufficient chromium to make them fluoresce bright red under a high-intensity white light. Another synthetic hydrothermal emerald, developed by the Crystals Research company of Melbourne, Australia, owes its colour to vanadium and not chromium. Vanadium was also used as the colouring element in the earlier productions of the Biron synthetic emerald, a product distinguished by its excellent colour and clarity.

The Biron manufacturing process was acquired in 1987 by Equity Finance of Perth, Australia. In an endeavour to avoid using the term 'synthetic' in advertising and selling its product, Excaliber Holdings Ltd (a subsidiary of Equity Finance) later launched the 'Pool Emerald', which was publicized as 'natural recrystallized emerald'. The name 'Pool' came from the Pool Emerald mine in Western Australia which supplied the low-grade natural emeralds used as source material in the hydrothermal synthesis of this product. In 1990 Biron International Limited also manufactured a pink beryl (similar to the morganite variety of beryl) using the hydrothermal process. However, the colour of this material was produced by titanium rather than by manganese as in the natural stone.

The skull-crucible process

Because the melting point of zirconia powder (used in the production of the diamond simulant *cubic zirconium oxide*) is well above 2000°C, it cannot be melted in any of the conventional refractory crucibles. Instead, crystals of this material are produced by means of a 'skull' melting technique which was originated in the Lebedev Physical Institute, Moscow.

The skull crucible consists of a circular arrangement of water-cooled copper pipes (Figure 15.12). The zirconia powder (plus a stabilizer whose purpose is explained later in this section) is packed into the crucible and melted by means of radio-frequency induction heating. As the zirconia powder is only electrically conductive at high temperatures, the melting process is initiated by placing a piece of zirconium

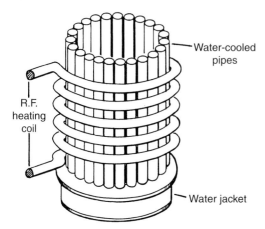

Water-cooled
pipes

R.F.
heating
coil

Water jacket

Figure 15.12 Sketch of a skull-crucible apparatus showing the vertical water-cooled pipes
and the RF induction coil

metal in the centre of the crucible. This oxidizes as it melts and becomes part of the
zirconium oxide source material.

When the bulk of the powder has become molten, its outer crust is kept below melting point by the cooled copper pipes and forms its own high-temperature crucible.
After several hours, the RF heating power is slowly reduced, and the transparent cubic
zirconium oxide crystals form as the melt cools (Figure 15.13). To relieve stresses in
the cooling crystals, they are held at 1400°C for an annealing period of up to 12 hours.

Zirconium oxide is cubic in its molten state, but normally becomes monoclinic (and
opaque) when it cools to room temperature. In order to maintain the molten zirconium
oxide in its cubic and transparent state as it solidifies, a 'stabilizer' is mixed with the
source material before heating. Suitable stabilizers for this purpose include magnesium, calcium and yttrium oxide (in industry, unstabilized opaque zirconium oxide is
used as a high-temperature refractory material).

Cubic zirconium oxide (CZ for short) is marketed as 'Phainite' in the former USSR,
as 'Djevalite' by Djevahirdjian of Switzerland, and as 'Diamonesque' by the Ceres

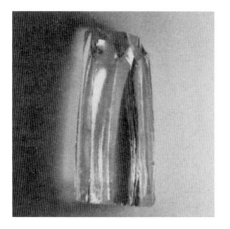

Figure 15.13 (left) A single crystal of CZ (weighing about 600 carats) as broken out of the mass of crystals in the skull crucible. (right) An 880 carat CZ replica of the Cullinan I diamond

Corporation in the USA. Swarovski in Austria is one of many other companies world-wide who produce large quantities of this popular diamond simulant. Coloured cubic zirconia is also manufactured by several of these companies, the colours being produced by rare earth and transition element additives.

Diamond synthesis

Although the event was not authenticated by an independent observer, the Swedish ASEA company claim to have been the first to develop a method of repeatedly synthesizing diamond in 1953. The reason given for the non-disclosure of their success at the time was the need to improve on the size and quality of the diamonds while keeping the process secret. It was also believed that no other teams were engaged at that time on synthesis attempts. The ASEA equipment consisted of a complex high-pressure sphere which used a thermite mixture for heating a carbon-iron matrix. The sphere was patented but only used experimentally. By 1955 the equipment had been replaced by a piston-and-cylinder apparatus which became the basis of ASEA's later production plant.

In 1955, however, General Electric of America filed a patent application for the synthesis of diamond and made public the successful conclusion of their work. Four years later, De Beers also successfully synthesized grit size diamonds, and using ASEA high-pressure presses developed their method into a commercially viable process (Figure 15.14). Since then, the synthesis of industrial grade grit-size synthetic diamond has been achieved in many countries including the USSR, Japan and the People's Republic of China.

The basic technique used in the large-scale synthesis of industrial diamond involves the dissolving of graphite in molten iron, nickel, manganese, or cobalt at high temperatures and pressures (HPHT synthesis). In this process the metal acts as a *catalyst* to greatly reduce the temperature and pressure necessary to convert the hexagonal atomic structure of graphite into the more tightly bonded cubic structure of diamond.

Figure 15.14 (left) Very small blocky crystals of synthetic industrial diamond. The habit of the crystals can be modified by adjustment of the process temperature and pressure. (right) A production unit containing a group of high-pressure presses used in the De Beers synthetic industrial diamond process. (De Beers)

The conversion is effected by applying pressure in the region of 110 000 atmospheres (around 1 600 000 psi or 11 030 MPa) to a pyrophyllite cartridge containing alternate discs of graphite and the metal catalyst, and at the same time heating this to around 3300°C. At this temperature the metal discs melt, dissolving the graphite. The temperature is then allowed to fall and the graphitic carbon recrystallizes as clusters of tiny synthetic diamonds, the process only taking two or three minutes. After the solidified metallic mass of the cartridge and its contents is removed from the press it is crushed, and the diamonds are extracted by dissolving away the non-diamond content with acids.

In 1970, synthetic gem quality and carat-size diamonds were grown under high-pressure/high-temperature (HPHT) laboratory conditions by General Electric of America. These diamonds were produced by a diffusion technique in which free carbon atoms were persuaded to crystallize on synthetic diamond seeds in the cooler section of a molten metal catalyst 'bath'. The carbon source material consisted of small diamond crystals placed in the hot section of the bath. The resulting tabular shaped diamonds were very expensive to produce in comparison with the price of the natural stone, and were therefore not economically viable as a commercial product. A year later, research workers in the former USSR announced that they too had synthesized gem-quality synthetic diamonds, but had also decided that at the time they were not an economic proposition.

In 1986, Sumitomo Electric Industries of Japan announced the commercial production of HPHT carat-size transparent yellow synthetic diamond crystals of gem quality (Figure 15.15). These diamonds, much superior in quality to the normal industrial synthetic diamond, were developed for various industrial purposes. The crystals, weighing up to 1.2 carats, were marketed only as sawn, laser-cut and partly polished rectangular pieces weighing in their finished form up to 0.40 carat. Although production costs of the Sumitomo product were said to be close to the value of natural stones of similar size, colour and quality, the company stated that it had no plans to release the product into the jewellery market. Not long after this announcement, Sumitomo samples were sent to the major gem laboratories, and a faceted yellow synthetic diamond similar to the Sumitomo product was seen in London!

In 1987, De Beers sent samples of their own HPHT yellow carat-size gem-quality synthetic diamonds to the GIA for inspection (Figure 15.15). Large gem-quality diamonds are claimed by the company to have been synthesized on an experimental basis in their Johannesburg-based Diamond Research Laboratory since the early 1970s. By 1988, the largest synthetic diamond crystal grown by De Beers (which the author inspected during a visit to the Diamond Research Laboratory that year) weighed around 11 carats. The Company stated that the purpose of their research programme to grow large gem-quality transparent synthetic diamonds is (like Sumitomo) to investigate possible high-technology industrial applications, and to understand the diamond synthesis process better. At that time they had no plans for the future distribution of these crystals to the jewellery or gem trade.

The De Beers synthetic diamond crystals are modified octahedrons. They are made by a flux method. Several crystals can be grown in the reaction vessel simultaneously (on synthetic diamond seeds). High-quality transparent colourless or blue synthetic diamonds are reportedly difficult to grow. Like all yellow synthetic diamonds (both gem and industrial quality) produced during this period by any organization, the De Beers stones were 100% Type Ib material. This type of diamond only contains nitrogen atoms, which are dispersed throughout the crystal lattice, and is rare in nature (natural yellow diamonds are usually a mixture of Type Ia and Type Ib material).

Figure 15.15 (top) Crystals of gem-quality synthetic diamonds produced by Sumitomo with, in front, examples of heat sinks cut from them. (bottom) Crystals of gem-quality synthetic diamonds produced by De Beers

In 1993, Thomas Chatham (son of Carroll Chatham) announced that his US company was about to market HPHT synthetic gem-quality diamonds produced in the former USSR at a price substantially lower than that of natural diamonds. Production and other problems, however, prevented this from happening. More recently, a company called Supersprings in Los Angeles began mounting yellow and light blue synthetic diamonds in their range of jewellery. These stones were being synthesized in Russia in the size range of 0.15–0.25 carats.

In 1996, no doubt because of the increasing problem of gem-quality synthetic diamonds entering the jewellery trade, De Beers unveiled two new identification instruments developed in its UK-based DTC Research Centre specifically to differentiate between natural and synthetic diamonds. These instruments, together with a third more recent model, are described in Chapter 16.

In 2004, Chatham Created Gems in the USA began marketing a series of new HPHT synthetic diamonds in a range of colours, some of which closely resembled the more muted hues and saturations of natural diamonds. These new synthetic diamonds are reportedly from a source in Asia and are being facetted in China in sizes from a few points (there are 100 points to the carat) up to 2 carats. (See Chapter 16 for the identification of these and other synthetic diamonds.)

Synthetic diamonds have also been made in the past by means of an explosive charge which is used to generate briefly the temperatures and pressures necessary to change graphite or amorphous carbon to diamond. Another method, developed by Linde of the USA, consists of passing methane over very small diamond seed crystals at temperatures between 600 and 1600°C. The carbon atoms in the methane are deposited as diamond on the seeds, but the growth rate is very slow. Physicists at the Harwell Atomic Energy Research Centre in the UK produced a similar growth by beaming high energy carbon ions onto diamond seed crystals. These latter processes would appear to have been the forerunners of the thin diamond film deposition technique described in the next section of this chapter.

Thin diamond films

Although only just within the specification of this book, some mention must be made of a relatively new technology which might just become a problem to gemmologists in the future. Research started around 1985 in the former USSR and in Japan into the *low*-pressure/high-temperature deposition of a thin layer of synthetic diamond onto various substrates such as silicon. This deposition method involves passing a mixture of methane and hydrogen through a microwave 'bath' which breaks the gas molecules into hydrogen and carbon atoms, and allows the carbon to reactively crystallize as diamond on the substrate surface. The main applications for such coatings are in the development of long-life tools and bearings, scratch-proof lenses, windows for scientific equipment, and the provision of specialized heat sinks. These thin film synthetic diamond products are now manufactured worldwide by companies such as De Beers, General Electric and Sumitomo.

Diamond-like carbon (DLC) films have also been developed as these can be produced more easily than diamond films. In DLC coatings, the carbon atoms are present in both diamond and graphite-type bonds, and the properties of the coating differ from those of diamond.

Although it is technically possible to coat thin diamond films onto gemstones, with many gem materials adhesion of the film often presents a problem. If a diamond simulant (such as CZ) were to be successfully coated with a film of diamond, this could create an identification problem as the reflectance meter might measure the high lustre of the surface layer and give a diamond reading. However, as the thickness of the coating is usually limited to one-thousandth of a millimetre (even this may take around an hour to deposit), it should not affect the results of a thermal conductance tester, even if the coating is in contact with the claws of a metal mount. Synthetic moissanites have been successfully coated with a thin film of diamond which would probably mask its electrical semi-conductivity test, and would show as diamond on a reflectance meter.

Chemical vapour deposition (CVD)

Virtually a spin-off from the industrial uses of thin diamond film technology, various forms of chemical vapour deposition (CVD) of synthetic diamond have been under development for many years, but were restricted by very slow growth rates. It was not until the early 1980s that a breakthrough in CVD diamond growth rates in Japan by researchers Sato, Matsumoto and Kamo made the technique a viable proposition. Unlike the now well-known high pressure, high temperature (HPHT) processes used to grow both industrial and gem-quality synthetic diamonds, the deposition of layers of gem-quality diamond uses low pressures (around one tenth of an atmosphere) and high temperatures.

Using hydrogen and methane as starting gases in the CVD process, the gas molecules are torn apart in a high-temperature plasma generated by microwaves in a reactor chamber. The liberated carbon atoms react at the surface of heated seed crystal substrates mounted at the base of the chamber to form CVD synthetic diamond. To prevent the production of polycrystalline material, single-crystal seed substrates are used. Hydrogen atoms liberated by the plasma play an important role in suppressing the formation of carbon. Growth rates can be increased by adding nitrogen to the feed gases, the resulting yellow colour of the diamond being reduced by subsequent HPHT annealing. The CVD diamonds produced by this method are now being grown to a sufficient thickness to be faceted into diamond brilliants. In 2003, Apollo Diamond Inc. of the USA announced plans to market CVD single-crystal synthetic diamonds. Identification of CVD diamonds is by means of DTC's DiamondView and DiamondPlus instruments (see also Chapter 16 under 'Diamond').

Synthetic jadeite

In 1984, General Electric of America produced white, green, black and lavender samples of synthetic jadeite. The original synthesis experiments were started only a few years after they had successfully synthesized gem-quality diamond. The starting material for synthetic jadeite included crushed silica glass as well as alumina and sodium carbonate, and was processed in a high-pressure apparatus at a temperature of around 1400°C. These synthetic jadeites were said to be the product of an experimental study and the company has no plans for commercial production.

'Synthetic' lapis lazuli, turquoise and opal

Although the Gilson man-made imitations of lapis lazuli, turquoise and opal (Figure 15.16) were introduced and advertised as synthetic versions of these gemstones, subsequent analysis of the materials by leading gemmologists has indicated that they should more correctly be described as simulants. The reason for this is that these products all contain some substances which are not present in the natural gemstones. For example, Gilson lapis lazuli has been found to be composed mainly of synthetic ultramarine and two hydrous zinc phosphates, constituents which are absent from natural lapis.

Figure 15.16 Gilson 'synthetics'. From left to right: lapis lazuli, turquoise and opal cabochons, with rough samples at the rear

A similar situation exists for Gilson turquoise and opal which are also said to contain materials not found in the natural stones. Some Gilson turquoise (see Figure 15.16) has unconvincing surface markings which are intended to imitate the limonite banding on the natural stone. Because these products are not truly synthetic, their main identification features are described in Chapter 17 which covers inorganic gemstones and their simulants.

The Gilson 'synthetic' opals are produced in both white and black versions. Of these, the white opal is the most realistic imitation. Although the Gilson process has remained a secret, the opals are thought to be made from a *silicon ester* solution. The silica particles, similar to the cristobalite spheres in natural opal, are allowed to settle in close-packed arrays by sedimentation over a period of several weeks, after which any remaining water is removed. The highly porous and brittle product is then impregnated and stabilized with an opaline silica.

Composite gemstones

With the exception of opal doublets and triplets, composite stones are fabricated with the intent to deceive. Because they form a separate and less obvious form of simulant, both their description and detection are covered in the following notes.

Composite stones consist of *doublets* and *triplets*, the components of which can vary from sections of the same mineral cemented together to make a larger whole, to stones having a crown of the gem being faked and a pavilion of glass or synthetic (e.g. a diamond crown and a synthetic corundum pavilion – see Figure 15.17).

Opal doublets are formed from a thin layer of precious opal backed with plastic or a layer of common or potch opal. Sometimes the opal 'doublet' may be cut from of a layer of opal backed by its own ironstone matrix. Opal triplets are made by cementing a dome of clear quartz (or synthetic spinel or corundum) over the top of an opal doublet. In this case, the top surface of the composite gem shows no sign of iridescence, and a 'distant vision' RI reading for the dome can be obtained from the surface of the triplet.

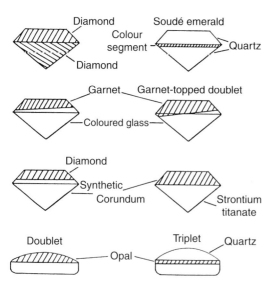

Figure 15.17 Cross-section of a selection of composite gems

Dating from Victorian times, a range of doublets has been produced using coloured glass to simulate the gem being imitated, and fusing on to the crown area a thin section of almandine garnet. These composite stones are described as garnet-topped doublets (GTDs for short). The high lustre and refractive index (1.77) of the garnet covered table makes them suitable for simulating corundum, and they are usually made to imitate ruby or blue sapphire, although emerald, peridot, yellow sapphire and topaz versions have also been fabricated.

As can be seen in Figure 15.17, the garnet layer is seldom symmetrically joined to the glass. This is because the two sections (consisting of a thin plate of garnet and a blob of glass) are placed in a heated mould and fused together before being cut and polished. Once a stone is suspected of being a GTD, this can easily be confirmed by first placing it table facet down on a piece of white paper. If it is then inspected under a strong light source, the pink rim of the garnet top can usually be detected around the join in all but red GTDs. Alternatively, the line of join between the top and base of a doublet can be made more visible by immersing the stone in a liquid having an RI close to that of the glass (e.g. bromoform). Even water will help to increase the contrast between the two components of a doublet (Figure 15.18).

Under the hand lens or microscope, careful examination will reveal not only the line of the join (which can be quite irregular – Figure 15.19) but also the presence of bubbles. With doublets having large accessible pavilion facets, measuring the RI of both

Figure 15.18 The join between the glass pavilion and the almandine crown of a garnet-top doublet becomes more apparent when the gem is immersed in water

Figure 15.19 The section of garnet on the crown of a GTD is often unsymmetrical

the table and a pavilion facet will also reveal the true nature of the stone (except in the more unusual case where the two sections are the same material, e.g. a crown of natural corundum and a pavilion of synthetic corundum, simulating ruby or sapphire).

Triplets such as the soudé emerald consist of a colourless quartz, synthetic spinel, or beryl top and a similar base. The colouring element completing the triplet can be a thin coloured layer of gelatine or a sintered spinel segment cemented or fused between the crown and pavilion sections. Unlike the GTD, the soudé emerald type triplet is a symmetrical product with the coloured layer inserted in the girdle area (see Figure 15.20).

Symmetrically made quartz/quartz composites have also been made with a central layer of colour filter to simulate amethyst. When immersed in a suitable liquid and inspected in line with the girdle, the crown and pavilion sections of these triplets can be seen to be colourless. Both sections of the soudé emerald may contain misleading

Figure 15.20 Immersion view of the three components of a soudé emerald. The top and bottom sections are colourless, the emerald green being produced by the thin central segment

inclusions, but under magnification a plane of bubbles is often visible at the colour layer in both types of composite.

Although it is not always possible to bear in mind the possibility of a composite product when examining a gemstone, with an unmounted stone the non-compatibility of an RI reading with other identification details (such as SG) should initiate the close inspection of the crown/girdle area for indications of more than one component (*in a practical exam always look for signs of a composite during the initial hand lens check of a faceted stone*).

Perhaps the most potentially dangerous composite stone is the diamond-topped doublet. This is made from a thin crown section of diamond cemented to a pavilion which can be a variety of materials from quartz to one of the man-made simulants of diamond (such as synthetic corundum, YAG, GGG, CZ or even the more recent diamond simulant synthetic moissanite). Occasionally, one sees a doublet in which a crown of YAG or synthetic corundum has been cemented to a pavilion of strontium titanate to 'quench' the excessive fire of that material and provide a harder wearing top surface. The resultant gem has more the appearance of a diamond than a stone made completely from either material, but with the low-cost availability of CZ, such a substitute is probably no longer a worthwhile proposition in modern jewellery.

The diamond/diamond doublet, although rarely seen, can also be a problem. The reason for this seemingly strange combination is the non-linear pricing structure of diamond (the comparative rarity of larger sizes makes the price of a 2-carat stone around four times that of a comparable 1-carat stone). The composite diamond/diamond doublet could therefore command (undetected) more than the total value of the individual pieces. Its detection relies on the affect the joined area has on the appearance of the stone. Total internal reflection is reduced and the join produces a mirror effect when viewed through the table. Bubbles can usually be seen in the plane of the join. A more dramatic test, although one *not* to be recommended, is to heat the stone just enough to soften any cement layer, when the top section can be carefully slid to one side!

Chapter 16

Identifying inorganic gemstones and their synthetic counterparts

Because of the large price differential that exists between natural gems and their synthetic counterparts, it is important that the gemmologist is not only able to identify a stone but is also able to determine whether it is a natural or a synthetic material. Gemstone *simulants* can in general be readily identified as their physical constants differ from those of the gemstone they imitate. In this respect, synthetic spinels (with the exception of the red variety) can also be regarded as simulants as they are mainly used to imitate other gemstones and not natural spinels (these stones are covered in the next chapter which deals with the identification of simulants).

However, the synthetic versions of gemstones such as diamond, emerald, quartz, ruby, sapphire and alexandrite pose a more difficult problem as their constants are (by definition) very close to those of the natural stone. Identification of these stones is mainly based on tell-tale indications within the gem of the very different conditions and timescales under which the natural and synthetic crystals have grown. These indications include growth lines, colour zoning, and features called *inclusions*, all of which can be used to help separate the natural from the synthetic gemstone.

Growth lines and colour zoning

Curved lines of growth and curved colour zoning are commonly seen only in coloured synthetic Verneuil corundums and in Verneuil-produced red spinel (i.e. flame-fusion products). They are due to the intermittent fall of droplets of molten alumina onto the boule's upper surface, and to the greater volatility of some of the colouring oxides compared to alumina. Growth lines are quite difficult to detect in all but the Verneuil ruby, but curved colour zones which are often broad enough to be seen with the naked eye (particularly when the stone is immersed) are easily detected in blue Verneuil sapphires and red Verneuil spinels.

As an indicator of natural origin, straight colour zoning (usually following the pattern of the lateral crystal axes) can be seen in many natural stones, including quartz, ruby, sapphire and emerald.

Twinning

Repeated or multiple twinning is a feature of some gemstones such as quartz, corundum and chrysoberyl, and can sometimes help to distinguish them from the synthetic product (as described later in this chapter). Twinning of a crystal can occur both during and after its formation, the latter being due to subsequent deformation. These forms, together with interpenetrant twins and contact twins are illustrated under 'Twinned crystals' in Chapter 4.

Types of inclusion

Perhaps the most useful internal features of a gemstone are inclusions, as these are directly related to the immediate environment in which the crystal grew. Inclusions in natural gemstones can be divided into three basic groups:

1. **Protogenic** or **pre-existing** inclusions. These consist of minerals (sometimes in the form of minute well-shaped crystals) which were present before the host crystal began to form. Apatite in corundum and hessonite garnet, mica in corundum, quartz and emerald, rutile in quartz, diamond in diamond, and pyrite in corundum and emerald are commonly found inclusions of this type.
2. **Syngenetic** or **contemporary** inclusions. These consist of materials which were present at the same time as the host crystal (they may have grown from the same solution as the host crystal, or have a similar atomic structure to the host). They can be present as crystals, as trapped liquid inclusions or as liquid trapped in a fracture which has then become sealed by the host (termed a 'healing feather'). Such a feather usually consists of a finely scattered group of separate liquid droplets, and can be seen in corundum, peridot, spinel and topaz. In a peridot, a similar disc-like feature is called a 'lily pad'.

   When trapped liquid inclusions incorporate a gas bubble and/or a miniature crystal, they are called two and three-phase inclusions respectively. Two and three-phase inclusions occur in emerald and topaz. Two-phase inclusions can also be seen in aquamarine. Common crystal inclusions in this category are rutile needles in corundum and quartz, olivine in diamond, spinel and zircon in corundum, and apatite in corundum, garnet and spinel. Minute spinel octahedra can produce a feather-like feature in some red spinel.

   Cavities in the host crystal may be the result of interrupted development. In this case the hollow so formed often has a regular profile and is filled with a liquid or a gas. Because of their distinctive shape, which may mimic that of the host crystal, these cavities are called 'negative' crystals. They can usually be distinguished from a solid crystal by the presence of gas bubbles. Negative crystals can be seen in corundum and spinel.
3. **Epigenetic** or **post-contemporary** inclusions. These occurred after the formation of the host crystal. They include the recrystallization in fractures of foreign materials, the development of asterism by exsolution of titanium dioxide as in corundum, the development of internal cleavages as in topaz and moonstone, and irradiation damage to the crystal lattice caused by radioactive materials in the host crystal as occurs in some green and brown zircons. In corundums from Sri Lanka, included crystals of zircon (containing the radioactive elements uranium and thorium) have expanded slightly as a result of internal alpha-particle irradiation, and produced stress fractures called 'zircon haloes'.

Identification features of natural and synthetic gemstones

The following section is confined to those natural stones having commercially produced synthetic counterparts (diagnostic inclusions in other natural gemstones are included in Appendix C). Coverage in this section is further restricted to identification features which are detectable using the gem test instruments described in previous chapters. In the case of synthetic stones, many of these features are due to gas bubbles, residual amounts of flux or aqueous solutions, or to metallic fragments from the synthesis apparatus. A later section in this chapter deals with more sophisticated test methods using the type of equipment which is only found in the larger gem testing laboratory.

Alexandrite (colour change variety of chrysoberyl)

The natural stone often contains feathers and two-phase inclusions. Stepped twinning planes may also be detectable under polarized light.

Synthetic alexandrite is grown by the flux-melt process (Creative Crystals), by the Czochralski crystal pulling method (Kyocera-produced 'Crescent Vert' or 'Inamori' alexandrite) and by the floating zone method (Seiko). The stones grown by the first two methods are characterized by swarms of dust-like inclusions (possibly parallel to the seed face). Randomly oriented needles, paralleled elongate crystals, and thin tri-angular platinum crystals may also be present. Earlier flux-melt stones contain liquid-filled healing feathers (Figure 16.1). Seiko synthetic alexandrites have an overall swirled appearance and contain tadpole-shaped gas bubbles similar to those seen in Verneuil synthetic rubies.

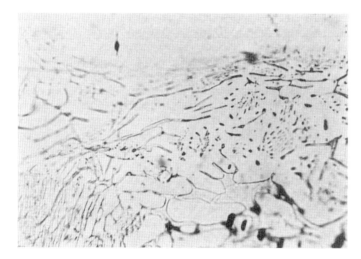

Figure 16.1 A typical liquid-filled healing feather in this flux-melt synthetic alexandrite. (Eppler)

Diamond

Natural diamond is host to a wide range of inclusions which include crystals of dia-mond, garnet, olivine, chrome diopside, chrome enstatite and pyrite. The surface of a diamond crystal (or a 'natural', i.e. part of the crystal surface on the girdle of a cut stone) may contain projecting triangular growth or 'trigon' etch features (Figure 16.2). The majority of these are negative 'trigons' which are orientated in the opposite direction to the triangular faces of the octahedron crystal. While all natural diamonds are virtually inert to SW UV, around 20% of them fluoresce strongly under LW UV (all diamonds, natural and synthetic, fluoresce under X-rays). Natural diamonds which fluoresce blue

Figure 16.2 Trigon etch pit features on diamond. (De Beers)

under LW UV, phosphoresce a faint yellow (this is *diagnostic* for diamond as no other blue-fluorescing mineral under LW UV has a yellow phosphorescence).

Synthetic gem-quality diamonds produced by General Electric (GE), Sumitomo and De Beers have several distinguishing features. The GE stones exhibit magnetic properties (owing to the metal catalyst used in the synthesis process), but this can only be detected using a strong magnet to attract an unmounted stone suspended from a thread or placed on a cork floating on water. Yellow GE stones show no reaction to LW UV, but fluoresce and phosphoresce greenish-yellow or yellow under SW UV, and (like colourless or brown GE stones) do not have a 414.5 nm absorption band. On crystals, an occasional large trigon may be present, but not in the numbers seen in natural crystals. Some yellow Russian synthetic diamonds, which have been subjected to a post-growth high-pressure/high-temperature (HPHT) treatment fluoresce more strongly to LW UV than SW UV (as do natural diamonds). On its own, this test is therefore no longer diagnostic for yellow synthetic diamonds. However, the presence of sharp absorption bands around 460, 560, 637 and 658 nm (many of which can be seen with a hand-held spectroscope) are distinctive of Russian synthetic diamonds.

Like the GE stones, Sumitomo gem-quality synthetic diamonds are inert to LW UV, and fluoresce under SW UV (but with no phosphorescence). Vein-like colourless zones have been seen in Sumitomo crystals, but these may be removed if the stone is cut. Whitish dust-size inclusions, visible under the microscope, are often randomly spread through the stone. Colour zoning and prominent graining may be present. Trigons are absent. Magnetic properties are variable.

De Beers gem-quality synthetic diamonds are also inert to LW UV, but not all stones fluoresce under SW UV. Greenish-yellow stones phosphoresce after SW UV exposure. Inclusions consist of relatively large black particles of flux having a metallic lustre (these are tabular, elongate or needle-like). Internal graining and distinctive colour zoning may be present. Only projecting pyramidal growth features are seen on the faces of De Beers crystals, not (recessed) trigons. Magnetic properties are variable.

While it is possible to identify most synthetic diamonds by their strong fluorescence under SW UV, and by their magnetic properties, the growing threat of lower-cost gem-quality synthetic diamonds was no doubt why, in 1996, De Beers' DTC Research Centre in the UK unveiled two instruments it had developed to positively distinguish natural from synthetic diamonds.

The first of these instruments 'DiamondSure' (Figure 16.3, top), detects the presence of the 415 nm absorption line found in all natural diamonds but absent from synthetic diamonds. Depending on the result of the test, the instrument displays the message 'PASS' or 'REFER FOR FURTHER TESTS' on its front panel. If the latter message is displayed, use is then made of the second instrument 'DiamondView' (Figure 16.3, below). This utilizes two high-intensity UV lamps (with a wavelength less than 230 nm) to produce a fluorescent image, which is processed by a specially configured computer and associated display to reveal the surface growth structures of the diamond. Synthetic diamonds show strong fluorescent yellow/blue/green patterns with major octahedral and cube growth features compared with natural diamonds, which show

Figure 16.3 View of the De Beers DiamondSure (top) with its fibre-optic probe mounted vertically, ready to test the 415 nm absorption response of an unmounted diamond on the black test platform. When testing mounted diamonds, the test platform is removed and the tip of the probe is hand held against the diamond's table facet. The De Beers DiamondView (below) uses a video camera, placed vertically between two high-intensity SW UV light sources, and a specially configured computer/display (here showing an image of the fluorescent growth structure produced by a synthetic diamond). Special stone holders are available for unmounted and ring-set diamonds

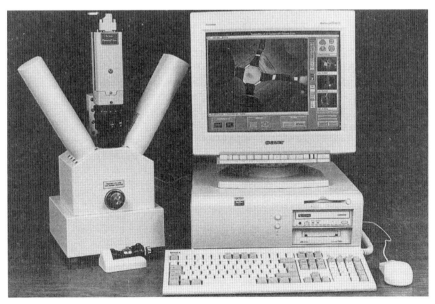

less dramatic and mainly blue octahedral or hummocky cuboid growth structures, slip bands and re-entrant features.

In 2003, both of these instruments were updated and a third model 'DiamondPlus' introduced. This additional instrument is housed in a case similar to that of the 'DiamondSure' and contains two solid-state lasers and two miniature spectrometers. It was designed to help with the detection of HPHT-treated Type II diamond (see 'HPHT enhancement of diamond' in Chapter 14) and to reveal the luminescent and spectrometer features associated with CVD synthetic diamonds (see 'Chemical vapour deposition (CVD)' in Chapter 15). Using the 'DiamondView, CVD diamonds fluoresce a strong orange. In some samples, the microscope may reveal parts of the diamond seed substrate.

Emerald (natural)

The distinguishing internal features of natural emeralds vary with the locality, and are listed here country by country:

Brazil	Biotite mica and thin liquid films resembling paving stones.
Colombia	Three-phase inclusions with jagged ends (Figure 16.4), which contain a gas bubble and a salt (halite) crystal (all mines); albite and pyrite crystals (Chivor mines); rhombs of calcite, and yellow/brown rhombs of parisite (Muzo mines).
India	Hexagonal negative crystals comprising two-phase inclusions resembling 'commas' (Figure 16.4), and mica.
Pakistan	Flakes of mica, crystals of phenakite and thin liquid films resembling the veil-type inclusions in flux-melt synthetics.
South Africa (Transvaal)	Green chrome mica flakes of fuchsite.
Russia (Siberian)	Flakes of mica and green actinolite crystals in blade form (Figure 16.5).
Zambia	Tourmaline crystals, mica flakes, elongate two-phase inclusions, fibrous and acicular tremolite crystals.
Zimbabwe (Sandawana)	Hair-like tremolite fibres (Figure 16.6), mica.

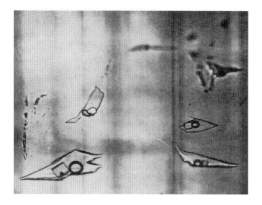 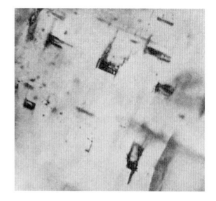

Figure 16.4 A Colombian Chivor mine emerald (left) showing three-phase inclusions. An Indian emerald (right) with comma-like two-phase inclusions

Figure 16.5 Large actinolite crystal in a Siberian emerald

Figure 16.6 Tremolite fibres in a Sandawana (Zimbabwe) emerald

Emerald (synthetic)

Synthetic emeralds grown by the flux-melt process (i.e. Chatham, Gilson, Lennix, Crescent Vert, Seiko, Zerfass and Russian products) generally have a lower RI, DR and SG than natural emerald (RI = 1.560, 1.563; DR = 0.003; SG = 2.65; compared with 1.577, 1.583; 0.006; 2.71 respectively). If bromoform is diluted to an SG of 2.65 using rock crystal (quartz) as an indicator, most of these synthetics will float or suspend in the liquid, but natural emeralds will sink. Separation can also be made with a refractometer.

The lower RI, DR and SG figures for the synthetic product are generally due to an absence of iron oxides which, when originally introduced, resulted in an unacceptable erosion of platinum in the synthesis apparatus. This absence of iron causes the emeralds to appear a more intense red under the Chelsea filter and under crossed filters (and to fluoresce more strongly under LW and SW UV) than do the majority of natural emeralds. For the same reason, the synthetic products are more transparent to SW UV. These tests should not be taken as conclusive evidence of a synthetic origin, however, as a few natural emeralds, particularly those from the Colombian Chivor mines, also show a strong red through the filter and behave similarly under UV.

In these respects, Crescent Vert stones and later versions of both Chatham and Gilson emeralds are much closer in their characteristics to the natural stone. This is thought to be due to the addition of some iron oxide to the melt which raises the constants, inhibits fluorescence and reduces SW UV transparency.

The constants of synthetic emeralds grown by the hydrothermal process (i.e. Lechleitner, Linde, Regency, Biron and Pool, and Russian products) are also much closer to those of natural emeralds. While the majority of these synthetics (coloured by chromium) appear pink to red through the Chelsea filter, some versions (such as the Regency) contain sufficient chromium to make them show bright red. These stones also fluoresce a brighter red than the natural gem when under LW and SW UV, and even fluoresce red when illuminated by a high-intensity white light source.

Some synthetic emeralds, such as early samples of the Biron, have been coloured by vanadium rather than chromium, and do not fluoresce at all (in the UK and Europe, only green beryls showing a chromium spectrum can be described as emerald).

There is also a significant difference between the thermal conductance of synthetic and natural emeralds, and this can be used as a method of identification by employing the type of thermal tester which has been designed for the identification of coloured gems (see under 'Thermal conductance testers' in Chapter 12).

Inclusions in synthetic emeralds are as follows:

Flux-melt stones	Twisted wispy veils or curved lace-like feathers (resembling thinly dispersed cigarette smoke – Figure 16.7), and transparent colourless crystals of phenakite (common to most synthetic emeralds – see Figure 16.8). Seiko stones contain dust-like particles near the surface, together with twisted two-phase feathers, and coloured growth bands parallel to the table facet. The liquid droplets making up the feathers in Lennix emeralds may also consist of two-phase inclusions.
Hydrothermal stones	The Lechleitner and Linde 'coated' emerald products have a network of cracks on the surface of the stone (Figure 16.9). Parts of the pavilion may also be left in a matt unpolished condition to retain maximum colour. If the stone is immersed in bromoform, the dark rim of the emerald-coloured synthetic beryl coating can be seen, together with dust-like crystals at the overgrowth junctions.

In the 100% hydrothermal stone it may be possible to see the seed plate, which is often lighter in colour. Linde stones, in addition to the curved lace-like feathers seen in other synthetic emeralds, sometimes contain nail-like inclusions consisting of conical tubes capped with phenakite crystals (Figure 16.10). Both the Linde and Regency emeralds are also highly fluorescent under LW/SW UV and intense white light, and appear bright red under crossed filters and through the Chelsea filter.

Russian hydrothermal stones contain characteristic phenakite 'daggers', partly healed cracks and the occasional brass-like metallic needles.

Morganite

Biron International Limited began producing hydrothermally grown synthetic pink beryl (i.e. the morganite variety of beryl) in 1990. This material is coloured by titanium (the natural stone is coloured by manganese), but should it be encountered as a facetted gemstone it can easily be identified as a synthetic by its lower constants (RI 1.578, 1.571; SG 2.685 compared with RI 1.586, 1.594; SG 2.80 for the natural beryl) and by the two broad absorption bands centred on 495 and 550 nm due to titanium (there is no discernable spectrum for the natural stone).

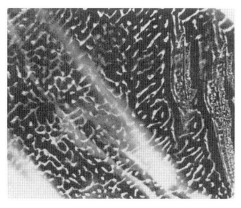

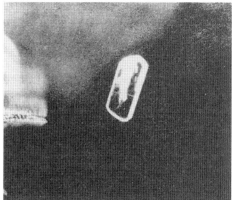

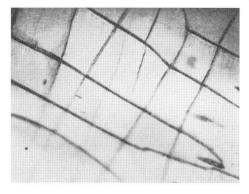

Figure 16.7 (top left) A twisted wisp-like feather in a Chatham flux-melt synthetic emerald. (Gübelin)

Figure 16.8 (above) Typical phenakite crystal as seen in many synthetic emeralds. (Gübelin)

Figure 16.9 (left) A network of fine surface cracks on a Lechleitner synthetic emerald-coated beryl

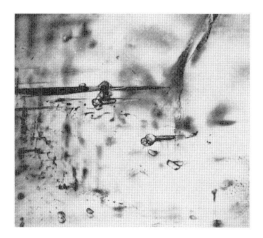

Figure 16.10 Nail-like inclusions formed by a phenakite crystal capping a needle/conical tube in a Linde hydrothermal synthetic emerald

Quartz

Synthetic quartz is grown hydrothermally, and has been produced as high-quality untwinned material for many years for the optical and electronic industries. Some years ago crystal growers in the USA and in the former USSR began producing quantities of synthetic quartz in gem colours (particularly amethyst, and in Russia, transparent pink quartz).

The detection of twinning planes in natural quartz has been one of the main methods of separating natural from synthetic amethyst and citrine (as described under 'Use of the polariscope' later in this chapter). However, this is no longer considered to be a completely secure technique as twinned synthetic quartz can be grown hydrothermally from twinned seeds. Identification is therefore more safely based on the presence of typical inclusions. In natural amethyst, a distinctive inclusion called a thumb print or 'tiger stripe' may be present (Figure 16.11), and although also due to twinning, it is unlikely to occur in this form in synthetic twinned material. Other natural inclusions in quartz are groups of prismatic crystals (Figure 16.12).

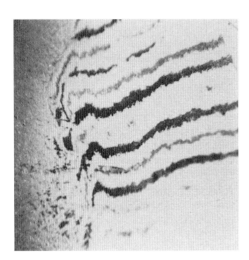

Figure 16.11 A typical tiger stripe inclusion in a natural amethyst variety of quartz

Figure 16.12 Groups of prismatic crystals in quartz

Synthetic amethyst sometimes contains breadcrumb-like inclusions, and in all synthetic quartz, traces of the colourless seed plate may be visible, accompanied by strong colour banding parallel to the seed plate.

Ruby (natural)

Hexagonal colour zoning is usually present in natural rubies, but is also present in some flux-melt synthetics although less distinctly. Other more distinguishing internal features of natural rubies vary with the locality, and are listed here country by country:

Myanmar (*Burma*)	Zircon, spinel, and rounded colourless crystals (Figure 16.13), rhombs of calcite and yellowish sphene crystals. Rutile needles (silk and star-forming – Figure 16.14). Wisps and swirls of colour (called 'treacle').
Sri Lanka	Long sparse rutile needles (silk and star-forming), zircon crystals with 'haloes', pyrite and biotite mica.
Thailand	Less inclusions than Myanmar (Burma) stones. Reddish-brown opaque almandine crystals. Partly healed cracks or feathers surrounding crystals, yellowish apatite platelets. Very little silk (rutile needles).
Tanzania	Rutile needles, apatite, zircon and calcite crystals. Whitish boehmite particles along intersecting twin lamellae planes.

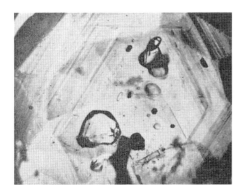

Figure 16.13 Crystal inclusions and colour zoning in Myanmar (Burma) ruby

Figure 16.14 Rutile needles forming a patch of silk in a Myanmar (Burma) ruby

Ruby (synthetic)

These have the same physical properties and constants as the natural stone, and the main identification features are those of growth lines, colour zoning and inclusions. The bulk of these stones are grown by the Verneuil flame-fusion process, although a number are now produced by the floating zone (Seiko), flux-melt (Chatham, Kashan, Knischka and Ramaura) and hydrothermal processes (Lechleitner). Identifying features in synthetic rubies are as follows (*note*: rutile, in the form of silk, is virtually absent from any of these products):

Verneuil stones
Curved growth lines (best seen under immersion with appropriate lighting such as diffused or 'shadowing' – see Chapter 13), clouds of minute gas bubbles and (rarely in modern productions) tadpole-like gas bubbles (Figure 16.15). Because stones, particularly large ones are usually cut (for maximum yield) with the table facet in line with the plane of the split boule (i.e. in line with the '*c*' axis), the orange-red pleochroic colour of ruby is often visible on a dichroscope through the crown facets. With most natural rubies the table facet is cut at right-angles to the '*c*' axis to avoid this colour.

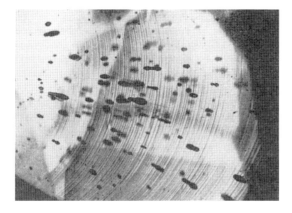

Figure 16.15 Curved growth lines and bubbles in synthetic Verneuil ruby

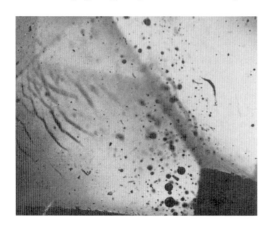

Figure 16.16 Fire marks (surface cracks caused by over-pressure polishing) and bubbles in a synthetic ruby

Fire marks are sometimes seen near facet junctions of the Verneuil stones. These are parallel crack-like markings caused by overheating during rapid polishing of this low-cost material (Figure 16.16). Verneuil rubies are very transparent to SW UV as they contain no iron oxides (oxides introduced into the source powder migrate out into the skin of the finished boule). Because of this and their chromium content they fluoresce a bright red under LW UV and under crossed filters.

With clean stones containing no obvious identification features, a Verneuil corundum can be identified by means of the Plato test described under 'Use of the polariscope' later in this chapter.

Star stones The asterism effect is usually sharper and located more on the surface of the stone than with the natural gem. In earlier productions of the synthetic Verneuil product, the arms of the star did not reach as far as the base line of the cabochon (see also under 'Surface diffusion' in Chapter 14).

Floating zone stones Seiko synthetic rubies grown by this process are relatively free from inclusions, but contain swirled growth features similar to the curved growth lines in Verneuil ruby. Clouds of bubbles may also be present.

Flux-melt stones In addition to producing the following types of synthetic ruby for the gem market, this process has been used (together with the Czochralski crystal pulling method) for the growth of high-quality ruby crystals for use in lasers.

Kashan rubies contain a variety of features including small elongated forms called 'paint splash' inclusions (Figure 16.17). At the lower quality end, these stones have coarse mesh-like networks of flux-filled whitish negative crystals (Figure 16.18). The Kashan was probably the first synthetic ruby to contain variable amounts of iron oxide. This invalidated the SW UV transparency test, and produced LW UV fluorescent and Chelsea filter results similar to those of the natural stone, (i.e. muted or inert). In these respects, and in appearance, the Kashan product resembles the Thai ruby.

Ramaura rubies contain feathers of orange-yellow and white flux, and comet-like inclusions with tails. Under LW and SW UV, chalky-white and bluish-white zones are visible. The addition of a fluorescent dopant to the source material results in a shift in the LW UV fluorescent

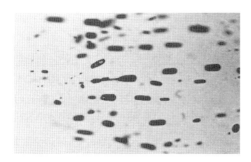

Figure 16.17 'Paint splash' inclusions in Kashan synthetic ruby. (Eppler)

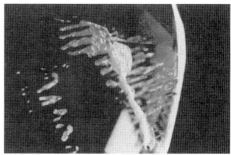

Figure 16.18 Network of flux-filled negative crystals in a Kashan synthetic ruby. (R.W. Hughes)

colour towards the yellow-orange (this is not easily detected in some stones and is best compared with the red fluorescent colour of a natural ruby in a darkened room).

Knischka rubies have negative crystals terminating in long crystalline tubes, black distorted hexagonal platelets of platinum and silver (mainly eliminated from later productions), and two-phase inclusions, the latter being regarded as diagnostic for this product. The stones fluoresce strongly under UV and under crossed filters.

Hydrothermal stones The production of synthetic ruby by this method has been mainly experimental. However, Lechleitner has used the process to develop an overgrowth of ruby (and blue sapphire) on Verneuil-produced seeds or preforms. The inclusions listed under 'Verneuil' in this section are present, and the overgrowth has a crazed surface similar to that illustrated in Figure 16.9. White flux residues producing feathers and wispy veils may also be present.

Sapphire (natural)

Strong hexagonal colour zoning is usually present in the natural stone, particularly with blue sapphires (Figure 16.19), although this also occurs in a less marked degree

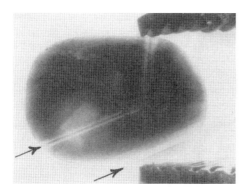

Figure 16.19 (left) Straight colour zoning in natural sapphire; (right) Curved colour zoning in synthetic Verneuil sapphire

in a few flux-melt corundums. Distinguishing inclusions vary with the source and are listed here country by country:

Australia	Strong colour zoning, zircon crystals with haloes (as in Sri Lankan sapphires), crystals of plagioclase feldspar.
Cambodia	Plagioclase feldspar and red pyrochlore.
India (*Kashmir*)	Milky zoning or cloudiness caused by layers of liquid inclusions; feathers and zircon crystals with haloes (the latter two features as in Sri Lankan sapphires).
Myanmar (*Burma*)	Convoluted healing feathers (looking like crumpled flags), short thick rutile needles, apatite crystals.
Sri Lanka	Rutile needles forming silk (and sometimes asterism), three-phase inclusions, zircon crystals with haloes and feathers, 'lines' of spinel octahedra (all seen in Figure 16.20). Elongate negative crystals.

Sapphire (synthetic)

Like synthetic ruby, synthetic sapphire has similar physical characteristics and constants to the natural stone. Identification, therefore, is mainly reliant on distinguishing internal features such as colour zoning and inclusions. The bulk of synthetic sapphire is grown by the Verneuil flame-fusion method, although a number are now also produced by the floating zone (Seiko), flux-melt (Chatham and Kyocera) and hydrothermal (Lechleitner) processes. Inclusions in synthetic sapphires are as follows:

Verneuil stones	Curved colour zoning (easily visible in blue stones – Figure 16.19) and small gas bubbles. With yellow and orange sapphires, the use of diffused transmitted light and a blue colour contrast filter helps to make curved colour banding more visible. Curved growth stria, similar to those seen in ruby, are present in most sapphires but are difficult to detect.
	Because stones, particularly large ones, are usually cut with the table facet in line with the plane of the split boule (i.e. in line with the 'c' axis) the bluish-green pleochroic colour of blue sapphire is often visible on a dichroscope through the crown facets. With most natural blue sapphires, the table facet is cut at right angles to the 'c' axis to avoid this colour. (When viewing a synthetic blue sapphire on the dichroscope it may also be possible to see the curved colour zoning at the same time).
	For stones having no obvious distinguishing features, Verneuil corundum can be identified by means of the Plato test (see under 'Use of the polariscope' later in this chapter).
Star stones	The asterism effect is usually sharper and located more on the surface of the stone than with the natural gem. In earlier productions of the synthetic Verneuil product, the arms of the star did not reach as far as the base line of the cabochon (see also under 'Surface diffusion' in Chapter 14).
Floating zone stones	Orange and pink Seiko sapphires have curved or swirled growth structures which are best seen with the stones in an immersion liquid such as di-iodomethane (methylene iodide).
Flux-melt stones	Chatham blue and orange sapphires contain strong hexagonal colour zoning. The orange product has solid flux inclusions in the form of

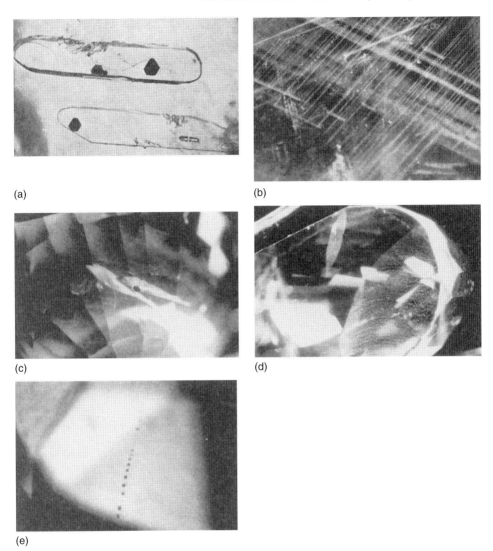

(a)

(b)

(c)

(d)

(e)

Figure 16.20 Features in Sri Lankan sapphires: (a) three-phase inclusions; (b) rutile needles forming long silk; (c) zircon crystals with stress haloes; (d) feathers; (e) 'chain' of spinel octahedra

Hydrothermal stones

wispy and lace-like feathers, and curled 'flags'. Hexagonal, triangular and needle-like platinum inclusions are also present. A Kyocera orange synthetic sapphire has less flux and platinum inclusions than the Chatham version, but contains a fine distribution of thinner inclusions.

Lechleitner blue sapphires are produced by hydrothermally coating Verneuil corundum seeds or preforms. The seed section contains gas bubbles and may also have white flux residues producing feathers and wispy veils. The overgrowth has a crazed surface similar to that shown in Figure 16.9.

Spinel

The natural stone is usually host to small spinel octahedra, often in lines, plus negative octahedral crystals, iron-stained feathers and zircon haloes. Most synthetic spinels are produced by the flame-fusion Verneuil method and are used as simulants of other gemstones with colours which are quite easily distinguishable from natural spinels. They can also be distinguished from natural spinel by means of their greater RI (1.727 compared with 1.717). The synthetic stone usually contains anomalous double refraction (seen as a 'cross-hatch' effect on the polariscope, and known as 'tabby extinction'). Blue and green stones coloured by cobalt show the characteristic three absorption bands at the yellow/green end of the spectrum; however, it should be remembered that there are some rare blue natural spinels which also contain some cobalt.

The only synthetic which resembles its natural counterpart is the rarely seen red synthetic spinel produced by the Verneuil method (see comments on its manufacturing problems in Chapter 15). However, red synthetic spinel has more recently been produced in the USSR by the flux-melt process. Because of its 1:1 ratio of magnesia and alumina, both Verneuil and flux-melt synthetic red spinel have the same SG value as natural spinel, but due to their extra chromium content the RI is usually slightly higher (1.73). The synthetic Verneuil stone often contains many small gas bubbles and has pronounced curved colour zoning which produces a 'shuttered' or 'Venetian blind' appearance. The flux-melt Russian version contains tension cracks (surrounded by strong anomalous double refraction) and small black flux particles.

Colour zone diffusion and induced feathers/fingerprints

If Verneuil corundum is heated to very high temperatures for long periods, the distinctive curved colour zoning becomes less visible, making the stone appear more like its natural counterpart.

Synthetic corundum can also have quite convincing feather 'inclusions' induced into the body of the stone by means of a heat treatment technique. The stones to be treated are first heated unevenly using an oxy-acetylene gas torch in order to produce internal fractures by way of differential expansion. The stones are then reheated to encourage partial healing of the cracks, which take on an appearance similar to the healing feathers in natural corundum.

When induced feathers are combined with colour zone diffusion, the Verneuil corundum loses two of its identifying characteristics, and discrimination then depends on a careful assessment of other features such as bubbles, SW UV transparency and use of the Plato test as described in the following section.

Use of the polariscope (identifying synthetic quartz and Verneuil corundum)

Identification of natural/synthetic quartz

The quartz synthesis technique was originally developed for the production of untwinned material (which is a prime requirement for electronic oscillator applications). Because of this, the absence of twinning has until recently been the basis of a test for synthetic quartz as suggested by Dr Karl Schmetzer.

In a simplified version of this test developed in the GIA Gem Trade Laboratory, the direction of the stone's optic axis (i.e. the '*c*' axis) is found by using the conoscope adaptation of the polariscope (see 'The conoscope (interference figures)' section in

Chapter 10), and this direction is marked on the stone using a felt tip pen or a spot of white typing correction fluid. The stone is then positioned and viewed in this orientation between the crossed polars of a polariscope (or a polarizing microscope) and preferably immersed in a liquid having an RI close to that of quartz – although water is adequate. If twinning exists, this will be evident as a series of coloured and paralleled interference bands (Figure 16.21).

Figure 16.21 Representation of the coloured interference bands caused by twinning in natural quartz (as seen in the direction of the optic axis under crossed polars)

However, as there is no need for synthetic quartz grown specifically for use in jewellery to be untwinned, it is possible that the producers of this material will begin to use twinned seed material, and the polariscope test will no longer be valid.

Other differences between the two materials are in the distribution of the amethyst colour. This is straight or angular in the natural stone with distinctly separate zones of purple and violetish blue or colourless, but is more diffuse in the synthetic stone. However, this feature may also be due to a difference between twinned and untwinned material. Additional discriminating differences between synthetic and natural quartz are described in an earlier section of this chapter under 'Quartz'.

The Plato test for Verneuil corundum

This is a method, developed by Dr W. Plato, for identifying synthetic Verneuil corundums which contain no detectable inclusions, growth lines or colour zoning. First, the direction of the stone's optic axis (i.e. the 'c' axis) is found by using the conoscope adaptation of the polariscope (see 'The conoscope (interference figures)' section in Chapter 10), and this direction is marked on the stone using a felt tip pen or a spot of white typing correction fluid. The stone is then positioned and viewed in this orientation under crossed polars (preferably at 20–30× magnification) while in an immersion liquid such as di-iodomethane (methylene iodide), a process made easier by using a horizontal-type immersion microscope. If two sets of bands intersecting at 60° are visible (Figure 16.22), then the stone is a synthetic Verneuil corundum.

Laboratory equipment and methods

Although outside the budget of most gemmologists/jewellers and small gem testing laboratories, there is a growing range of sophisticated analysis equipment that is being

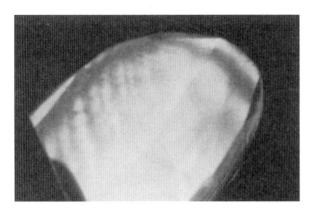

Figure 16.22 Plato lines as seen in a yellow synthetic sapphire when viewed parallel to the '*c*' axis under crossed polars

used to identify the nature of potentially high-value gemstones. The use of a liquid nitrogen cooling technique for revealing faint absorption lines diagnostic of irradiation and heat treatment in fancy coloured natural diamonds has already been mentioned in Chapter 14 under 'Irradiation methods'. The following is a brief description of some of these equipments and methods.

Electron microprobe

This is used for the non-destructive analysis of gemstone constituents and surface-reaching inclusions. It consists of a vacuum chamber for the specimen, an electron gun, X-ray spectrometers and an inspection microscope. The gem to be analysed is positioned under the focused beam of electrons using the microscope. As the electrons bombard the surface of the gem, X-rays are emitted, each constituent element in the bombarded area emitting X-rays at a characteristic wavelength. This X-radiation is detected and its wavelengths measured by the spectrometers. Tables of emission spectra can then be used to identify the elements responsible for these wavelengths of emission. For the quantitative analysis of the chemical composition of the sample, the X-ray data is usually analysed by computer, and a comparison check can also be made against calibration specimens.

A related equipment using EDXRF (energy-dispersive X-ray fluorescence) spectrometry, which has become standard equipment in many gemmological laboratories, has shown that while synthetic rubies contain few trace elements, the presence of certain elements (i.e. Mo, La, W, Pt, Pb, Bi) are found only in flux-melt synthetic rubies, while Ni and Cu are associated with hydrothermal synthetic rubies. If these elements are not detected, the presence of Ti, V, Fe and Ga are diagnostic for a natural ruby.

In addition, trace-element analysis of this type can help determine the mining locality of the sample. Rubies from a basaltic deposit (i.e. Cambodia and Thailand) have a relatively high Fe and low V content, while stones from ruby-bearing marbles (i.e. Myanmar, Nepal and China) have a relatively low Fe and high V content.

Scanning electron microscope (SEM)

For gemmological purposes, one of the most important applications of the SEM is its use as an energy dispersive spectrometer. In this mode it operates in the same way as the electron microprobe (described in the previous section) and is used to evaluate the X-rays produced when the electron beam strikes the gemstone under test. In its primary function as a microscope, the SEM uses a focused beam of electrons to bombard the sample under test and thereby cause the emission of secondary electrons from the sample's surface. The varying intensity and distribution pattern of these secondary electrons are detected and used to produce a picture of the sample's surface on a video display.

The advantage of an SEM over an optical microscope is that electrons have a much shorter wavelength than visible light, and can therefore reveal much smaller surface features. Magnifications of up to $250\,000\times$ are attainable with this technique, which made it possible to discover the cristobalite spheres responsible for opal's iridescence (see Figure 8.9).

Spectrophotometers

These instruments are now widely used in gemstone analysis. They consist of a light source (which may cover the UV and IR as well as the visible section of the spectrum), a monochromator which can be tuned over the spectral range, a detector unit for sampling the transmitted or reflected light from the specimen and a control unit to amplify and display the output of the detector unit.

Spectrophotometers often contain a means of automatically scanning and recording absorption or transmission spectra. When they are used in the visible range to measure the colour of a sample, they usually have the computing ability to translate the recorded spectrum into tristimulus values and CIE colour co-ordinates (Figure 16.23).

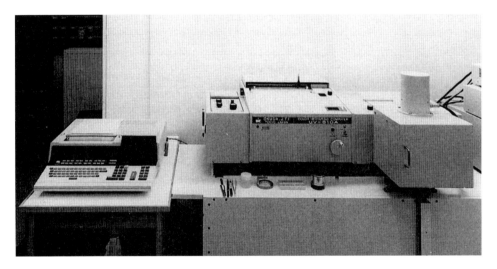

Figure 16.23 A spectrophotometer designed to measure the colour grade of cut stones. On the left is the control unit, in the centre the sample chamber and detector, and on the right the light source and monochromator

A UV spectrophotometer analysis technique for discriminating between natural and synthetic rubies by means of their characteristic UV bandwidth/wavelength was developed by the Swiss gemmologist Bosshart. This was made less positive by the Ramaura synthetic ruby (now no longer in production) whose characteristics in the UV overlapped those of natural stones. Other similarities between some Chatham rubies and a few natural rubies from Myanmar (Burma) and Sri Lanka, and between some Kashan rubies and natural stones from Kenya and Myanmar, made it necessary to corroborate the results with microscopic examination of inclusions.

For gemmological purposes, the most useful area of spectroscopic analysis lies in the infrared. Using an instrument such as the Nicolet 60SX Fourier transform infrared spectrometer, the water content of natural emeralds can be used to distinguish them from synthetic stones (both flux-melt and hydrothermal). This is done by comparing their water absorption bands centred around $3600\,cm^{-1}$ (2778 nm) in the infrared. A similar technique is used to distinguish natural alexandrite from most flux-melt synthetic alexandrites. Infrared analysis is also used to verify the tints in fancy coloured diamonds as natural or artificially induced (see under 'Irradiation methods' in Chapter 14).

Nuclear magnetic resonance (NMR) is another technique used in the non-destructive determination of a gem's chemical constituents. The basic principle of NMR is the selective absorption of externally applied electromagnetic radiations by systems of atoms or molecules in the sample under test. This absorbed energy is caused by the interaction between the sample's nuclear magnetic moments and a high-intensity external magnetic field.

The magnets used by an NMR spectrophotometer to produce the intense magnetic fields necessary for these measurements are either iron-cored electromagnets weighing several tons, or cyromagnets operating at liquid helium temperature. The instrument's detector probe is fed from a radio-frequency transmitter and receiver which monitor the energy absorption of the various nuclei in the sample. The instrument is controlled by a computer which analyses the resulting data.

NMR can be used as a quantitative method for determining most gem constituents including hydrogen, aluminium, silicon, beryllium, lithium, sodium and phosphorous. A full NMR analysis can form a 'fingerprint' for that material, and can be used to discriminate between natural and synthetic stones.

A similar technique, called neutron activation analysis (NAA) has been used to distinguish natural from synthetic stones by means of the detection of the element gallium, which while present in all natural stones is not normally found in synthetics. However, following disclosure of the technique, several synthetic emeralds, a synthetic alexandrite and one synthetic ruby have since been found to contain a similar small quantity of gallium! Although this invalidates the test method, a complete absence of gallium in a gemstone would still indicate that it is a synthetic.

Raman spectroscopy

When a high-intensity light (such as a laser beam) impinges on a surface, a small amount of the light undergoes *Raman scattering*. This scattered light has components which are longer or shorter in wavelength than the incident beam (described as *Stokes* and *anti-Stokes* wavelengths – see 'Photoluminescence and Stokes' law' in Chapter 12). The effect is caused by molecular vibrations in the surface of the material under test. The degree of wavelength difference, or *Raman shift*, between the incident beam and the scattered light depends on the nature of the surface atoms and their structural bonding. This spectral shift enables diagnostic absorption spectra in the infrared to be seen in the visible region of the spectrum by means of a standard spectrophotometer. The use of Raman spectroscopy in gemmological research is increasing as the databank of Raman spectra produced for a range of materials is built up.

X-ray equipment

At the simplest level, an X-ray source can be used to discriminate between natural and cultured pearls by means of a contact X-ray picture. A more sophisticated piece of equipment (Figure 16.24) makes use of the Laue diffraction pattern of dots generated on

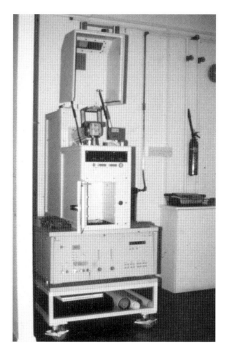

Figure 16.24 A specially commissioned X-ray unit used mainly for pearl testing (see pearl lauegrams in Chapter 18, Figures 18.11 and 18.12). (Courtesy of the Gem Testing Laboratory of Great Britain)

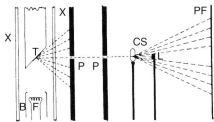

Figure 16.25 (left) Diagram showing method of producing a lauegram. X = X-ray tube (B, F, T are bias shield, filament, target); P = pinhole collimator; CS = crystal specimen, L = lead disk to absorb main X-ray beam; PF = photographic film

Figure 16.26 (below right) Laue diffraction pattern (taken parallel to the '*c*' axis of a beryl crystal) showing six-fold symmetry: (below left) Lauegram taken parallel to the trigonal axis of symmetry of a crystal

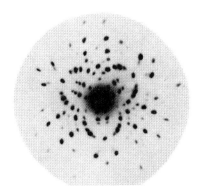

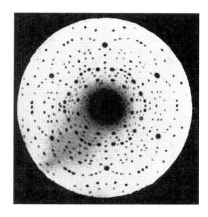

a photographic film when a narrow beam of X-rays are scattered by the atomic planes in the pearl's crystalline aragonite layers or (in the case of cultured pearls) by the mother-of-pearl nucleus (Figure 16.25). Laue diffraction patterns (lauegrams) are also used in this way to detect the crystallographic symmetry of a specimen (Figure 16.26).

Mineral identification can be achieved by means of X-ray diffraction using a powder sample of the specimen (i.e. a scraping). Because of the random orientation of the individual crystals in the powder, the resulting lauegram consists of a series of concentric rings instead of a pattern of dots (Figure 16.27). By using a reference file of pictures recorded in this way from a range of substances, it is possible to make an identification match.

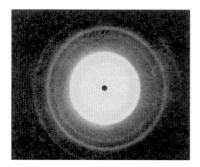

Figure 16.27 Powder diffraction photograph of diamond

X-rays can also be used to distinguish between diamond and its simulants. Diamonds are transparent to X-rays, but all diamond simulants absorb X-rays in varying degrees (Figure 16.28). (Also see under 'Diamond' earlier in this chapter).

Another technique associated with the identification of diamond is X-ray topography, which has been used in the experimental fingerprinting of the stone. A vertical ribbon-like beam of collimated X-rays is used to scan the diamond, which is positioned so that the beam is diffracted by the atomic layers in the crystal lattice. The emerging X-rays fall on a photographic plate to produce a projection topograph showing a pattern of distinguishing crystal defects within the stone (Figure 16.29).

Figure 16.28 A low-energy X-ray tester for distinguishing diamond from its simulants

Figure 16.29 A projection X-ray topograph of a brilliant-cut diamond showing various crystal defects

Quantitative cathodoluminescence

Cathodoluminescence is the fluorescent effect displayed by some materials when they are bombarded with a beam of electrons in a vacuum. In the quantitative cathodo-luminescence (CL) method developed by Dr J. Ponahlo, the photocurrent, I, of the luminescence sensor is plotted against the power of electron excitation, W. Natural emeralds show only a very small linear increase of I against W, but synthetic emeralds show a large increase. Natural Thai rubies produce CL intensities which are lower by a factor of 10–100 compared with those of synthetic rubies under the same excitation. Quantitative CL therefore appears to offer a relatively safe method of discriminating between natural and synthetic rubies and emeralds.

In a later extension to this method, Dr Ponahlo has employed microspectrophotometry techniques to reveal even more distinctive differences between natural and synthetic gemstones by means of their CL spectra. Using this method, it is now possible to compare the luminescing spectra of ornamental stones and to distinguish between lapis lazuli from Afghanistan, Chile and the former USSR. Sumitomo and De Beers gem-quality synthetic diamonds can also be distinguished from natural diamonds by means of their CL spectra. (Also see under 'Diamond' earlier in this chapter).

Chapter 17

Identifying inorganic gemstone simulants

In contrast to synthetic gemstones, which have essentially the same chemical composition, crystal system and physical constants as their natural counterparts, a simulant need only have a superficial resemblance to the gemstone it imitates. Because of this, the simulant's constants are usually quite different from those of the genuine stone. Although this provides the main means of identification, the presence of diagnostic inclusions should not be overlooked when checking a suspect stone (see under 'Identification features of natural and synthetic gemstones' in the previous chapter). Materials used to simulate the more valuable gem minerals range from natural gemstones (sometimes dyed) to various man-made products.

Glass (paste) is perhaps the most common of the less expensive imitations, and can be identified by its non-crystalline character, by the presence of strain bands as visible on the polariscope (see under 'Using the polariscope' in Chapter 10), its poor thermal conductance (it feels warm to the touch), its relative softness (evidenced by rounded and worn facet edges), and the presence of conchoidal fractures, uneven colour distribution (colour swirls) and gas bubbles (Figure 17.1). Composite gemstones and their

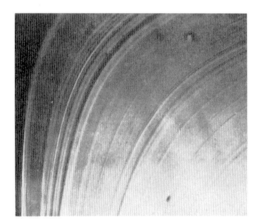
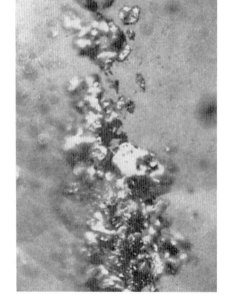

Figure 17.1 (above) Colour swirls/stria in glass: (right) Typical bubbles in glass

detection have already been covered in Chapter 15 as not all of these materials are truly simulants (e.g. opal and diamond doublets).

As already mentioned, the detection of both glass and the other gemstone simulants is relatively simple, as their constants are rarely the same as those of the gemstone they imitate. Mounted diamond simulants are probably the most troublesome because their refractive indices are usually above the range of the standard refractometer. For this reason, they have been dealt with in some depth in the following notes. (The identification of organic gem materials and their simulants is covered in the next chapter.)

Alexandrite

Both synthetic corundum and synthetic spinel have been produced to imitate the characteristic colour change of the alexandrite variety of chrysoberyl. However, the colour of synthetic corundum (doped with vanadium) changes from an amethyst purple in filament light to a greyish blue in daylight and cannot be mistaken for the brownish-red to green colour change of genuine alexandrite. Although the colour change of the synthetic spinel simulant is more convincing, today this stone is rarely encountered. The strong pleochroism of alexandrite (red, orange, green in filament light) also serves to distinguish it from that of the corundum simulant (brownish-yellow, purple), as does a diagnostic absorption line in the corundum simulant (due to vanadium) at 475 nm. However, it is just possible that the even stronger pleochroism of the rare andalusite gem (green, yellow, red) may cause it to be confused with alexandrite. Table 17.1 lists the identifying constants of these stones (in this and the following tables, H = hardness on the Mohs scale).

Table 17.1

Gemstone	RI	DR	SG	H
Alexandrite	1.746, 1.755	0.009	3.72	8.5
Synthetic corundum	1.762, 1.770	0.008	3.99	9.0
Synthetic spinel	1.727	–	3.64	8.0
Andalusite	1.63, 1.64	0.01	3.18	7.5

Aquamarine

With the exception of paste simulants and composite stones (detection of the latter is covered in Chapter 15), the most common aquamarine simulant is synthetic blue spinel. A simple test on the polariscope (taking care not to be mislead by the strain patterns often seen in synthetic spinel) will separate the doubly-refracting aquamarine from the singly-refracting spinel. On the spectroscope, a faint cobalt spectrum will be visible in aquamarine-colour spinel, and the stone will appear pink through the Chelsea filter. Other possible simulants of aquamarine are blue zircon (whose high double refraction, 0.058, shows as a 'doubling' of the pavilion facet edges; see Figure 17.2), topaz and beryl glass (i.e. fused beryl coloured by cobalt). The latter two are distinguished by their constants as listed in Table 17.2.

Figure 17.2 The high DR of zircon causes a doubling of the pavilion facet edges as viewed through the table facet

Table 17.2

Gemstone	RI	DR	SG	H
Aquamarine	1.574, 1.580	0.006	2.71	7.5
Synthetic spinel	1.727	–	3.64	8.0
Topaz	1.61, 1.62	0.01	3.56	8.0
Zircon	1.93, 1.99	0.058	4.68	7.5
Beryl glass	1.52	–	2.44	7.0

Diamond

Because of its value, diamond is probably the most imitated of all the gemstones. Like emerald, ruby and alexandrite, whose value sometimes even exceeds that of diamond, there are many simulants of diamond including natural gemstones such as the colourless varieties of quartz, topaz, corundum and zircon, all of which can easily be distinguished from diamond by virtue of their double refraction. With the exception of zircon, it is possible to identify each of these stones on the refractometer. Measurement of RI will also identify colourless synthetic spinels, synthetic corundum and paste.

Flint glass was probably the first man-made diamond simulant, although its use in antique jewellery probably owed more to its pleasing degree of 'fire' (its dispersion of 0.04 is close to that of diamond) than to its superficial resemblance to diamond. With the exception of synthetic colourless corundums and spinels, which were introduced as diamond simulants almost 75 years ago, the majority of man-made simulants of diamond are spin-offs from specialized crystals grown for the electronics, laser and space industries.

Of these, YAG (yttrium aluminium garnet), GGG (gadolinium gallium garnet), CZ (cubic zirconium oxide) and lithium niobate have no counterpart in nature and are correctly described as man-made or artificial products rather than synthetics. Until 1987, strontium titanate, another diamond simulant, was in the same category. However, in that year natural grains of the material were found in the former USSR, and named 'tausonite'. Therefore strontium titanate should now be described as a synthetic rather than a man-made product. Cubic zirconium oxide (CZ), whose manufacture is described in Chapter 15, is the most convincing and widely manufactured of all of the simulants to date. In 1996, a new diamond simulant, synthetic moissanite (a yellow to colourless silicon carbide with a hexagonal crystal system) was introduced by C3 Incorporated of the USA. However, this material has a large double refraction (0.043), a high dispersion (0.104), is an electrical semiconductor (see under 'Electrical properties' and 'Near-UV transparency tester' in Chapter 12) and should not be difficult to distinguish from diamond. The constants, including dispersion, of all these diamond simulants are listed in Table 17.3.

Table 17.3

Gemstone	RI	DR	Dispersion	SG	H
Diamond	2.417	–	0.044	3.52	10.0
Flint glass	1.6–1.7	–	0.04	3.0–4.0	5.0
Quartz	1.54, 1.55	0.009	0.013	2.65	7.0
Topaz	1.61, 1.62	0.01	0.014	3.56	8.0
Zircon	1.93, 1.99	0.058	0.039	4.68	7.0
Synthetic spinel	1.727	–	0.02	3.64	8.0
Corundum	1.76, 1.77	0.008	0.018	3.99	9.0
Synthetic rutile	2.61, 2.897	0.287	0.280	4.2–4.3	6.5
YAG	1.83	–	0.028	4.58	8.5
Strontium titanate	2.41	–	0.190	5.13	5.5
Lithium niobate	2.21, 2.30	0.09	0.120	4.64	5.5
GGG	1.97	–	0.045	7.05	6.0
CZ	2.15–2.18	–	0.065	5.6–6.0	8.0
Synthetic moissanite†	2.65, 2.69	0.043*	0.104	3.22	9.25

* Usually cut with optic axis at right angles to table facet – DR may not show on polariscope unless sample is tested in various positions.
† (See also under 'Thin diamond films' in Chapter 15).

Additional tests for diamond

The 'light spill' or 'tilt' test

If a correctly proportioned brilliant-cut stone is illuminated and viewed against a dark background, with its table facet at right angles to the line of vision, the stone will appear uniformly bright. This is because its pavilion facets are acting as mirrors and reflect the light back through the table facet by means of total internal reflection.

    If the stone is a diamond (and is correctly proportioned) it will be possible to tilt the top edge of the stone away from the line of vision without losing the overall brilliant appearance. If, however, the stone is a diamond simulant (with a refractive index lower than that of diamond) the optics of the stone start to fail as it is tilted. The result is that the pavilion facets furthest from the eye begin to look black because they are no longer acting as mirrors (i.e. the light 'spills' out of these facets instead of being reflected back through the table facet – see Figure 17.3). The lower the refractive index of the simulant, the more marked is this effect (Figure 17.4).

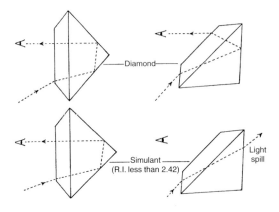

Figure 17.3 The 'light spill' test. Sketch comparing the total internal reflection of light in an ideal-cut diamond with the loss of light through the back (pavilion) facets of a tilted brilliant-cut diamond simulant

Figure 17.4 Five diamond simulants photographed at an angle to illustrate the 'light spill' test. The centre stone is a YAG (RI = 1.83); the top left is a GGG (RI = 1.97); top right is a lithium niobate (RI = 2.25); bottom left is a CZ (RI = 2.16); bottom right is a strontium titanate (RI = 2.42) which, like diamond, shows no light leakage through the pavilion facets

The exceptions to this test (i.e. stones that react optically like diamond) are strontium titanate (also known now as synthetic tausonite; see Appendix D) synthetic moissanite and synthetic rutile, which have RIs similar to or greater than that of diamond. Strontium titanate and synthetic rutile can be identified by their excessive 'fire' (their dispersion is several times that of diamond). Synthetic moissanite has a high double refraction and can be identified (like zircon) by the 'doubling' of its pavilion facets when viewed obliquely through the table facet. The other possible exception to this test is a simulant (such as CZ) whose pavilion has been cut deeper than diamond's ideal proportions in order to compensate for its lower RI. In this case the stone produces an overall total internal reflection even when tilted.

Diamonds with the small table facet and deep pavilion of the 'Old English' cut will fail this check, so before making a tilt test it is advisable to ensure that the stone's profile approximates that of diamond's ideal cut (see under 'Diamond cuts' in Chapter 19).

The 'dot-ring' test

This is more suited to unmounted stones than is the light spill test. However, like the previous test it also depends on the RI of the stone and its proportions, and so has the same possibility of the errors and exceptions mentioned previously. The test is carried out by first making a small black dot on a piece of white paper. The stone under test is then placed table facet down on the paper and centred over the dot. If the stone is a simulant (with an RI less than that of diamond) the dot will appear as a ring around the culet (the point of the pavilion). This effect is due to the failure of the pavilion facets to act as 'internal' mirrors, with the result that the dot is seen though the sides of each pavilion facet and forms a ring (*note*: shallow-cut diamonds will fail this test).

The 'transmission' test

This depends on the same optical situation as the dot-ring test, but instead of placing the stone over a dot, it is positioned table facet down on any strongly coloured surface. If the colour does not show through the rear of the stone it must be either a diamond, a strontium titanate, a rutile, a synthetic moissanite or a simulant with a deep cut pavilion (*note*: shallow-cut diamonds will also fail this test).

The facet condition test

Diamond is the hardest of all natural or man-made substances, and this makes it possible to achieve a very high quality of polish on its facets. Because of its hardness, it is also possible to polish the facets so that they are perfectly flat and meet each other at sharply defined edges. Softer stones will not have the same degree of polish and their facet edges will be slightly more rounded. If a diamond simulant (even a colourless sapphire) has been worn for several years, it may also be possible to detect signs of damage or abrasion at the facet edges.

Reflectance and thermal conductance tests

Because of diamond's high refractive index (and that of several of its simulants) it is not possible to measure this on a refractometer. However, as there is a direct relationship between reflectivity and refractive index, it is possible to identify diamond and its simulants using an electronic reflectance meter (see Chapter 9).

Diamond has a very much greater thermal conductance than any of its simulants (with the one exception of synthetic moissanite – which like zircon can be identified by its large double refraction). One of the most widely used checks for diamond is made with the aid of a thermal conductance tester. Because the advantages and disadvantages of the reflectivity and thermal tester techniques nicely complement each other, some instruments have been marketed which combined the two methods in one instrument (see Figure 12.12).

The weight/girdle diameter test

An unmounted stone can be identified by checking its weight against its girdle diameter. This identifies the stone by, in effect, estimating its SG. An indication of the relationship between these dimensions for diamond and a few of its simulants is shown in Table 17.4 which assumes that the stones are cut to the ideal diamond proportions (maximum/minimum acceptable tolerances in cut will cause the weight to vary by plus or minus 10%. Because of this, synthetic moissanite, with an SG of 3.22, could overlap into diamond's values and because of this has been omitted from the table).

Table 17.4

Girdle diameter (mm)	Carat weight (to two decimal places)				
	Diamond	CZ	YAG	GGG	Strontium titanate
2.0	0.03	0.05	0.04	0.06	0.04
4.0	0.23	0.38	0.30	0.47	0.34
6.5	1.00	1.65	1.30	2.00	1.46
8.0	1.87	3.07	2.43	3.74	2.27
10.0	3.64	6.00	4.74	7.30	5.31

Emerald

Many emerald simulants (but not *synthetic* emeralds) can be detected by means of the Chelsea filter (see Chapter 13). The test is best carried out under strong filament lighting while holding the filter close to the eye. Emeralds (which are coloured by chromium) appear pink to red through the filter, while emerald simulants containing no chromium appear green. However, there are some emeralds (e.g. from South Africa) which do not appear pink through the filter, while there are simulants such as beryl glass (i.e. fused beryl coloured by chromic oxide), demantoid garnet and green zircon which do. The latter two simulants produce 'negative' readings on the refractometer (i.e. they are above its range), and, unlike emeralds whose SG is lower than 2.86, an unmounted green garnet and green zircon will sink in bromoform.

The identifying constants of emerald and its simulants are listed in Table 17.5.

Table 17.5

Gemstone	RI	DR	SG	H
Emerald	1.57, 1.58	0.006	2.71	7.5
Demantoid	1.89	–	3.85	6.5
Zircon	1.78–1.99	0–0.03	3.9–4.68	6.0–7.5
Sapphire	1.76, 1.77	0.009	3.99	9.0
Peridot	1.654, 1.690	0.036	3.34	6.5
Jadeite	1.66–1.68	Polycrystalline	3.30–3.36	6.5–7.0
Tourmaline	1.62, 1.64	0.018	3.05	7.0
Beryl glass	1.52	–	2.44	7.0

Jade

There are many simulants of the two jade minerals *jadeite* and *nephrite*. Both of these gem minerals exhibit a polycrystalline character under the polariscope (i.e. they pass light through the analyser filter in any rotational position), and this serves to distinguish them from some imitations. Green jadeite can sometimes be identified under the spectroscope from its intense absorption band at 437 nm in the violet. Jadeite, which is generally more valuable than nephrite, often has a characteristic dimpled or 'orange peel' surface when polished due to its uneven hardness. However, the modern practice of polishing with diamond grit eliminates this feature. Both jadeite and nephrite are porous and can be dyed green to enhance their appearance. Dyed material may appear pinkish under the Chelsea filter and has an indistinct absorption band in the red; dyed jadeite has no diagnostic absorption band at 437 nm. The identifying constants of jadeite, nephrite and a selection of their simulants are listed in Table 17.6.

Table 17.6

Gemstone	RI	SG	H
Jadeite	1.66–1.68	3.30–3.36	6.5–7.0
Nephrite	1.60–1.641	2.90–3.02	6.5
Hydrogrossular	1.74–1.75	3.6–3.67	7.0–7.5
Bowenite	1.56	2.58–2.62	4.0–5.0
Amazonite	1.52–1.54	2.56	6.0
Verdite	1.58	2.80–2.99	3.0
Prehnite	1.61, 1.64	2.88–2.89	6.0
Saussurite	1.57–1.70	3.00–3.40	6.5

Lapis lazuli

'Swiss lapis', a fairly common simulant of lapis lazuli, is made by colouring jasper with 'Prussian blue' pigment. It can be identified by its lack of pyrite inclusions and its slightly higher refractive index (1.54 compared with 1.50 for the genuine material). Another simulant is sodalite, which can be distinguished by its specific gravity (lapis, 2.8; sodalite, 2.28).

A more expensive lapis lazuli simulant was made by the German Degussa company in 1954. This is a sintered form of synthetic spinel, made by heating a mixture of alumina, magnesia and cobalt to a temperature just below the melting point of spinel. The result is a coarse-textured material closely resembling lapis lazuli in colour. Gold flakes are said to have been included in the mix to imitate the pyrite inclusions in the natural rock. This may have been because the process was not compatible with the introduction of pyrite flakes.

Gilson 'synthetic' lapis lazuli is included here because it contains constituents not present in the natural stone, and this classifies it as a simulant. This material has as one of its ingredients the mineral lazurite, another component of the natural stone, plus pyrite flakes. The Gilson product is much more porous than lapis, has a lower SG (2.36 compared with 2.8) and reacts more rapidly to acids. It has a higher lustre than the natural stone, and unlike lapis lazuli is transparent to X-rays.

Turquoise

Turquoise simulants include dyed chalcedony and howlite, odontolite (also called 'bone turquoise'), lazulite, amazonite and chrysocolla. Identifying constants of these imitations are listed in Table 17.7.

Table 17.7

Gemstone	RI	DR	SG	H
Turquoise	1.61–1.65	–	2.6–2.9	5.5–6.0
Chalcedony	1.53–1.54	–	2.58–2.64	6.5
Howlite	1.59	–	2.58	3.5
Odontolite	1.57–1.63	–	3.0–3.25	5.0
Lazulite	1.615, 1.645	0.03	3.1–3.2	5.5
Amazonite	1.52–1.54	0.008	2.56	6.0
Chrysocolla	1.50	–	2.00–2.45	2–4

'Bonded' turquoise has been produced from poor colour or friable turquoise. The source material is crushed and then bonded with polystyrene resins or sodium silicate. Imitations made from coloured plaster of Paris also exist. All of these materials lack the surface structure of the genuine stone which under the microscope can be seen to consist of crystalline particles interspersed with amorphous whitish powder.

Gilson man-made turquoise is also classified as a simulant as it contains material not found in the natural stone. Under the microscope (at 50× magnification) it is seen to be composed of small dark blue angular particles in a white background 'matrix', a structure quite unlike that seen in the genuine stone.

Opal

The Gilson man-made opal, like the Gilson lapis lazuli and turquoise, contains materials not found in the natural gem and is therefore classified as a simulant. Both white and black versions have been produced. Of these, the white product is the most convincing. The black Gilson opal should not pose a problem to anyone familiar with the natural gem.

Under high magnification (60×), the yellow segments of the older white Gilson opals resemble a series of closely spaced lines reaching down from a central spine. The structure within an orange segment of the later white product, however, has a crazy-paving appearance. Many of the white opals have a pronounced columnar structure when viewed sideways on. The colour segments in later Gilson white and black opals have a 'lizard skin' appearance (Figure 17.5). Many Gilson opals have a high porosity (this causes them to stick to the tongue, an effect not so marked in most natural opals).

All samples of Gilson opal simulants fluoresce a dusty green colour under SW UV, and a few fluoresce under LW UV. Any opals showing no signs of fluorescence are natural. When a natural opal does fluoresce under UV (usually with a white or cream tint) it also phosphoresces. This phosphorescence, when viewed under dark conditions, continues for up to 12 seconds. Signs of phosphorescence are virtually non-existent in darker Gilson specimens.

Opal simulants made from a monodisperse polystyrene latex have been produced in Japan. Because of the friability of the product, the stones are encased in an acrylic coating which gives them a 'glassy' appearance. These stones have a much lower SG than natural opal (1.2 compared with 2.1), and a higher RI (1.51 against 1.45). They are also hydrophobic (a drop of water placed on the surface forms a hemispheric bead – natural opals are hydrophilic and water spreads rapidly across their surface).

Figure 17.5 The 'lizard skin' effect seen in a Gilson imitation white opal

The 'Slocum Stone', introduced in the late 1970s, is a glass imitation of opal made in a range of colours including black, white, amber and green. Its SG (2.4–2.5) is higher than that of opal, as is its RI (1.49–1.51). The Slocum Stone's play of colour appears to be due to very thin tinsel-like flakes which produce interference colours. Also visible within the material are odd-shaped bubbles and swirl marks.

Quartz

Because of its greater value, amethyst is the variety of quartz which is most likely to be simulated either by coloured glass, by a symmetrical triplet (similar to a soudé emerald) made from a crown and pavilion of colourless quartz and a central section of colour filter material, or by purple synthetic corundum.

A coloured glass simulant can be identified by its strong anomalous double refraction on the polariscope, by included bubbles and by its constants. The components of the quartz triplet can be made more visible by immersing the stone in water. If bromoform (with an RI of 1.59) is used as the immersion fluid, both crown and pavilion sections are dramatically revealed as colourless, with the dark colouring layer clearly visible. A plane of bubbles is often visible in the area of the join. The purple synthetic corundum simulant can be identified by its constants (RI 1.76, 1.77 and SG 3.99 compared with 1.544, 1.553 and 2.65 for quartz).

Ruby

Ruby simulants include paste and composite stones (see Figure 15.17). Natural stones used to imitate ruby include pyrope and almandine garnet, spinel, tourmaline, zircon and possibly pink topaz. Table 17.8 lists the identifying constants of these simulants together with those of ruby.

Table 17.8

Gemstone	RI	DR	SG	H
Ruby	1.77, 1.76	0.008	3.99	9.0
Pyrope	1.74–1.77	–	3.7–3.8	7–7.5
Almandine	1.77–1.81	–	3.8–4.2	7.5
Natural spinel	1.717	–	3.6	8.0
Synthetic spinel	1.73	–	3.6	8.0
Tourmaline	1.62, 1.64	0.018	3.05	7–7.5
Topaz	1.63, 1.64	0.008	3.53	8.0
Zircon	1.93–1.99	0.059	4.68	7.25

Pyrope and almandine garnet, and red spinel (natural and synthetic), particularly in small sizes, may prove difficult to distinguish from ruby by eye alone. However, the lack of dichroism in these simulants, together with their distinctive absorption spectra and dissimilar character under the polariscope will serve to identify and separate them. Unlike most rubies, the two garnets do not fluoresce under LW UV because of their iron content.

Sapphire

Although sapphire is imitated by both paste and composite stones (see Chapter 15, and Figure 15.17), the most convincing simulants are blue synthetic spinel and a few natural blue gemstones. Blue synthetic spinel is easily identified by its low refractive index, its isotropic character on the polariscope, its typical cobalt absorption spectrum, and the fact that it appears pink through the Chelsea filter. Table 17.9 lists the identifying constants of sapphire and its simulants.

Table 17.9

Gemstone	RI	DR	SG	H
Sapphire	1.76, 1.77	0.008	3.99	9.0
Synthetic spinel	1.727	–	3.64	8.0
Topaz	1.61, 1.62	0.01	3.56	8.0
Tourmaline	1.62, 1.64	0.018	3.10	7.0–7.5
Kyanite	1.715, 1.732	0.017	3.65–3.69	5.0–7.0
Beryl	1.570, 1.575	0.005	2.69	7.5
Iolite (cordierite)	1.54, 1.55	0.009	2.57–2.6	7.5
Zoisite	1.69, 1.70	0.009	3.35	6.5

Chapter 18

Identifying organic gem materials and their simulants

Organic gem materials differ from the inorganic variety in that they are the products of living organisms. With the exception of pearl and coral, these materials are all *non-crystalline*, which further distinguishes them from the majority of inorganic gemstones.

Natural pearls

Despite their softness and their vulnerability to acid attack, perspiration and cosmetics, pearls have managed to maintain their popularity. Like other valuable gems, natural or 'native' pearls have many simulants. These can be divided into two broad groups: *cultured* pearls and *imitation* pearls. Although both natural and cultured pearls can easily be distinguished from the imitations, one of the more difficult tasks facing the gemmologist and jeweller has been the separation of the natural pearl from the cultured product.

The natural pearl is formed inside a pearl oyster (i.e. a *mollusc*) probably as the result of a parasitic larvae penetrating the creature's shell and causing an irritation or infection. The earlier theory that this was caused by a grain of sand is now largely discounted. The mollusc reacts to this local inflammation by surrounding the site of the irritation with a 'sac' of cells from a protective inner flap of tissue called the *mantle* or *pallium*. This sac of mantle tissue in turn secretes a deposit which encapsulates the intruding larvae with concentric layers of *nacre*.

The nacreous deposit consists of a very thin cellular network of *conchiolin* (a brown organic substance) whose interstices are filled with minute crystals of *aragonite*. These crystals are an orthorhombic form of calcite and are orientated with their principal axes at right angles to the layer. At the beginning of the mollusc's life, nacre is also deposited on the inside walls of its shell to form the mother-of-pearl material which is dealt with later in this chapter. This material is secreted on the shell by the entire surface of the mantle and is similar in structure to the layers which form on the growing pearl.

Once a mollusc has enclosed a source of irritation with nacre, it continues to deposit further layers for up to seven years, the thousands of thin translucent films of nacre eventually forming a pearl (Figure 18.1). If the site of the originating irritation occurs within the mollusc's mantle tissue, a rounded *cyst* pearl is formed. However, if the site is next to the nacreous inner surface of the mollusc's shell, a *blister* pearl is produced which becomes attached to the shell (Figure 18.2). While the cyst pearl is a completely formed and usually spherical gem, the blister pearl has to be cut out of the mollusc's shell, and is therefore incompletely formed (usually hemispherical to three-quarters spherical). With blister pearls, the part which was joined to the shell is devoid of

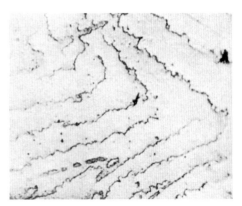

Figure 18.1 (left) The concentric structure in a section of a natural pearl: (right) Ridges of the overlapping nacreous layers on the surface of a natural pearl at 90× magnification. (V.G. Hinton)

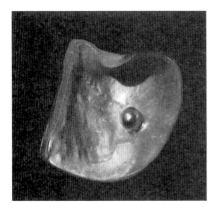

Figure 18.2 A natural blister pearl still attached to the nacreous surface of the mollusc's shell

nacre. Because of this, it is either smoothed off and then hidden by the jewellery setting, or the bare patch is covered by a section of mother-of-pearl (similar to the larger 'Mabe' pearl described in the 'Cultured pearl' section of this chapter).

Although most molluscs have the ability to produce pearls in this manner, the main source of these gems is the *Pinctada* salt-water species. These molluscs live in warm shallow waters, and are harvested for their pearl content in many locations including the Persian (Kuwait–Bahrain) Gulf, the Gulf of Mannar (between India and the north-west tip of Sri Lanka), the South Sea Islands, the Pacific coast of Mexico, the Gulf of California and the north-west coast of Australia.

Salt-water pearl fishing around the islands and inlets of Japan was traditionally carried out by girl divers called 'Ama'. Diving is still performed by the Amas, but mainly for the benefit of tourists (Figure 18.3). The once-naked girl divers are now clad from head to foot in white overalls (which are worn not so much for modesty but to frighten off sharks!).

The principal attraction of the pearl lies in its iridescent sheen. This is the result of a combination of interference and diffraction effects in the thin surface layers of the gem, and is termed the pearl's 'orient'. Although the majority of pearls have a silver-white

Figure 18.3 Japanese 'Amas' preparing to dive for pearl-bearing molluscs. (Mikimoto)

orient, pearls found off the Pacific coast of Mexico and in the Gulf of California have black metallic hues (said to be caused by the content of the sea water). Pink or rosée pearls are rarer and command a higher price. Another type of pearl having this colour is classified as a 'Pink Pearl' by CIBJO because of its non-nacreous surface. These Pink Pearls have a porcelain-like exterior coloured with flame-like markings. They are found in a univalve mollusc called the 'great conch' (*Strombus gigas*) off the coasts of Florida and the West Indies.

Freshwater pearls are produced by various clam and mussel species, and are found mainly in the rivers of the UK, the USA, the former USSR and China, the latter country harvesting the majority of these pearls. In general, freshwater pearls have a more subdued orient than oyster pearls and are less important commercially. The famous Abernethy freshwater pearl, weighing 43.6 grains (10.9 carats), was found in Scotland's River Tay in 1967. Although the refractive index of a pearl is difficult to measure, its value ranges from 1.52 to 1.66 (1.53 to 1.69 for black pearls). The specific gravity (SG) of a natural pearl varies from 2.68 to 2.78.

Cultured pearls

The first known attempts to culture this organic gem date back to the thirteenth century, when the Chinese produced poor quality and misshapen pearls by introducing irritant material between the mollusc's shell and its mantle. For centuries the Chinese used this technique to coat small images of Buddha with a thin layer of nacre by inserting them in the mantle of freshwater mussels.

The early history of the cultured pearl in Japan is confused by conflicts and jealousies between the various rival pioneers of the art. However, it is now generally accepted that Tatsuhei Mise produced the first cultured spherical pearls around 1900. More successful methods, based on scientific experiments, were later developed by both Tokichi Nishikawa and Kokichi Mikimoto. Although the name of Mikimoto is synonymous today with cultured pearls, his 'whole wrap' method of culturing proved to be too difficult and delicate (the bead nucleus was wrapped in the mantle tissue and tied with a silk thread), and it is the simpler Nishikawa 'piece' technique that is now used by Mikimoto and others to produce spherical cultured pearls.

Sadly, the waters surrounding the Mikimoto Pearl Island in Toba bay have become too polluted for pearl production, although the island still attracts tourists to its pearl museum and Mikimoto shop. Another hazard for the salt-water mollusc is the annual scourge of the red plankton micro-organism, (the 'red tide') which in a bad year can wipe out the molluscs of an entire fishery.

The various methods and techniques of cultured pearl production can be summarized as follows.

Basic method

This is the 'piece' method devised originally by Tokichi Nishikawa. It consists of inserting a 5–7 mm diameter mother-of-pearl bead, plus a small cube of mantle tissue (one of many prepared from another 'sacrificial' oyster), into a 3-year-old mollusc. The valves of the mollusc are wedged open and the bead and cube inserted in an incision made in the soft tissue. The wedge is then removed and the mollusc placed in a cage and returned to the sea (Figure 18.4). Three years later, the mollusc is removed from the sea again, and the mother-of-pearl bead, now transformed into a cultured pearl with a 1–2 mm layer of nacre, is extracted (Figure 18.5), the unfortunate mollusc dying in the process.

Figure 18.4 A Japanese salt-water cultured pearl fishery. The cages of molluscs are suspended beneath wooden rafts. (R.V. Huddlestone)

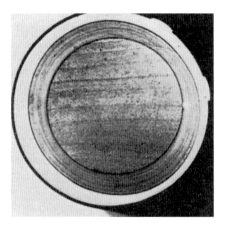

Figure 18.5 Thin section of a cultured pearl showing the parallel layers of the large bead nucleus surrounded by concentric layers of nacre

Salt-water Mabe pearls and cultured blister pearls

Cultured blister-type pearls, known as *Mabe* pearls, are relatively large (10–20 mm in diameter) and hemispherical in shape. They are produced by cementing the base of a hemispherical pellet of soapstone (steatite) to the inside surface of the mollusc's shell. This material is chosen because the subsequent coating of nacre can be easily separated from its surface. The mollusc is then returned to the sea where the pellet becomes covered with layers of nacre. After 2 or 3 years, the mollusc is recovered and the resulting pearl is sawn out of its shell. The soapstone bead is then removed from the pearl, the cavity cleaned out (and sometimes tinted) and either a smaller mother-of-pearl bead is cemented in its place, or the void is filled with a resinous compound. Finally, the base of the Mabe pearl is covered with a polished piece of mother-of-pearl.

Cultured blister pearls are produced by a similar method, except that a mother-of-pearl nucleus is used. However, unlike the Mabe pearl, this nucleus is left in the pearl after it is cut from the mollusc. Usually these pearls are hemispherical to three-quarters spherical in shape and have a sawn-off base which exposes both the nucleus and the nacreous skin.

To avoid confusion between Mabe pearls, cultured blister pearls, and the use of the word 'Mabe' by the Japanese for a particular type of oyster, the Mabe pearl (which consists of two or more components) is more accurately described as a *Mabe composite cultured pearl*.

Freshwater Biwa pearls

In later experiments with pearl nucleation using a freshwater clam, it was found that pearl formation can occur if just a fragment of mantle is inserted into the mollusc. However, the size of the resulting pearl is smaller as it is almost entirely composed of deposited layers of nacre. Because of the absence of a solid bead nucleus, these pearls are classified as *non-nucleated* (and sometimes described as *tissue nucleated*). They are called *Biwa* pearls as they are mainly farmed around the shores of Lake Biwa in Japan. Biwa pearls are produced in the large freshwater clam *Hyriopsis schlegeli*.

The original reason for omitting the bead nucleus was that it was unsuccessful when used with this variety of mollusc. Another difference between freshwater farming on Lake Biwa and salt-water culturing is the simultaneous production of several pearls in the large Schlegeli clam (Figure 18.6). Also, after the first crop is harvested the clam is

Figure 18.6 Shell of the *Hyriopsis schlegeli* with a sample of the Biwa pearls produced in this freshwater clam

returned to the water, when a second crop may sometimes occur naturally. The first harvest usually yields pearls of an oval or baroque shape, and because of the absence of a bead nucleus, these are small. Poorly shaped first-harvest or second-harvest pearls are often re-inserted as 'seeds' into another clam to improve their size and shape.

Distinguishing natural from cultured pearls

Except to the expert dealer in pearls, the identification of natural and cultured pearls can be difficult as both have surfaces of the same material. However, subtle differences between these surfaces will usually provide enough clues to a knowledgeable pearl merchant for a positive identification. For example, the more translucent outer coating of the cultured pearl gives it a waxy lustre. With drilled pearls, inspection of the drill hole will often reveal the sharp boundary between the bead nucleus and the outer coating of a cultured pearl. Beyond this boundary there is also a complete absence of growth bands. Inspection of the drill hole is aided by illuminating it with a thin glass-fibre light guide coupled to a strong light source (Figure 18.7).

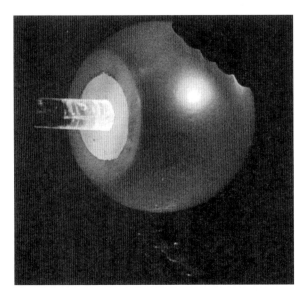

Figure 18.7 The bead and nacreous layers in a cultured blister pearl illuminated by a fibre-optic light guide

A technique called 'candling' has been used to reveal the difference between cultured and natural pearls. To make the test, the pearl is rotated slowly in front of a strong light which is completely masked except for a 1 mm square test aperture. With a cultured pearl, the tell-tale stripes caused by the structure of the mother-of-pearl bead will be projected onto its surface (no such pattern is seen with a natural pearl), and as the pearl is turned it produces two flashes of light per rotation. This principle of detection was made use of in the *lucidoscope* which consisted of an illuminated immersion cell fitted with an iris aperture. To improve the transmission of light through the parallel-banded layers of the bead nucleus, the pearl was immersed in pure benzine which leaves no residue on evaporation.

The mother-of-pearl bead in the cultured pearl has a greater specific gravity than the natural pearl, and this makes possible another simple test. If the pearls to be identified are placed in a heavy liquid which has been adjusted to the SG of the Iceland spar variety of calcite (i.e. bromoform or di-iodomethane diluted with 1-bromonaphthalene to an SG of 2.71), the majority of natural pearls, having an SG between 2.68 and 2.7, will float, while the majority of cultured pearls will sink (some Mabe composite cultured pearls filled with a resinous compound rather than mother-of-pearl have a much lower

SG of around 2.2 and will also float in the 2.71 liquid and in the standard quartz test liquid of 2.65).

The two following tests give a more reliable identification than those so far described.

The endoscope

This older method, though still in use in some laboratories, uses an instrument called an *endoscope* (Figure 18.8). The instrument was originally manufactured in France, but is now no longer available. It consists of a hollow needle over which the pearl is threaded.

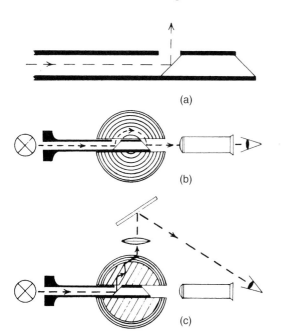

Figure 18.8 The endoscope in use testing pearls

The needle contains at its far end a short metal rod whose ends are polished at 45° to form two mirror surfaces (Figure 18.9(a)). The needle has a small aperture cut in it

(a)

(b)

(c)

Figure 18.9 (a) Construction of the needle section of the endoscope. (b) A natural pearl is identified by its concentric layers which channel light rays into the viewing microscope. (c) In a nucleated pearl, light escapes through the mother-of-pearl bead and can be seen as a line of light on the pearl's surface

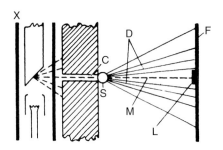

Figure 18.10 Diagram illustrating the production of a pearl lauegram: X = X-ray tube; C = collimator; S = pearl; D = diffracted X-rays; M = main X-ray beam; L = lead disk to absorb main beam (usually omitted when testing pearls); F = photographic film

immediately in front of the first mirror surface, and this allows light (which is injected along the needle) to be reflected out of the hole and into the pearl.

The pearl under test is moved carefully to and fro over the aperture in the needle by means of a lever arrangement, while the output end of the needle is viewed through a microscope. If the pearl is a natural one, a position can be found where light entering the pearl through the aperture is channelled round its concentric layers and is then reflected into the microscope by the needle's second mirror as a flash of light (Figure 18.9(b)).

If, however, the pearl is cultured, no flash of light will be seen in any position. In this case, the pearl is rotated on the needle until a position is found where the light rays pass through the parallel layers of mother-of-pearl and appear as a line of light on the outside surface of the pearl. A lens and a mirror are fitted to the endoscope to enable the operator to view this effect from the microscope position (Figure 18.9(c)).

X-ray methods

The other confirmatory pearl test uses an X-ray generator set. In natural pearls, the aragonite crystals in the many layers of nacre are aligned radially round the pearl with their principal axes at right angles to the surface. If the pearl is placed in a narrow beam of X-rays (Figure 18.10), these crystals scatter some of the rays and produce a *Laue* diffraction pattern of spots which can be recorded on photographic film. Because of the atomic structure within the aragonite crystals, this diffraction pattern will have a *hexagonal* symmetry for *any* orientation of the natural pearl (Figure 18.11).

With the nucleated cultured pearl the nacreous layer is relatively thin, and this hexagonal pattern is only produced in *two* positions (see Figure 18.12(a)) where the crystals in the mother-of-pearl bead are parallel to the line of the X-ray beam. In other positions of the pearl, either a *four-point* symmetry pattern, (Figure 18.12(b)) or a merging of the four-point and six-point pattern is produced (having the appearance of a continuous ring – Figure 18.12(b)). Because of this a pearl must be checked in at least two positions, at 90° to each other, to avoid an incorrect identification.

One problem with the X-ray diffraction test is that if the pearls are part of a necklace they must be unstrung so that the individual pearls can be tested in several different orientations. However, with careful adjustment of exposure levels and times, it is possible to make a contact X-ray picture, or *radiograph*, of a complete string of pearls by broadening the X-ray beam. This produces a clear picture of the difference in X-ray transparency between the outer layers of a cultured pearl and its mother-of-pearl or glass bead nucleus (Figure 18.13). With a natural pearl, the transparency varies more evenly with the thickness of the pearl, and fine concentric growth lines are also visible. This method of detection with non-nucleated cultured pearls is more difficult as there is no bead nucleus to show the difference, but only small irregular hollows or

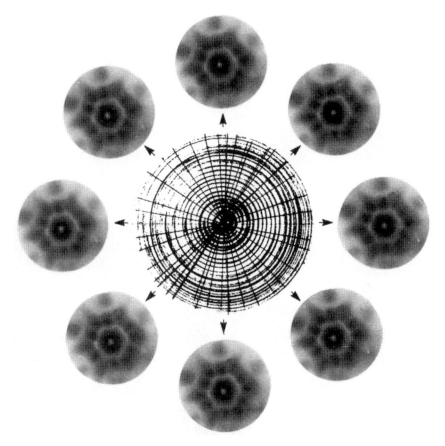

Figure 18.11 Diagram showing the six-spot hexagonal X-ray diffraction pattern obtained in any direction through a natural pearl. (A.E. Farn)

patches near the centres. However, Biwa pearls fluoresce strongly under X-rays, and this can provide the extra evidence needed to distinguish them from natural pearls.

Imitation pearls and their identification

Because of the popularity of pearls (and the scarcity of natural pearls), imitations have been made of them throughout history. These imitations have consisted mainly of the following three types: hollow glass spheres filled with wax, solid spheres of glass or plastic, and solid mother-of-pearl spheres. The first type, invented in France in the seventeenth century and called 'Roman pearl', had its interior painted with a pearl 'essence' before being filled with wax. The two more recent versions are given their lustrous outer appearance by being dipped in a special preparation.

This preparation, called 'essence d'orient', once consisted of *guanine* crystallites extracted from the scales of a fish called the *bleak*. These crystallites are now obtained from the scales of the herring in fisheries in the Bay of Fundy between Nova Scotia and New Brunswick in Canada. In both cases the guanine extract is suspended in a special solvent before being added to a cellulose lacquer.

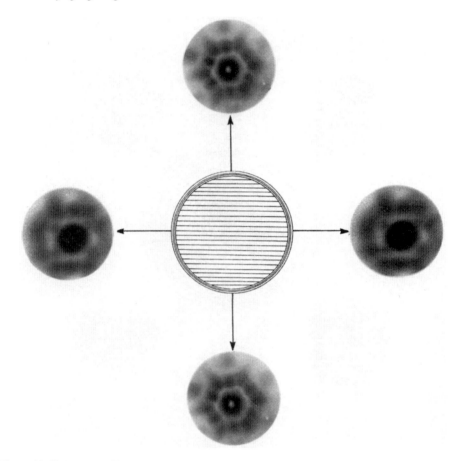

Figure 18.12(a) X-ray diffraction patterns obtained with a cultured nucleated pearl. The six-spot pattern is obtained perpendicular to the mother-of-pearl layers, and a four-spot pattern in line with the layers. (A.E. Farn)

In the Perlas Manacor factory in Majorca, the manufacturing process of solid glass imitation pearls starts with a rod of opalescent white glass. The end of this rod is held in a gas flame and molten globules of the glass are then deposited on to a wire that is rotated to produce a spherical bead. The wire is pre-coated with a thin layer of clay-like material which enables the beads to be easily removed from the wire when cool (this also provides a ready-made drill hole).

In the next operation, the beads are mounted in banks of a hundred on the metal stems of a pin board. The pin board is then loaded into a rotary dipping chamber which contains the pearl essence. During the coating process, the beads receive at least five successive layers of the pearl essence. After coating, the banks of beads are passed slowly through a tubular furnace to bake on the surface layers. In a final operation, the beads are given a coating of cellulose acetate and cellulose nitrate to enhance the iridescent effect between the various layers (Figure 18.14).

A slightly more sophisticated imitation, marketed as the 'Angelo' pearl, uses the same type of bead nucleus that is implanted in molluscs for pearl culture. This product has

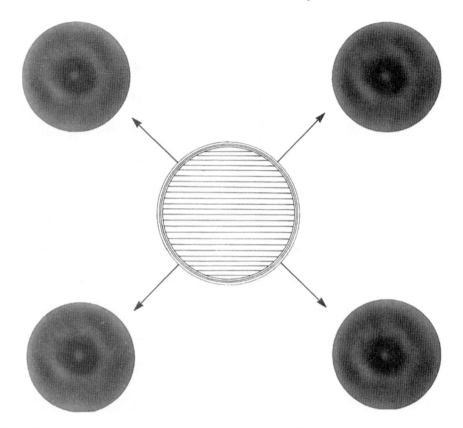

Figure 18.12(b) Diagram showing the merged four- and six-spot diffraction pattern obtained in directions diagonal to the mother-of-pearl layers. (A.E. Farn)

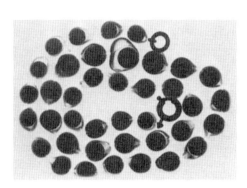

Figure 18.13 Contact X-ray photograph of a string of cultured nucleated pearls

Figure 18.14 Imitation pearls in a necklace and brooch

three coatings of a plastic-like material applied to it, the second one giving a discrete play of colour.

The difference between the slightly rougher layers of natural and cultured pearls and the smoother surface of imitations can often be detected by rubbing them against the teeth. However, a more reliable method is to inspect the edges of the drill holes. With imitation pearls these are usually ragged, with chips visible in the coating together with some indication of the underlying glass surface (Figure 18.15). Under the microscope (at 40 to 80× magnification), the typical edges of overlapping layers of nacre are absent from the

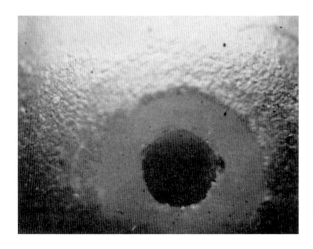

Figure 18.15 Surface of an imitation pearl showing underlying glass bead near drill hole

imitation gem (see Figure 18.1). A needle inserted at an angle into the drill hole can be used to detect any soft wax filling of the rarer hollow-sphere variety.

Black pearls from the Gulf of California and the Pacific coast of Mexico often have their colour enhanced by being stained with *silver nitrate*. This tends to produce an unnaturally even and intense black coating, which is noticeably different from the more bronze or greyish iridescent shades of the untreated pearl. Black pearls fluoresce a faint red under crossed filters or LW UV, but this effect is completely inhibited when they are stained with silver nitrate. Imitation black pearls in the form of polished beads of hematite lack the iridescent lustre typical of the genuine article. Hematite imitations can also be distinguished by their high SG (4.95 to 5.16).

An X-ray radiograph picture can also be used to detect imitation pearls, which unlike the natural pearl and the cultured pearl are completely opaque to X-rays. A radiograph of a pearl necklace can sometimes reveal a mix of imitation and cultured pearls (Figure 18.16).

Shell

Mother-of-pearl material is obtained from the iridescent nacreous layers deposited on the inside surfaces of mollusc shells. Like the pearl, this iridescence is caused by the interference and diffraction of light in the thin overlapping layers of translucent nacre. The shell of the *abalone* or *Haliotis* mollusc is particularly colourful in this respect with banded growth contours (the mollusc also produces baroque pearls with the same iridescence as the shell lining). Sections of abalone shell (also called *paua* shell) are widely used in costume jewellery. Mother-of-pearl is also used as an inlay material and for making buttons, knife handles and other ornaments.

Cameos carved from the *Giant Conch* shell and the *Helmet* shell (and from agate) make use of two or more contrasting layers of shade or colour. The uppermost layer is carved into a relief figure or design, while the underlying layer is exposed to form the background (Figure 18.17). Cameo shells with a brown underlying layer come from

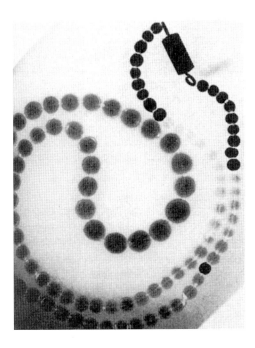

Figure 18.16 Contact X-ray picture showing a mixture of imitation (opaque) and cultured (translucent) pearls in a necklace

Figure 18.17 Example of a shell cameo

waters around the Bahamas, and shells with an underlying orange colour come from around the coastline of the Malagasy Republic. (See cameos and intaglios under 'Gemstone carving' in Chapter 19.)

Identification of shell is mainly by means of the hand lens or microscope which reveals the overlapping layers of the surface structure. The flame structure of the conch pearl may also be visible. Plastic imitations lack these features.

Amber

This is a fossilized form of resin consisting of a mixture of hydrocarbons plus succinic acid. It was originally extruded from the trunks of various types of tree, including the conifer, some 20–60 million years ago in the *Miocene* to *Palaeocene* period.

The Greeks gave amber the name 'electrum' (i.e. 'sun made'). Subsequently, amber was found to produce a negative electrostatic charge when rubbed, and to be capable of picking up small fragments of tissue (an ability shared by several of its plastic imitations). Because of its initial association with amber, this phenomenon gave rise to the word 'electricity' which was coined from the Greek 'electrum'.

Amber is usually a pale yellow-brown translucent to opaque material, but reddish-brown, greenish, bluish-grey and black varieties are also found. *Sea amber* (also called 'sea stone' and 'strand amber') is recovered from the Black Sea, the coastline of Sicily, and from the shores of the Baltic, where it is known as the 'gold of the Baltic'. *Pit amber* is mined principally in Palmnicken near Kaliningrad (formerly Konigsberg) in the former USSR. Less important sources exist in Mexico, Romania, Myanmar (Burma) and the Dominican Republic.

Pieces of amber that are large enough to be fashioned into gems or ornaments are called *block amber*. Smaller pieces of acceptable quality are heated to about 180°C and compressed together (or extruded through a sieve) to form *reconstructed* or *pressed* amber (also known as 'ambroid'). The majority of amber from northern Europe has been treated in this way so that it can be moulded easily into beads of various shapes.

Cloudy amber is sometimes clarified by heating it in rape seed oil. The oil penetrates the bubbles, which are the cause of the cloudiness, and makes them transparent. If the amber is then cooled too rapidly, this results in the production of stress cracks in the form of iridescent spangles (Figure 18.18). These internal 'enhancements' soon became

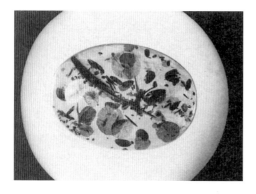

Figure 18.18 The 'sun spangle' stress fractures in clarified amber

popular enough to be produced intentionally, and for a while were even dyed red or green. Other rather more macabre inclusions of an organic nature are insects (Figure 18.19).

Among the many amber imitations, perhaps the most convincing is *copal resin*. This is a 'young' resin compared to amber, and is derived from various tropical trees. It is mainly used as the basis for copal varnish. Because it has a similar SG to amber (1.08), it cannot be distinguished from it by hydrostatic weighing or immersion techniques. Unlike amber, however, copal softens in ether, and a small drop of the liquid will leave a dull spot on its surface. Copal also crumbles easily under a knife blade (or even under thumbnail pressure). When fractured, its surface has a characteristic crazed network of fine cracks.

Small dead insects have been inserted into molten copal resin in an attempt to deceive, but in addition to the above test indications, there is no sign of the faint blur produced by the dying struggles of a live insect trapped in amber!

Another amber simulant, similar to copal resin, is kauri gum (from the kauri tree in New Zealand's North Island). Like copal resin, it is generally a younger material than

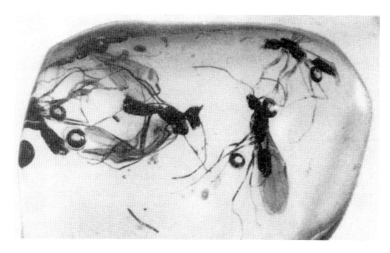

Figure 18.19 Insects encapsulated alive in amber

amber, although fossil kauri gum also exists. The younger gum is more soluble in ether than amber, but the fossil gum may pass this test.

Glass imitations of amber are easy to detect because of their hardness. This ranges from 5.0–6.0 compared with amber's 2.0–2.5 (Burmese amber is the hardest, Dominican the softest). Glass can also be separated from amber because it is colder to the touch, and, like plastic imitations, because it has a greater SG than amber's 1.08.

A suitable 'heavy liquid' for testing amber can be made up by dissolving 10 level tea-spoons of salt (50 grams) in a half-pint (0.275 litre) of water. This produces a solution having an SG of around 1.13. Amber (and copal resin/kauri gum) has an SG of 1.08, and will float in the salt solution, but glass and plastic imitations, such as *Bakelite* and *casein* will sink (Bakelite and casein give off a smell of carbolic acid and burnt milk respectively if touched with a hot needle). Amber has a refractive index of 1.54 (as has copal resin/kauri gum), and this can sometimes be used as an identifying feature when checking a bead (particularly one with conveniently flat faces).

Ivory

Elephant tusks are, sadly for that animal, a well-exploited source of *dentine ivory*. However, ivory is also obtained from other sources including the tusks of the walrus and the hippopotamus, the front horn-like tooth of the narwhal (a dolphin-like arctic whale), the teeth of the sperm whale, and more rarely, from the tusks of fossilized mammoths.

Dentine ivory consists almost entirely of dentine material, but may sometimes include some enamel and other organic substances. Ivory is relatively soft (2.0–3.0) and has a splintery fracture. The best-quality material comes from the tusks of the Indian elephant which are smaller than those of the African elephant. Mammoth ivory from the Urals and Siberia in the former USSR is harder than other varieties of ivory, but is usually flawed by cracks. The refractive index of ivory produces a vague line on the refract-ometer around 1.54. Ivory's specific gravity lies between 1.7 and 2.0.

Ivory imitations include many plastics, the most convincing being *celluloid*. This is generally less dense than ivory (SG of 1.3–1.8), and unlike ivory can be pared with a knife. *Bone* is another material used to imitate ivory, but has a greater SG (1.9–2.1).

Under the microscope, bone peelings show a multitude of small tubes (called 'Haversian canals') which appear as dots in cross-section or as lines at right angles to the dots (Figure 18.20). At higher magnifications, the bone cells resemble the shape of an outspread spider.

Ivory, when viewed under the microscope, shows two sets of crossing parallel lines like the engine-turned pattern sometimes seen on watchbacks (Figure 18.21). This is an optical effect caused by the submicroscopic structure of ivory.

With *walrus* ivory there are contrasting inner and outer parts of the material. The inner part has a distinctive marbled appearance. The structure of *hippopotamus* ivory is similar to that of elephant ivory, but has a finer more closely spaced array of concentric lines. A cross-section of the protruding horn-like tooth of the *narwhal* (once sold as the 'horn of the unicorn') displays angular concentric rings and is hollow centred.

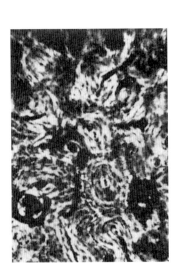 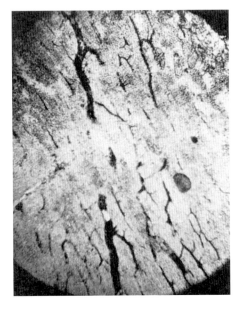

Figure 18.20 (left) Microphotograph of a traverse section of bone: (right) Microphotograph of a longitudinal section of bone

Figure 18.21 The 'engine-turned' effect seen on the surface of elephant ivory

Sperm whale teeth have a sharp boundary between the inner and outer areas (the latter may show concentric rings in cross-section and parallel lines following the curve of the tooth). The inner part shows two converging sets of parallel lines.

Ivory simulants described as 'vegetable ivory' are derived from the *doom palm nut* and the *corozo nut*. Products made from either nut have SGs below that of dentine ivory (1.39–1.40), and a peeling shows a pattern of interconnecting oval cells. Items carved from vegetable ivory are usually small, as the size of the nuts ranges from only 2–3 cm in diameter.

Ondontolite

Also known as *bone turquoise*, this is a bluish fossil bone or dentine ivory obtained from prehistoric animals such as the mammoth. It owes its colour to *vivianite*, an iron phosphate, and is used as a turquoise simulant. Under the microscope, its organic structure is clearly visible. It is much more dense than ivory (SG of 3.0–3.25), is harder (5.0) and has a higher refractive index (1.57–1.63).

Tortoiseshell

The main source of supply of tortoiseshell material comes not from the tortoise, but from the outer shell or scale of the *Hawksbill sea turtle*. The individual plates, or *blades*, of the turtle's shell are a mottled yellow-brown. The front blades are called *shoulder plates*, the centre ones *cross-backs*, the side ones *main plates*, and the rear ones *tail plates* (Figure 18.22). A clear yellow tortoiseshell is obtained from the turtle's under-shell.

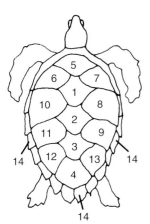

Figure 18.22 Diagram showing the blades of the Hawksbill turtle shell: 1, 2, 3, 4 = cross-backs; 5, 6, 7 = shoulder blades; 8, 9, 10, 11 = main plates; 12, 13 = tail plates; 14 = hoof

The RI of tortoiseshell is in the region of 1.55, and its SG is around 1.29. Like amber, small pieces can be softened by heating (to 100°C) and moulded together to form more usable sizes.

Various plastics, including *casein*, are used to imitate tortoiseshell. These can be distinguished from the genuine material by the nature of the body colour. In tortoiseshell, this can be seen to be made up of small patches of pigment when viewed under the microscope. The colour in plastic imitations is either more homogeneous or is distributed in the material as swathes of colour (Figure 18.23).

Figure 18.23 (top) Microphotograph of the colour patches in tortoiseshell: (bottom) The colour swathes in a plastic imitation of tortoiseshell

Jet

Once used in Victorian mourning jewellery, jet is a form of fossilized wood. Carboniferous in composition, it contains traces of mineral elements such as aluminium, silicon and sulphur, plus 12–19% of mineral oil. Jet is similar in this respect to lignite or brown coal, and is halfway in formation between peat and bituminous coal. It has a hardness of 2.5–4.0, an SG of 1.3–1.4, and an RI in the region of 1.64–1.68. During Victorian times the principal source of jet was from mines in northern Yorkshire. The town of Whitby became the main site of the jet fashioning industry, and this is where the jet carver's art flourished in the mid-nineteenth century (Figure 18.24). Today, the

Figure 18.24 A selection of jet bracelets made from sections strung on elastic

main sources of good quality 'hard' jet are Spain, France, Germany, the former USSR and Utah in the USA.

The five principal simulants of jet are a moulded glass known as 'French jet' or 'Vauxhall glass', vulcanite (a hard black vulcanized rubber also known as 'ebonite'), plastics such as Bakelite and epoxy resins, and black-stained chalcedony. French jet has a much higher SG (around 2.9) and is harder (6.0) than jet. It is also colder to the touch and may contain bubbles. Vulcanite gives off a rubbery sulphurous smell if touched with a hot needle or worked with a file, and does not show the conchoidal fractures typical of jet. Bakelite is a phenoloc resin (patented in 1906) and was used as both an amber and jet simulant in Edwardian times. Like the more recent epoxy resin simulants, it has a characteristic smell of carbolic acid when touched with a hot needle. Although Bakelite is as hard as jet, it is tougher and unlike jet shows no signs of conchoidal fractures round the drill holes of beads. Chalcedony can be identified by its greater hardness, its lower RI and its coldness to the touch.

Among several less common simulants of jet is anthracite, which is a compact non-bitumous variety of coal. It has a vitreous to metallic lustre which gives it a 'harder' look than does the softer lustre of jet. Small ornaments are still being made of it in the UK (Figure 18.25).

Figure 18.25 An anthracite carving

Coral

This is another popular Victorian gem material. Coral is composed mainly of calcium carbonate in the form of fibrous calcite. Its branching plant-like structure (Figure 18.26) is formed by the skeletal remains of various types of marine polyp that live in

Figure 18.26 Diagram showing the growth of precious coral and the way beads are cut from the 'tree'

colonies in shallow subtropical waters in the Mediterranean Sea and around the coast-line of Australia, Japan and Malaysia. Precious coral is mainly white, pink, red, dark red and black, and is used in necklaces, bracelets and occasionally cameos and carv-ings. The rose pink variety is called 'angel's skin', and the less valuable white coral is sometimes stained to imitate this more attractive tint. Black 'coral' is composed of an organic horn-like material and grows to a height of several feet around the Malaysian Archipelago, the coastline of northern Australia and in the Red Sea.

There are many coral simulants including plastics, pink-stained vegetable ivory and stained bone. Stained vegetable ivory and bone can be identified by their structures (as described in this chapter under 'Ivory'). The identifying features of coral include its SG (2.6–2.84 for white and pink varieties; 1.34–1.46 for the black variety) and the fact that it effervesces when a small drop of acid is applied to its surface. Stained white coral can be distinguished from natural pink coral by applying a swab of solvent (such as amyl acetate) to its surface to remove some of the colour. Pink coral usually has a higher SG than the stained white variety (2.84 compared with 2.6) as it is mainly composed of the aragonite form of calcite.

Chapter 19

The fashioning of gemstones

With the possible exception of a few gems, such as pearl* and occasionally amber, most of our organic and inorganic gem materials have their appearance enhanced by some form of shaping or polishing. With translucent or opaque stones this may be as basic as an abrasive tumbling operation to develop a smooth and lustrous surface while retaining the baroque profile of the original rough material. Alternatively, stones of this type may be cut in the rounded *cabochon* form (Figure 19.1) to bring out their surface colour or banding as with malachite and agate. Another use of the cabochon style is to emphasize sub-surface sheen features such as *chatoyancy* and *asterism* (i.e. in cat's-eye stones such as quartz and chrysoberyl, and star stones such as ruby and sapphire). If the stone is translucent but dark, the cabochon can be shallow cut to lighten the colour. Sometimes the back of a cabochon is hollow cut to produce the same effect.

Figure 19.1 Profile of the cabochon cut. Stones are sometimes hollow cut, as shown on the right, to lighten their colour

Although tumble polishing a rough stone improves its appearance, and the cabochon cut is appropriate for opaque and 'speciality' stones, most transparent gem minerals are faceted to enhance features such as *body colour*, *brilliance* (the surface and total internal reflection of light), *fire* (dispersion) and *scintillation* (the sparkles of reflected light produced when the light source or the gemstone is moved).

The surface component of a gemstone's brilliance (i.e. its lustre or reflectivity) is related to its refractive index by Fresnel's equation (see Chapter 9). However, the proportion of brilliance due to the light reflected from *within* the stone depends on the style of cutting, which in turn is partly dictated by the critical angle of the gem material.

Critical angle

In Chapter 9, reference was made to the way in which light rays, passing through an optically dense medium, are reflected back from the surface of a less dense medium provided that the angle between the incident rays and normal is greater than the critical angle of reflection of the two media.

---

* The rarely practised 'doctoring' or 'skinning' of pearls involves the removal of a bad-coloured or blemished layer by careful filing or abrasion.

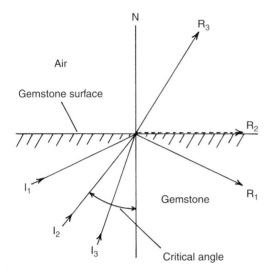

Figure 19.2 Light rays are reflected back from a gemstone's facet at angles to the normal (N) which are greater than the critical angle, and are refracted out of the gemstone at angles less than the critical angle

If the denser medium is a gemstone, and the less dense medium is air, as in Figure 19.2, the ray I_1 will be reflected back into the gemstone as R_1. At the critical angle of total reflection, however, ray I_2 will cease to obey the law of reflection, and will be refracted along the surface of the gemstone as R_2. At angles less than the critical angle, all rays (such as I_3) will be refracted out of the gemstone into the surrounding air (R_3).

In order to fully develop the potential brilliance of a transparent faceted gemstone, it is important that as many as possible of the rays entering the gem through its top (*crown*) facets are reflected back from the lower (*pavilion*) facets, and re-emerge from the crown of the stone as a result of *total internal reflection*.

To achieve this objective, both the lapidary and the diamond polisher must adjust the angles of the crown and pavilion facets so that the majority of rays entering the crown facets meet the interior faces of the pavilion facets at angles to the normal, which are *greater* than the critical angle of the stone. If the angles of the facets are wrong, many of these rays will pass out through the pavilion facets, and the stone will appear dark. It is equally important that the rays reflected back from the pavilion facets meet the crown facets at an angle *less* than the critical angle. If they fail to do this, they will be reflected back into the stone again.

The critical angle of a gemstone is dependent on both the RI of the gemstone and that of the surrounding medium, as shown in the following equation:

$$\text{Critical angle} = \text{angle whose sine equals} \quad \frac{\text{RI of surrounding medium}}{\text{RI of gemstone}}$$

As the surrounding medium is air (RI = 1),

$$\text{Critical angle} = \text{angle whose sine equals} \quad \frac{1}{\text{RI of gemstone}}$$

To determine the critical angle of a gem material, invert its RI value (i.e. divide it into 1.0). This will give the sine of the angle, which can then be looked up in a set of trigonometric tables (alternatively, the critical angle can be obtained directly from a gem's RI by using a computer program such as GEMDATA, available from Gem-A Instruments Ltd); for diamond, with an RI of 2.417, the critical angle is 24.43°; for quartz, with an RI of 1.545, the critical angle is 40.33°.

From this it can be seen that for maximum brilliance, the pavilion/girdle angle for a quartz gemstone will be significantly different from that for a diamond. In Figure 19.3 (top diagram), a single ray (1) is shown undergoing total internal reflection in a diamond (in either direction). A second ray (2) is shown entering the diamond's table facet at a shallow angle and being refracted out through the back of the stone via the 'cone' formed by the critical angle. If a quartz gem is cut to the same pavilion/girdle angle as the diamond, neither ray entering the table facet would be reflected back from the pavilion facets as they would meet them at an angle less than 40.33° (the critical angle for quartz). However, if the pavilion/girdle angle for quartz is increased from 40 to 43° (see lower diagram in Figure 19.3), then ray (1) is reflected back successfully through the table facet.

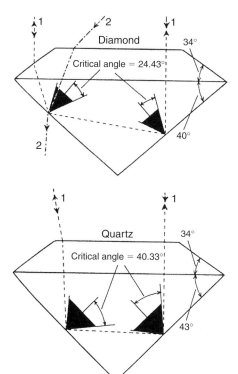

Figure 19.3 The top diagram shows the total internal reflection of ray 1, which meets the pavilion facets of diamond at an angle greater than the critical angle. Ray 2 meets the pavilion facets at an angle less than the critical angle and is refracted out of the pavilion. In the bottom diagram, the pavilion has been made deeper to maintain the total internal reflection of ray 1 in quartz, which has a much larger critical angle than diamond

The diagrams in Figure 19.3 are only simple ones intended to illustrate the importance of the critical angle in the design of a gemstone's profile. Ray path diagrams that take into account all of the rays entering the crown facets are much more complicated. Because of the complexity of such diagrams, computer programs have been developed to plot them, and this provides a means of rapidly determining how a gem's brilliance is affected by small variations in its crown and pavilion facet angles.

Critical angle and the jewellery owner

The importance of a gemstone's critical angle has an aspect not always appreciated by the owner of gem-set jewellery. If the pavilion facets of a gemstone are allowed to become coated with grease and soap, the result will be a reduction in the stone's overall brilliance. This is because the RI of grease and soap is greater than that of air, and this will *increase* the gem's critical angle (see equation). The effect is particularly noticeable in the case of a brilliant cut diamond, 83% of whose brilliance comes from total internal reflection, and is sufficient justification for cleaning jewellery occasionally in a mild grease solvent.

Cutting styles

With a colourless transparent stone, one of the main aims is to achieve a polished gem which has the maximum possible brilliance of appearance. As mentioned earlier, this brilliance is dependant both on the *surface reflectivity* or *lustre* of the stone (which with some stones is evident even in the rough uncut state) and on the *total internal reflection* of those rays entering the gem's crown facets. If the stone has an appreciable degree of dispersion, the cut must also exploit this to bring out the gem's 'fire'.

The dispersion of white light into its spectral colours which produces the effect of fire (Figure 19.4) also reduces the amount of undispersed white light being reflected back from the pavilion facets. For this reason, the design of the stone's cut has to strike a balance between fire and brilliance. Incident rays which meet the table facet at right angles (i.e. normal) to its surface are not refracted as they pass from the air into the stone, and are therefore not dispersed as is the ray shown in Figure 19.4.

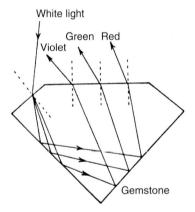

White light

Green Red

Violet

Gemstone

Figure 19.4 The white light entering a brilliant-cut diamond at an angle to its surface is dispersed into its spectral colours and totally internally reflected at multi-coloured 'fire'

If, for the sake of simplicity, we consider only those rays which meet the main table facet at right angles, the larger the area of this facet, the more brilliant will be the appearance of the stone. To achieve more fire, however, the table area must be reduced so that more rays can enter the stone at an angle through the crown side facets, and undergo dispersion. With coloured gemstones, a cutting style may be chosen which emphasizes or lightens the depth of colour viewed through the crown facets.

In the case of high-value gemstones, such as diamond, the style of the cut may also be influenced by the shape of the rough crystal. This is because of the general need to obtain the best 'yield' (i.e. percentage of polished to rough weight). In some coloured stones, such as sapphire, the colour may be unevenly distributed, and the symmetry of the cut is then modified to place the main area of colour low in the pavilion. This has the effect of making the colour appear to flood the stone when viewed through the crown. With stones having strong pleochroism, the crown will be carefully placed in relation to the crystal's optic axis so as to show only the best colour through the table facet (or, in the case of andalusite, to show all of its pleochroic colours). The following sections describe the main cuts used with both diamond and other gemstones.

Diamond cuts

One of the earliest cutting styles for diamond was the *point cut*. This simply consisted of polishing the octahedral diamond crystal while leaving the basic shape intact (as the octahedral faces of a diamond represent the hardest planes and cannot be worked, the stone must have been polished at a slight angle to these faces).

The next development was to polish a facet on one of the pyramidal peaks of the octahedron to form what is called the *table cut*. Over the next few hundred years, the forerunners of the present-day diamond polishers began to experiment with the simple table cut in order to improve the stone's brilliance and fire. By trial and error, the present-day *brilliant cut* slowly evolved as polishers began to understand the optical improvements that could be obtained by adjusting the angles and proportions of the stone's facets, and as advancing technology improved the precision with which stones could be cut. Legend has it that a seventeenth-century Venetian lapidary named Vincenzio Perruzzi was one of the earliest to fashion a diamond with a recognizable table, crown and pavilion, although this has never been substantiated.

In parallel with the slow development of the brilliant cut, another popular shape, the *rose cut*, came into being (Figure 19.5). This consisted of a flat base and a pyramid crown of triangular facets. The rose cut was by no means ideal optically (its lack of total internal reflection was sometimes improved by the addition of a mirror backing). A variant, the double rose cut, which had an identical crown and pavilion, was an improvement in terms of brilliance although the angles of the back facets were still not ideal for reflecting maximum light back through the crown. Other variants of the basic rose cut were the *Antwerp* or *Brabant rose* whose crown was cut at a shallower angle, and the *Dutch rose* which had a higher crown. Today, the rose cut is only occasionally used for small diamonds and zircons.

The *cushion* or *antique cut* (Figure 19.6) is an early variant of the brilliant cut, and was favoured for diamonds from Brazil as a better yield was obtained from the rough.

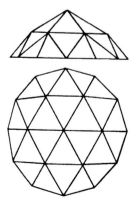

Figure 19.5 The basic profile of the rose cut

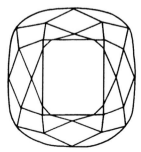

Figure 19.6 The antique or cushion cut

However, this cut suffers from some loss of brilliance. For small diamonds (below a 2 mm girdle diameter) the *single* or *eight cut* has been used in the past to reduce the cost of polishing a less valuable stone. The single/eight cut has only eight facets surrounding the table, and eight pavilion facets. However, with the development of semi-automatic faceting machines it is now economic to fully facet even much smaller stones.

When polishing a diamond from a rough crystal, the yield rarely exceeds 50%, and may be as low as 40%, depending on the shape of the crystal. To maximize the yield, an elongated octahedral crystal may, for example, be fashioned into an oblong *baguette*, an *oval*, a *marquise* or a *pear-shaped* profile rather than the more traditional round 'brilliant' (see Figures 19.7 and 19.8).

Figure 19.7 Four brilliant-cut diamonds, with at the top left an emerald-cut diamond

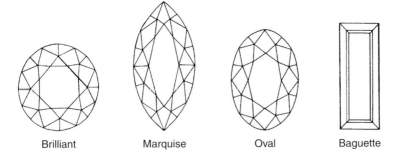

| Brilliant | Marquise | Oval | Baguette |

Figure 19.8 From left to right; the brilliant, marquise, oval and baguette cuts of diamond

The modern brilliant cut (Figure 19.9) consists of fifty-seven facets, of which thirty-three are above the girdle and form the crown, and twenty-four are below forming the pavilion. One additional facet, called the *culet*, is sometimes polished on the base of the pavilion. It is claimed that this is done to protect the point of the pavilion from damage, although it also conveniently disguises any asymmetry at the junction of the pavilion facets! As the culet is a facet through which light rays can escape from the stone, it should ideally be no larger than a pin head. It can be seen as a point of light when an illumination source is viewed through the stone's table facet.

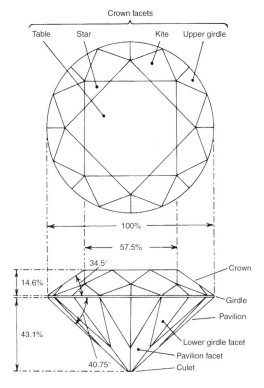

Crown facets

Table   Star   Kite   Upper girdle

100%
57.5%
34.5°
14.6%
Crown
Girdle
Pavilion
43.1%
Lower girdle facet
Pavilion facet
40.75°
Culet

Figure 19.9 The ideal Scan DN proportions and angles for a brilliant-cut diamond (kite facets on the crown are also called 'main crown facets'). The Eppler cut has a table facet width, crown depth and pavilion depth of 56%, 14.4%, and 43.2%, respectively. The crown angle is 33.17°, and the pavilion angle is 40.83°. The Tolkowsky ideal cut has a slightly smaller table (53%) and a deeper crown (16.2%). All percentages refer to a girdle diameter of 100%

If a diamond is cut with its pavilion deeper or shallower than the ideal profile, some of the light entering the crown facets will fail to be totally internally reflected. When viewed through the crown, shallow stones will appear dark in the central area of the table (creating a 'fish-eye' effect), while over-deep stones will appear dark around their upper girdle facets.

Several brilliant-type cuts have been designed with additional facets to increase the diamond's brilliance (Figure 19.10). Two of a series of more recent brilliant cuts

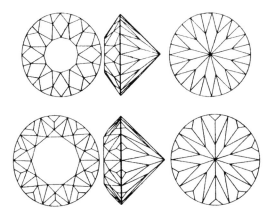

Figure 19.10 (top) The 85-facet King cut has 49 crown facets (including a 12-sided table facet) and 36 pavilion facets (plus a culet). (bottom) The 101-facet Magna cut has 61 crown facets (including a 10-sided table facet) and 40 pavilion facets (plus a culet)

Figure 19.11 (left) The hexagonal Fire-rose cut is also available as a round, pear-shape, marquise or heart-shape. The cut is designed to give a higher yield and increased brilliance in strongly coloured diamonds (De Beers). (right) The 12-sided oval-shaped Dahlia cut gives a higher yield than other more conventional cuts for elongated diamonds. It is also suitable for diamonds having a pronounced colour. (De Beers)

(designed by G. Tolkowsky to celebrate the De Beers centenary) are shown in Figure 19.11.

The zircon cut

This style, based on the ideal cut for diamond, is designed to improve the brilliance of a zircon (whose RI is somewhat lower than that of diamond). Although having the same crown/girdle and girdle/pavilion angles as diamond's brilliant cut, the zircon cut reduces the amount of light leakage from the rear of the stone by employing extra facets which are placed between the culet and the main pavilion facets (Figure 19.12).

The emerald cut

Designed specifically for emerald, this cut not only enhances the stone's colour, but allows for its shock sensitivity by omitting the otherwise vulnerable corners (Figure 19.13). The cut is also known as the 'step' or' 'trap cut', and is often used for diamonds and other gems. A variant, called the *radiant cut* (Figure 19.14), is used specifically for diamonds.

Crown   Pavilion

Figure 19.12 The zircon cut has additional facets placed around the culet to prevent the loss of light through the base of the pavilion

Figure 19.13 The emerald, trap or step cut

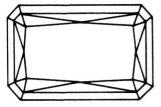
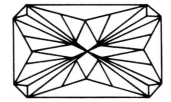

Figure 19.14 The 61-facet radiant cut used for diamond. It has emerald-cut facets on the crown, and triangular facets on the pavilion. The radiant cut has some of the fire and scintillation of a round brilliant in the centre, and in this respect is superior to the emerald cut for diamond

The scissors or cross cut

This is basically a modification of the emerald cut, but has a rectangular plan profile and triangular facets around the table (Figure 19.15). It is used, like the emerald cut, to enhance the colour of a stone while maintaining a reasonable degree of brilliance. There is, however, some loss of light through the centre area of the pavilion.

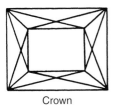
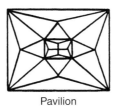

Crown   Pavilion

Figure 19.15 The scissors or cross cut

The mixed cut

This is mainly used for coloured gemstones. It consists of a brilliant-cut crown (oval or round) and a step- or trap cut-pavilion (Figure 19.16). The style is frequently used by native cutters and is often made with an over-deep pavilion in order to retain maximum weight in the stone.

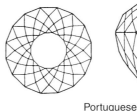
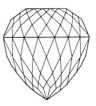

Mixed   Portuguese

Figure 19.16 (left) The mixed cut is mainly used for coloured stones. It has a brilliant-cut crown and a trap- or step-cut pavilion. (right) The Portuguese cut is occasionally used for large coloured stones

Other styles

There are a great variety of cutting styles in addition to the main traditional ones mentioned in the preceding paragraphs. Many of these are variants of the basic cuts and are designed to accommodate the final shape of the rough stone. Some of the simpler styles have descriptive names such as the hexagonal, square, triangular and keystone cuts. One of the more elaborate variants used for larger coloured stones is the Portuguese cut, which can have up to 177 facets (Figure 19.16). Laser equipment is increasingly being used to produce new cutting styles.

Gemstone polishing

In the very early days of gemstone fashioning, a polisher or lapidary would cut and polish both diamonds and other gemstones. However, over the years the gemstone world has grown into two distinct and separate industries ranging from mining to polishing and marketing. Diamonds are now almost exclusively polished by diamond cutting specialists, and all the other gemstones are cut and polished by lapidaries. Although very rarely a lapidary may cut a diamond, a diamond cutter is always most reluctant to risk his or her polishing equipment on anything other than diamond!

Lapidary techniques

The basic method of polishing a non-diamond gemstone is a combination of sawing and rough grinding by the cutter to produce the required profile or *preform* (with the surface of the stone left in a matt state), followed by a finishing operation by the polisher to achieve the final surface or facet lustre.

With a piece of rough having strong pleochroism, chatoyancy or asterism, it will first be necessary to orientate the gem material correctly. For cat's-eye and star stones the fibres, needles or cavities producing the sheen effect must be positioned parallel to the base of the cabochon. As mentioned in Chapter 5, the lapidary must also be aware of any planes of cleavage in his or her stone in order to avoid accidental breakage during polishing. The lapidary must ensure that any polished facet is tilted by at least 5° from such a plane. If an attempt is made, for example, to polish a facet at right angles to the length of a topaz prism, the result will be a pitted surface where layers of topaz have cleaved away from the stone (Figure 19.17).

Figure 19.17 Cleavage damage to topaz caused by the table facet being polished at right angles to the crystal's *c* axis. (J. Gemmell)

Next, the rough gemstone material is 'slabbed' (i.e. sawn) into suitable pieces using a rotary saw whose cutting edge is impregnated with diamond grit. These pieces are then marked with the profile of the required finished stone. If a cabochon is to be produced, the sawn 'blank' is passed to the cutter who develops the rounded base outline using a water-cooled vertical grinding wheel. Following this operation the stone is cemented to a *dop stick* for easier handling, and the dome of the cabochon formed on another finer grinding wheel. The final polish is achieved on a horizontal rotary *lap* using an appropriate polishing powder. For cabochon polishing, the lap may be of wood and contain grooves having the necessary radii.

To produce a faceted gem, the rough gemstone is sawn into a suitable shape and the resulting blank passed to the cutter who grinds the facets on the stone using an iron, copper or gunmetal lap (*glass* laps have been marketed in the USA – these have a flatter surface which is less scratch-prone than metal). The polishing lap may be powered from an electric motor or from a hand-cranked flywheel (this latter method is claimed to give better control). However, even the primitive bow-driven vertical polishing lap can still be seen in use in India and Sri Lanka (Figure 19.18).

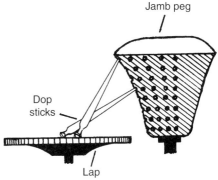

Figure 19.18 (top left) The bow-driven vertical polishing lap still used by some native lapidaries in India and Sri Lanka. (above) Sketch showing the use of the jamb peg. (left) A lapidary using a jamb peg (on the far side of the lap)

Just as traditional is the use of a *jamb peg*, a mushroom-shaped block of wood containing a series of holes which act as pivot sockets for the free end of the dop stick. The various angles of the gem's facets can then be ground by selecting the appropriate hole to anchor the end of the dop stick (Figure 19.18). A more up-to-date method uses a mechanical faceting head to hold and position the stone (Figure 19.19).

A wide variety of polishing powders and lap materials have been used over the years. Traditional polishing powders include *rottenstone* and *emery*. Rottenstone (also called *silicon dioxide* and *tripoli*) is a white to pale-brownish silica-based 'soft' abrasive powder

Figure 19.19 (left) A preformed and dopped stone inserted in a mechanical faceting head which can be set to preselected angles. (right) The stone, positioned to an appropriate angle by the mechanical faceting head, is brought into contact with the polishing lap

derived from decomposed limestone. Emery is a coarse carborundum powder (i.e. silicon carbide). Today, these have largely been replaced by metallic powders including the oxides of cerium and tin. Graded corundum and diamond powder are also widely used, the latter providing much faster cutting which partly compensates for its higher cost.

With the exception of diamond powder, which is often mixed with a lanolin cream, all of these polishing powders are used in a suspension of water. In addition to metal and glass, lap surface materials include wood, hard felt, leather and even plastic. The lapidary will usually have a variety of laps, each reserved for a particular gem or range of gems (in this way, the possible contamination of the lap surface by fragments of other harder gem materials is avoided). The diamond powder grade used for rough stock removal is 60–90 micrometres size. Grades then progress through 30–60 micrometres for shaping, 6–12 micrometres for the pre-polishing finish, to 1–6 micrometres for the final polish.

The value of the final polished stone depends on many factors including the depth of colour, freedom from flaws and quality of cut. Several organizations have developed grading standards for coloured gemstones. However, because of the complexity of assessing factors such as pleochroism, hue, colour saturation and colour zoning, none of these standards have become internationally recognized in the same way as have the diamond grades for colour and clarity, which are described later in this chapter.

Gemstone carving

Gemstone carving is also the province of the lapidary, and has been so for thousands of years. Chinese jade carving is perhaps the most well known, and is still carried on in large lapidary workshops in Beijing, which employ around forty lapidaries on each of several floors (Figure 19.20).

Historically, the bulk of early Chinese jade carvings made use of nephrite from eastern Turkistan. From the eighteenth century onwards, top-quality Chinese jade carvings began to be made from jadeite imported from Burma (now called Myanmar). In 1850, boulders of nephrite were discovered in rivers in Siberia, and these would have provided a more accessible source for the Chinese carvers. However, by this time Burmese jadeite, with its more intense colours, had become the preferred material with Chinese carvers.

Cameos and *intaglios* (the latter an incised carving sometimes used as a seal stone) have long been carved from shell, coral and harder gem materials such as agate in Italy and many other parts of the world. Another gemstone and cameo carving centre of some importance is the area around Idar-Oberstein in Germany (Figure 19.21).

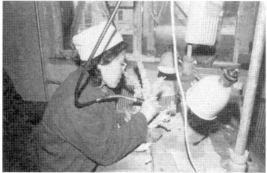

Figure 19.20 Gemstone carving of nephrite (top), coral (middle) and turquoise (bottom) in a large lapidary establishment in Beijing, China

With the advent of ultrasonically cut and rapidly reproduced copies of the works of master cameo carvers it has become important to distinguish hand cut cameos from the less valuable article. As the ultrasonic technique produces an effect described as 'freshly fallen snow', one method of detection is to closely inspect the surface of the cameo. This frosted appearance can be seen in the deep recesses of any cameo which has been produced using this technique, even if it has been subsequently hand finished.

The tools used for carving vary from steel blades and files to rotary burrs using silicon carbide bits. Because of its hardness, corundum is fashioned using diamond-tipped tools.

Figure 19.21 (left) A cameo carved from agate. (Emil Becker). (right) A pochard duck carved from blue chalcedony, with tiger's-eye head, hawk's-eye tail and silver legs and feet, mounted on an azurite/malachite base. (Emil Becker)

Diamond polishing methods

That diamond, the hardest of all gem minerals, can be fashioned at all is due to its differential hardness (see Chapter 6). Starting with the rough diamond crystal, an expert 'designer' (usually the owner of the diamond cutting firm) inspects the stone and decides how it should be cut to produce the best yield in polished stones. He or she then marks the surface of the stone with Indian ink to indicate the directions in which the crystal must be sawn or cleaved (with large stones, both methods may be employed). Depending on the quality of the stone (i.e. its colour and freedom from flaws) the designer will also decide on the 'make' of the final polished stone. A good make is one which closely approaches the proportions and angles of the ideal brilliant cut. However, if the stone is of low quality, he or she may decide to depart somewhat from the ideal cut to produce a larger more valuable stone.

Cleaving is performed by first carefully identifying the stone's cleavage plane, and then scratching a shallow groove called a *kerf* parallel to this plane in the surface of the diamond. This is done using a sharp fragment of diamond held in a dop (in some diamond cutting establishments, the kerf groove is now cut with a laser beam). Next, a thin-bladed chisel is placed in the groove and given a sharp blow to part the stone along the line of the cleavage plane (in effect, the cleaving blade acts as a splitting wedge in the kerf – see Figure 19.22). At this stage, cleaving may also be used to remove a section of flawed stone.

Sawing is carried out using a thin rigidly clamped phosphor bronze blade which is impregnated with fine oil-bound diamond dust and rotated at 5000 r.p.m. (revolutions per minute) (Figure 19.23). Sawing can only take place in two directions across the crystal grain, one of these directions being parallel to the natural girdle of the octahedral

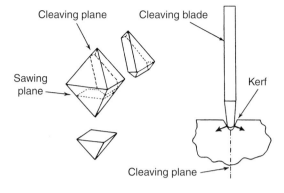

Cleaving plane

Cleaving blade

Sawing plane

Kerf

Cleaving plane

Figure 19.22 (left) One of the four cleaving planes and one of the two sawing directions in a diamond. (right) The edge of the diamond cleaver's blade acts as a wedge to part the crystal

Figure 19.23 (top) The diamond crystal, cemented in a double-sided dop is lowered under pressure on to the phosphor bronze saw blade which is dressed with diamond dust. (Monnikendam). (below) A double bank of sawing machines attended by the diamond sawyer. (Monnikendam)

crystal; the other is at right angles to this through the tips of the bi-pyramids and the centres of their common baselines.

Laser 'sawing' has also been developed for diamond work. Because of the difficulty of maintaining the focus of the laser beam at depth, this technique is usually limited to stones below 3 carats. The cut produced by the laser is wedge shaped and therefore more wasteful than conventional sawing, but this is offset by the greater speed of cutting and the fact that the sawing direction is independent of the diamond's *grain* (i.e. the planes of cleavage and minimum/maximum hardness related to the crystal axes).

The sawn or cleaved stone is then *bruted* or *rondisted* (i.e. rounded) by rotating it on a lathe-like spindle and bringing another diamond into contact with it as a cutting tool (the other diamond often being the remaining section of the sawn stone – Figure 19.24). The bruter may sometimes leave a small fragment of the outer surface or 'skin' of the diamond crystal on the girdle. This is called a 'natural', and indicates to the designer that the maximum possible width of the crystal has been utilized (such a natural may include a *trigon* which can subsequently serve as an identification feature for diamond).

Figure 19.24 The bruter rondisting a sawn octahedral diamond. A corner of the other section of the sawn stone is clamped in a dop stick and used as the cutting tool. (Monnikendam)

Next, the crown and pavilion facets are ground by mounting the bruted stone in a holder called a *dop* and bringing it into contact with the surface of a 12-inch (30-cm) diameter iron lap called a *scaife* (Figure 19.25). This has a porous surface and is dressed with a mixture of olive oil and diamond dust and rotated at 3000 r.p.m.

Figure 19.25 Mechanical clamp-type dops holding diamonds on a scaife. The dops are mounted on the end of tangs, which are weighted to achieve the necessary polishing pressure. (De Beers)

Two types of dop are used, the older version employing a low-melting-point solder to secure the diamond, and the more modern type using a mechanical clamping device. Both types of dop are mounted on the end of a short copper rod which is bent by hand to present the diamond to the scaife's surface at the appropriate facet angle. The rod is clamped in a saddle-shaped *tang* (Figure 19.25), which is weighted to achieve the correct grinding pressure. In more elaborate mechanical dops, the angles relative to the surface of the scaife are preset and can be selected by an indexing arrangement.

Using the dop, the polisher grinds the crown and pavilion facets in a fixed sequence, checking each one with a hand lens before proceeding to the next. Two facets are first ground on the crown of the stone placing them on opposite sides of the girdle. Next, two more facets are placed between these and positioned exactly opposite each other. In this way a symmetrical pattern is built up, and this is used as a guide to position the pavilion facets.

Several fully automatic diamond bruting and faceting machines have been developed, and are in use throughout diamond cutting centres where labour is expensive. These faceting machines are designed mainly for polishing small diamonds where costs must be kept low in accord with the value of the stone.

The work of producing a faceted diamond is usually divided (in all but the smaller cutting establishments) between some six experts. The *designer* marks the stone for cleaving or sawing. The *cleaver* and/or *sawyer* performs the initial 'breaking up' of the stone. The *bruter* then produces the basic rounded profile (or, by off-setting the stone, an oval or marquise shape). The *cross-cutter* takes the bruted stone and grinds the table, the sixteen main crown and pavilion facets, and the culet (see Figure 19.9). If the cross-cutter has found it necessary to polish his or her sequence of facets slightly off the centre line of the stone (in order to remove a surface inclusion or flaw), they will return it to the bruter to regrind the girdle so that it is symmetrical with the facets. The stone is finally passed to the *brillianteerer*, who grinds the last sequence of twenty-four crown and sixteen pavilion facets, and gives all the facets their finishing polish.

Diamond grading

At the end of the polishing operation, the final value of the polished diamond will depend on four basic parameters. These are popularly known as the 'four C's' of diamond grading. These four C's and their grading criteria are as follows.

Colour

This is perhaps the most important single factor. Polished yellow Cape series diamonds are subjectively graded for colour by placing them in a white grooved grading tray and comparing their colour with a set of *master stones* (i.e. carefully selected diamonds or diamond simulants – Figure 19.26). Grading is effected by comparing the stone's body colour with that of the master stones by viewing them through the side of the pavilion under 'daylight' type illumination. Alternatively, a diamond's colour can be measured objectively with a specially adapted spectrophotometer (see Figure 16.24). International colour grades for Cape series polished diamonds are listed in Table 19.1.

Clarity

This is the freedom of the stone, as inspected at 10× magnification, from any internal feature (such as an inclusion) which will reduce the brilliance of the stone. The grading

Figure 19.26 A diamond master stone set used for comparison colour grading

Table 19.1 International colour grades for polished diamonds

Traditional UK	German RAL Scan DN (0.5 carats upwards)	Scan DN (under 0.5 carats)	GIA	CIBJO	IDC
Finest white	River	Rarest white	D	Exceptional white +	Exceptional white +
			E	Exceptional white	Exceptional white
Fine white	Top Wesselton	Rare white	F	Rare white +	Rare white +
			G	Rare white	Rare white
White	Wesselton	White	H	White	White
Commercial white	Top crystal	Slightly tinted white	I	Slightly tinted white	Slightly tinted white
Top silver Cape	Crystal		J		
Silver Cape	Top Cape	Tinted white	K	Tinted white	Tinted white
			L		
Light Cape	Cape	Slightly yellowish	M	Tinted colour	Tinted colour 1
			N		
Cape	Light yellow	Yellowish	O		Tinted colour 2
			P		
			Q		Tinted colour 3
			R		
Dark Cape	Yellow	Yellow	S–Z		Tinted colour 4

system takes into account the position of the flaw within the stone. Flaws visible in the table area produce a lower grade than those of the same size only visible through the girdle area of the crown facets. Table 19.2 lists the international clarity grades, and Tables 19.3 and 19.4 list the UK and CIBJO clarity grading standards for polished diamonds.

Cut

Until recently this has simply been an assessment of how closely the proportions and angles of a diamond follow those of the ideal brilliant-cut diamond (of which there are

Table 19.2 International clarity grades for polished diamonds

Traditional UK	RAL	Scan DN	GIA	CIBJO	IDC
Flawless	IF	FL	IF	Loupe-clean	Loupe-clean
		IF			
		VVSI 1	VVSI 1	VVS 1	VVS 1
VVS	VVS	VVSI 2	VVSI 2	VVS 2	VVS 2
		VSI 1	VSI 1	VS 1	VS 1
VS	VS	VSI 2	VSI 2	VS 2	VS 2
		⎰SI 1	SI 1	SI 1⎱	
SI	SI	⎱SI 2	SI 2	SI 2⎰	SI
1st PK	PK 1	PK 1	I 1	PI	PI
2nd PK	PK 2	PK 2	I 2	PII	PII
3rd PK	PK 3	PK 3	I 3	PIII	PIII
Spotted					
Heavily spotted					
Rejection					

Table 19.3 Traditional UK clarity grading standards (10× magnification)

Flawless	Clean stone
VVS	Very fine white or black spots, not central
VS	Fine white or black spots, not central
SI	Small inclusions, some central
1st PK	Black or white spots, seen with difficulty, mainly central
2nd PK ⎱	
3rd PK ⎰	Several white or black spots, easily seen – can be central in 3rd PK
Spotted ⎱	
Heavily spotted ⎬	Obvious large inclusions, cracks, etc.
Rejection ⎰	

Table 19.4 CIBJO clarity grading standards (10× magnification)

Loupe-clean	A diamond is classified as loupe-clean if it is completely transparent and free from visible inclusions
VVS *(VVS 1, VVS 2)	Very, very small inclusion(s) which are hard to find with a 10× loupe
VS (VS 1, VS 2)	Very small inclusion(s) which can just be found with a 10× loupe
SI 1, SI 2	Small inclusion(s), easy to find with a 10× loupe, but not seen with the naked eye through the crown facets
PI	Inclusion(s) immediately evident with a 10× loupe, but difficult to find with the naked eye through the crown facets. Not impairing the brilliancy
PII	Large and/or numerous inclusion(s), easily visible to the naked eye through the crown facets, and which slightly reduce the brilliance of the diamond
PIII	Large and/or numerous inclusion(s), very easily visible with the naked eye through the crown facets, which reduce the brilliance of the diamond

* Subgrades VVS 1, VVS 2, VS 1 and VS 2 are used only for stones of 0.47 carats and larger.

several versions – see Figure 19.9). Also taken into account were any external flaws and additional or misaligned facets. The overall quality or 'make' of the cut depends on the balance of brilliance, fire and scintillation of the finished diamond, but there are other criteria (such as polish and optical symmetry) which also play a part.

Because of the complexity of producing a set of guidelines for accessing cut, grading standards had only been formulated so far for colour and clarity (as illustrated in Tables 19.1–19.4). In 2004, however, after 15 years of research including consultations with diamond cutters, dealers, retailers and potential customers, the GIA finally published details of their 'Diamond Cut Grading System'. This is specifically for round brilliant-cut diamonds, requires a standard viewing environment, and is based on carefully defined criteria which include brightness, fire, scintillation, polish and symmetry. The software used in the GIA Diamond Cut Grading System predicts the overall cut grade for values input by the user, who can also obtain a prediction of the cut grade for a polished diamond even when it is only in the planning stage.

Carat weight

This is perhaps the simplest of the four parameters, and in terms of value is only complicated by the non-linear pricing structure of the diamond (i.e. because of the rarity factor of large diamonds, the price of a 2-carat stone may be four times that of a 1-carat stone of identical grade).

The normal unit of weight for diamonds is the metric carat, which is also used for non-diamond gems (1 carat = 0.20 grams). A diamond's weight may also be expressed in two other ways; small stones are measured in 'points' (there are 100 points to the carat); larger stones (including uncut diamonds) may be referred to as one, two or three 'grainers' (1 grain, like the pearl grain, equals 0.25 carat).

Chapter 20

Practical gemstone identification

In Chapter 1 the science of gemmology was defined as the study of the technical aspects of gemstones and gem materials. However, the *application* of gemmology as a science ranges from the routine bench work identification of gemstones by the gem dealer, jeweller and jewellery valuer through to the research carried out in the world's major gemmological laboratories.

As a text book for students of gemmology, the main emphasis in this third edition has been placed on the coverage of subjects relevant to the theory syllabi of the GAGTL's Foundation and Diploma courses, and this has included the elements of geology, crystallography, chemistry, and the specialized instruments used to measure the various physical constants of gem materials (see the standard student's gem kit in Figure 20.1). As the book is also intended to be a work of reference, particular attention has been paid to text links between the chapters, and to an extensive subject index.

Knowing that readers of this book will include a large proportion of gemmological students who will be preparing for the GAGTL's Diploma examinations (which

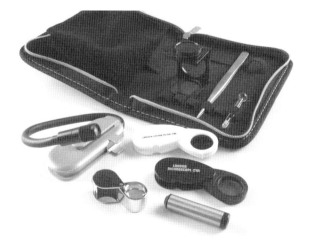

Figure 20.1 Gem-A Instruments Ltd standard gem kit consists of an 18 mm 10× hand lens, stainless steel tweezers, a folding polariscope with conoscope rod, Chelsea filter, Polaroid dichroscope, diffraction spectroscope, flexi torch, and a folding zip travel case. Gem-A Instruments Ltd professional gem kit has in addition a refractometer, a LW UV key ring, and an Osram flat light. The Polaroid dichroscope is replaced with a calcite version, and the hand lens and the spectroscope are replaced with wide-field versions

includes a 3½-hour practical identification session) the final chapters of this book have concentrated on the identification of gemstones and gem materials. The following notes are intended to guide the novice gemmologist in the correct sequence of tests to be taken in order to arrive at a positive identification of an unknown gemstone in the shortest time.

Observation

First impressions can be very important, and a close overall inspection of the gemstone under a suitable light source may reveal many features which will later aid in its identification (see Figure 20.2). For example, the presence of surface damage can indicate a stone's relative hardness.

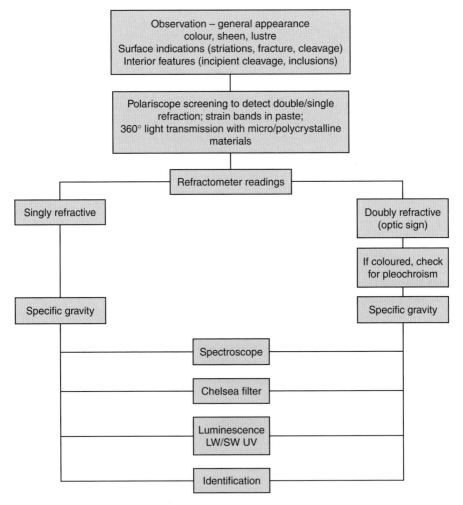

Figure 20.2 This diagnostic chart begins with a listing of the various internal and external features of a gemstone as observed first with the unaided eye and then with a 10× hand lens, and suggests a logical sequence of tests leading to a final identification. Test limitations make SG measurements practical only with unmounted gemstones, and make RI readings on crystals not normally possible at all (if transparent or translucent a crystal's optical character can be verified with the polariscope)

With faceted stones the condition and sharpness of the edges formed by the facet junctions can also indicate the hardness of the material. The moulding ridge around a stone's girdle may warn of a possible paste (glass) specimen. Colour can form a useful guide if only to rule out several of the possible identifications (see Appendix C).

With opaque ornamental gems their colour appearance can often immediately identify the material (e.g. the swirls of light and dark green in malachite; the concentric bands of pink and white in rhodochrosite).

If the gem is transparent, inspection with a $10\times$ hand lens may also reveal a diagnostic inclusion (e.g. a 'horsetail' inclusion in a green demantoid garnet) or signs of incipient cleavage (e.g. as in topaz). While examining the stone, look closely in the area of the crown and upper pavilion facets for signs of a composite gem. If found this could prevent an incorrect identification later on. The presence of gas bubbles or colour swirls may indicate a paste (glass) material, or curved colour zoning may reveal a Verneuil synthetic sapphire.

Exploring the stone further may reveal signs of chatoyancy and asterism, and these features will form additional clues when making an identification. The hand lens can also be used to gauge the size of a gem's double refraction by looking obliquely through the table facet for any sign of 'doubling' of the stone's pavilion facet edges (as in zircon). When inspecting a transparent colourless gem, take note of the presence and amount of 'fire' visible, as this will be relevant information when checking for diamond and its simulants. Take particular note of the lustre of a stone as this can also provide a subjective clue to its identity (a high adamantine lustre, for instance, will indicate diamond, and a vitreous lustre may point to a colourless sapphire). With experience, all of these initial observations can be made quickly and can then determine the sequence and number of objective tests necessary to confirm a positive identification.

Selecting the appropriate tests

Choosing the tests which will lead to a speedy identification may not seem a priority when time is not a factor, but in a retail working environment (and in the GAGTL's Diploma practical exam!) it is important to make just enough *objective* tests to arrive at a positive identification. Ideally, two such tests (such as an RI reading and an SG measurement) will be sufficient, but if the objective tests available are limited to one, as can happen with mounted stones, then identification must be supported by other more *subjective* evidence (such as signs of cleavage, pleochroism and luminescence and, where relevant, the results of a Chelsea filter examination).

One the most important of the objective tests is that involving the refractometer, as this instrument can yield precise figures to 2 decimal places (and estimated to 3 places) for RI and DR, as well as verifying an anisotropic gem's optical character and optic sign. However, it is recommended that *all* transparent and translucent gemstones are first screened on a polariscope. This will usually provide a non-ambiguous doubly-refractive (or isotropic) indication and forecast the presence of single or double shadow edges in subsequent refractometer readings. In addition, the polariscope may reveal the dark strain bands typical of paste (glass), and can identify a micro/polycrystalline gem by transmitting light through the crossed polars in any orientation of the gemstone (transparent to translucent *crystal* specimens can also be checked on the polariscope for optical character). After this initial screening the testing sequence as outlined in Figure 20.2 can continue, as appropriate.

Despite the fact that SG measurements constitute a useful objective test for gemstones, this is only possible with unmounted stones. However, the measurement of SG is a useful way of obtaining extra diagnostic information when identifying a crystal specimen. Using the hydrostatic method of 'in air' and 'in water' weighing, even the SG of large carvings can be measured with a simple spring balance and a container of water (the low sensitivity of the spring balance is compensated by the large displacement of the carving – this method is sufficiently accurate to separate jadeite from nephrite).

Although the spectroscope can provide a diagnostic spectrum of a faceted gemstone, the instrument is also useful when identifying a crystal specimen, particularly as flat enough faces on a crystal are seldom available and this usually rules out the use of the refractometer for crystal identification.

After selecting the necessary minimum of objective and subjective tests, the procedure outlined in Figure 20.2 is finally concluded by making a positive identification from the accumulated results (see also 'Examination notes' in Appendix E).

Appendices

Appendix A Bibliography

Anderson, B.W. (1976) *Gemstones for Everyman*, Faber and Faber, London

Anderson, B.W. (1990) *Gem Testing*, 10th edn, Butterworth-Heinemann, Oxford

Arem, J.E. (1987) *Color Encyclopedia of Gemstones*, 2nd edn, Van Nostrand, New York

Bank, H. (1973) *From the World of Gemstones*, Penguin, Innsbruck

Barnard, H. (2000) *The Diamond Formula*, Butterworth-Heinemann, Oxford

Bruton, E. (1978) *Diamonds*, 2nd edn, NAG, London

Campbell, M.C. (2004) *Gem and Ornamental Materials of Organic Origin,* Elsevier, Oxford

Church, Sir A. (1883) *Precious Stones*, London

Cressey, G. and Mercer, I. (1999) *Crystals*, 2nd edn, Natural History Museum, London

Davies, G. (1984) *Diamond*, Hilger, Bristol

Elwell, D. (1979) *Man-made Gemstones*, Horwood, New York

Farn, A.E. (1986) *Pearls: natural, cultured and imitation*, Butterworth-Heinemann, Oxford

Fraquet, H.R. (1987) *Amber*, Butterworth-Heinemann, Oxford

GIA (1993) *The GIA Diamond Dictionary,* 3rd edn, GIA, Carlsbad, CA

Gill, J.O. (1978) *Gill's Index to Journals, Articles and Books relating to Gems and Jewelry*, GIA, Carlsbad, CA

Gübelin, E.J. (1979) *Internal World of Gemstones*, 2nd edn, Butterworth-Heinemann, Oxford

Gübelin, E.J. and Koivula, J. (1992) *Photoatlas of Inclusions in Gemstones*, ABC, Zurich

Günther, B. (1981) *Tables of Gemstone Identification*, Elizabeth Lenzen, Kirschweiler

Hecht, E. and Zajac, A. (1974) *Optics*, Addison-Wesley, Reading, MA

Holmes, A. (1978) *Holmes' Principles of Physical Geology*, 3rd edn, Nelson, London

Hoover, D.B. (1992) *Topaz,* Butterworth-Heinemann, Oxford

Hughes, R.W. (1977) *Ruby and Sapphire,* RWH Publishing, USA

Hughes, R.W. (1990) *Corundum*, Butterworth-Heinemann, Oxford

Lenzen, G. (1983) *Diamonds and Diamond Grading*, Butterworth-Heinemann, Oxford

Leonardus, C. (1750) *The Mirror of Stones*, London

Liddicoat, R.T. (1989) *A handbook of gem identification*, 12th edn, GIA, Carlsbad, CA

Marrison, L.W. (1966) *Crystals, Diamonds and Transistors*, Penguin books, Harmondsworth

Matthews, W.H. (1976) *Introducing the Earth*, Bailey Brothers and Swinfen, Folkstone

Muller, H. (1987) *Jet*, Butterworth-Heinemann, Oxford

Nassau, K. (1980) *Gems Made by Man*, Chilton Book Co., Radnor, PA

Nassau, K. (1994) *Gemstone Enhancement*, 2nd edn, Butterworth-Heinemann, Oxford

Nassau, K. (2001) *The Physics and Chemistry of Color*, 2nd edn, J. Wiley & Sons, USA

Newman, R. (2005) *Diamond Handbook*, International Jewelry Publications, USA

Nicols, T. (1652) *A Lapidary; the history of precious stones*, Cambridge University Press, Cambridge

O'Donoghue, M. (1987) *Quartz*, Butterworth-Heinemann, Oxford

O'Donoghue, M. (1998) *Gemstones*, Chapman and Hall, London

O'Donoghue, M. (1997) *Synthetic, Imitation and Treated Gemstones*, Butterworth-Heinemann, Oxford

O'Donoghue, M. (2005) *Gems*, 6th edn, Butterworth-Heinemann, Oxford

O'Donoghue, M. and Joyner, L. (2003) *Identification of Gemstones*, Butterworth-Heinemann, Oxford

Pagel-Thiessen, V. (1980) *Diamond Grading ABC*, 7th edn, Pagel-Thiessen, Frankfurt

Read, P.G. (1983) *Gemmological Instruments*, 2nd edn, Butterworth-Heinemann, Oxford

Read, P.G. (1988) *Dictionary of Gemmology*, 2nd edn, Butterworth-Heinemann, Oxford

Rouse, J.D. (1986) *Garnet*, Butterworth-Heinemann, Oxford

Schumann, W. (2001) *Gemstones of the World*, 3rd edn, NAG, London

Sinkankas J. (1985) *Gem Cutting*, 3rd edn, Van Nostrand, New York

Sinkankas J., and Read, P.G. (1986) *Beryl*, Butterworth-Heinemann, Oxford

Smith, G.F.H. (1972) *Gemstones*, 14th edn, Chapman and Hall, London

Themelis, T. (1992) *The Heat Treatment of Ruby and Sapphire*, Gemlab Inc., USA

Theophrastus, (1774) *History of Stones*, London

Tolanski, S. (1962) *The History and Use of Diamond*, Methuen, London

Walton, Sir J. (1952) *Physical Gemmology*, Pitman, London

Watermeyer, B. (1980) *Diamond Cutting*, Purnell, Cape Town

Webster, R. (1976) *Practical Gemmology*, 6th edn, NAG, London

Webster, R. (1998) *The Gemmologist's Compendium*, 7th edn, NAG, London

Winter, C.H. (2003) *OPL   A Student's Guide to Spectroscopy*, OPL Press, UK

Yaverbaum, L.H. (1990) *Synthetic Gems – production techniques*, Noyes Data Corporation, Park Ridge, NJ

Zancanella, V. (2004) *Tanzanite*, Alcione, Trento, Italy

Appendix B The organic gems

Chemical composition, varieties, constants, characteristics and localities

Amber

Composition: Fossilized tree resin containing a mixture of hydrocarbons plus some succinic acid, and consisting mainly of carbon (79%), hydrogen (10%) and oxygen (11%)

Varieties: Opaque white, lemon, golden, red-brown, brown and black. Translucent to transparent colourless, pale yellow, deep yellow, light to deep red, and occasionally green and blue/grey

Refractive index: 1.54
Specific gravity: 1.05–1.10
Hardness: 2.5
Lustre: Resinous
Luminescence: Bluish-white (LW UV), Greenish (SW UV)
Inclusions: Plant particles and insects, bubbles
Localities: Sea amber from the Baltic and North Sea coasts. Pit amber from Sicily, Romania, Myanmar, the Dominican Republic and Kaliningrad in the former USSR

Copal
Composition: A younger fossil tree resin than amber (sometimes used as an amber simulant)
Constants: Copal resin has constants similar to those of amber, but unlike amber it is softened by ether, has a crazed surface, and crumbles easily under the knife blade
Inclusions: Occasionally contains insects
Localities: East Africa and South America

Coral
Composition: A branching tree-like structure formed by the skeleton remains of various types of marine polyp, and consisting either of a fibrous aragonite form of $CaCO_3$ (white and pink varieties) or of a horn-like material consisting of conchiolin (black and golden varieties – although these are not strictly corals in the gemmological sense)
Varieties: White, pink ('angel's skin) and red, golden and black
Refractive index: Around 1.48 for carbonate coral; 1.56 for conchiolin type
Specific gravity: 2.60–2.70 for carbonate coral; 1.37 for the conchiolin type
Hardness: 3.5 for carbonate coral; 2.5–3.0 for the conchiolin type
Lustre: Waxy
Localities: White, pink and red coral is found in shallow sub-tropical waters in the western Mediterranean and the Malay and Japanese seas. Black and golden coral is harvested off the Malaysian Archipelago, the coastline of northern Australia, the Red Sea and off the Hawaiian islands

Ivory
Composition: Ivory consists almost entirely of dentine, although enamel and other organic substances are associated with the complete tusk or tooth
Varieties: Ivory is obtained from elephant tusks, the tusks of the walrus and the hippopotamus, the teeth of the narwhale and the sperm whale and, more rarely, the tusks of fossilized mammoths
Refractive index: A vague refractometer reading around 1.5
Specific gravity: 1.7–2.0
Hardness: 2.0–3.0

Jet
Composition: Organic, carboniferous with traces of mineral elements. Jet is a form of fossilized wood (similar to lignite or brown coal) halfway in formation between peat and bituminous coal

Varieties: Hard jet and soft jet (black and dark brown)
Refractive index: 1.64 to 1.68
Specific gravity: 1.3–1.4
Hardness: 2.5–4.0
Lustre: Resinous
Fracture: Conchoidal
Localities: France, Germany, Spain, UK (on the Yorkshire coast around Whitby), USA and the former USSR

Odontolite

Composition: A bluish fossil bone or ivory obtained from prehistoric animals such as the mammoth. The bone contains calcite and owes its colour to vivianite, an iron phosphate. It is also known as bone or fossil turquoise, and is used as a turquoise simulant
Refractive index: 1.57 to 1.63
Specific gravity: 3.0–3.25
Hardness: 5.0
Locality: France

Pearl (natural)

Composition: Calcium carbonate (86%), conchiolin (12%), water (2%)
Varieties: White, black and pink blister, cyst and freshwater pearls; non-nacreous pink conch pearls
Refractive index: 1.52 to 1.66 (1.53 to 1.69 for black pearls)
Specific gravity: 2.68–2.78 (conch pearls, 2.83–2.86)
Hardness: 2.5–3.5
Luminescence: Most black pearls fluoresce a faint reddish-brown under LW UV
Localities: The Kuwait–Bahrain Gulf, the Gulf of Mannar, the South Sea Islands, the Pacific coast of Mexico (black pearls), the Gulf of California (black pearls), the north-west coast of Australia, and around the islands and inlets of Japan. Pink (non-nacreous) conch pearls are found off the coasts of Florida and the West Indies. Freshwater pearls from the rivers of the UK, the USA, the former USSR and the People's Republic of China

Shell

Composition: Overlapping layers of sub-microscopic crystals of calcium carbonate forming the inner nacreous layer of a mollusc's shell
Varieties: Mother-of-pearl from various shells having an iridescent lustre including the *Pinctada maxima*, *Pinctada margaritifera* and *Haliotis* (also called 'abalone', 'paua' or 'ear shell'). The shell of the Turbo sea snail is used to produce imitation pearls (called 'antilles' or 'oil' pearls). Cameos are cut from shells having layers of contrasting colour or shade, such as the Helmet shell (*Cassis madagascariensis*) and the shell of the giant conch (*Strombus gigas*)
Localities: The Pinctada and Haliotis shells are found in American, Australian, Japanese, Korean and New Zealand waters. The Helmet shell is found in West Indian waters, and the giant conch in both West Indian and Florida coast waters

Tortoiseshell
Composition: Derived mainly from the shell of the Hawksbill sea turtle, this material consists mostly of keratin (a protein forming the basis of horns, claws, nails, etc.)
Refractive index: 1.55
Specific gravity: 1.29
Hardness: 2.5
Luminescence: The yellowish parts of the shell fluoresce a bluish-white under LW UV
Localities: The Hawksbill turtle is found in tropical and sub-tropical seas, particularly around Malaysia, the West Indies and Brazil

Appendix C The inorganic gems

Chemical composition, varieties, constants, characteristics and localities

With doubly refracting stones the difference between the two RIs is indicated by 'x − y'; when these RI values are significantly greater than the figure for DR, this is shown as 'x to y', to indicate the possible range in which RI values can occur. With singly refractive gems, and poly/microcrystalline stones the possible RI range is also indicated by 'x to y'.

Andalusite
Composition: Al_2SiO_5
Crystal system: Orthorhombic (polymorphous with kyanite and fibrolite)
Habit: Prismatic (pseudo-tetragonal in outline)
Varieties: Transparent brownish-green and green. An impure variety, chiastolite, is an opaque yellow-white material containing carbonaceous inclusions in the form of a black cross
Refractive index: 1.63–1.64
Double refraction and optic sign: 0.01, negative
Dispersion: Low (0.016)
Specific gravity: 3.18
Hardness: 7.5
Cleavage: Distinct (prismatic)
Fracture: Sub-conchoidal
Lustre: Vitreous
Pleochroism: Strong (yellow, green, red)
Absorption spectrum: Brownish-green variety – band in violet, fainter band in blue. Green variety – band in violet plus six narrower bands in blue-green
Localities: Brazil, Sri Lanka, Spain (chiastolite from Australia, France, USSR, Zimbabwe)

Apatite
Composition: $Ca_5(F, Cl)(PO_4)_3$
Crystal system: Hexagonal
Habit: Prismatic (terminated by truncated bi-pyramids, see Figure 4.8)
Varieties: Transparent blue, green, yellow, violet
Refractive index: 1.63–1.64
Double refraction and optic sign: 0.003, negative

Dispersion: Low (0.013)
Specific gravity: 3.2
Hardness: 5.0
Cleavage: Poor (basal and prismatic)
Fracture: Conchoidal
Lustre: Vitreous
Pleochroism: Distinct in blue (blue, yellow)
Luminescence: Complex (see Table 12.1)
Absorption spectrum: Yellow-green stones show rare earth spectrum due to didymium, consisting of two groups of closely spaced lines in yellow (strong) and green (see Plate 10)
Localities: Brazil, Myanmar, Canada, Mexico, Sri Lanka, USA

Benitoite
Composition: $BaTiSi_3O_9$
Crystal system: Trigonal (maximum symmetry)
Habit: Trigonal bi-pyramid
Varieties: Transparent to translucent light blue, dark blue, colourless
Refractive index: 1.75–1.80
Double refraction and optic sign: 0.047, positive
Dispersion: High (0.046)
Specific gravity: 3.65–3.68
Hardness: 6.5
Cleavage: None
Fracture: Conchoidal
Lustre: Vitreous plus
Pleochroism: Strong (blue, colourless)
Luminescence: Strong blue fluorescence in SW UV (colourless stones may exhibit a dull red fluorescence under LW UV)
Locality: San Benito County, CA, USA

Beryl
Composition: $Be_3Al_2(SiO_3)_6$
Crystal system: Hexagonal
Habit: Prismatic (see Figure 4.8)
Varieties: Transparent to opaque; aquamarine, bixbite (red), emerald, vanadium beryl (green), goshenite (colourless), heliodor (yellow/golden), morganite (pink)
Refractive index: 1.56 to 1.59 depending on variety (see Appendix F) (most flux-melt synthetic emeralds have significantly lower RIs than natural emeralds, 1.560–1.563)
Double refraction: 0.005–0.008, negative, DR value depends on variety (see Appendix F) (most flux-melt synthetic emeralds have lower DR values than natural emeralds – 0.003)
Dispersion: Low (0.014)
Specific gravity: 2.7–2.8 depending on variety (see Appendix F)
Hardness: 7.5–8.0
Cleavage: Poor (basal)
Fracture: Conchoidal to uneven
Lustre: Vitreous

Pleochroism: Medium to weak (emerald, morganite, bixbite, vanadium beryl – shades of body colour; aquamarine – body colour and colourless)

Luminescence: Emerald – medium red to pink fluorescence under LW UV, SW UV and X-rays, but inhibited by iron oxide content. Morganite – crimson fluorescence under X-rays

Absorption spectrum: Emerald shows significant differences in spectrum between the ordinary and the extraordinary ray. The ordinary ray has a doublet in the deep red (680/683 nm), a line at 637 nm, a broad weak absorption band centered on 600 nm in the yellow, and in chrome-rich stones a line in the blue at 477 nm. In the extraordinary ray, the doublet is stronger but the 637 nm line is missing. In its place are two diffuse lines at 646 and 662 nm; the broad absorption band is much weaker and there are no lines in the blue (see Plate 10). Blue aquamarine has weak bands in the blue (456 nm) and violet (427 nm). Green aquamarine has a 537 nm band in the green in the extraordinary ray. Maxixe and maxixe-type dark-blue beryls have bands in the red (695, 654 nm) with weaker bands in the orange, yellow and yellow-green (628, 615, 581, 550 nm)

Inclusions: Emerald (see Chapter 16). Aquamarine – two-phase inclusions, 'rain' and mica (most stones are free of inclusions)

Localities: Aquamarine – Brazil, Myanmar, the Malagasy Republic, Namibia, Nigeria, USA, the former USSR. Bixbite – USA. Emerald and vanadium beryl – Afghanistan, Brazil, Colombia, India, Pakistan, South Africa, the former USSR, Zambia, Zimbabwe. Goshenite – USA. Heliodor – Brazil, the Malagasy Republic, Namibia, USA. Maxixe and maxixe-type – Brazil. Morganite – Brazil, the Malagasy Republic, USA, Zimbabwe

Bowenite

Composition: A hydrated magnesium silicate (a hard variety of serpentine)
Crystal system: Monoclinic
Habit: Massive, polycrystalline
Varieties: Translucent, yellow-green to blue-green often containing whitish patches
Refractive index: 1.56
Specific gravity: 2.58–2.59
Hardness: 4.0–5.0
Cleavage: None
Fracture: Even to hackly
Lustre: Waxy
Localities: Afghanistan, China, New Zealand, South Africa

Calcite

Composition: $CaCO_3$
Crystal system: Trigonal
Habit: Prismatic (profuse)
Varieties: Transparent colourless Iceland spar. Transparent brown and yellow gem quality calcite. Satin spar is a fibrous form of calcite. A banded stalagmatic dimorphous variety called aragonite is used for carving. Calcite is the basis of all true marbles and limestones. In its orthorhombic aragonite form it is a major constituent of pearls
Refractive index: 1.486–1.658

Double refraction and optic sign: 0.172, negative
Dispersion: Low (0.02)
Specific gravity: 2.71 (Iceland spar variety)
Hardness: 3.0
Cleavage: Perfect, rhombohedral (three directions)
Fracture: Conchoidal (rarely seen because of perfect cleavage)
Lustre: Vitreous
Luminescence: Variety of fluorescent effects
Inclusions: In Iceland spar variety, incipient cleavage planes
Localities: Worldwide

Chalcedony

Composition: SiO_2 (microcrystalline quartz)
Crystal system: Trigonal
Habit: Massive
Varieties: Agate (all colours with curved concentric bands). Aventurine (opaque with mica flake inclusions, green, golden brown). Bloodstone (opaque green with spots of red/brown jasper). Chalcedony (translucent unbanded grey and blue). Chrome chalcedony (translucent green). Chrysoprase (translucent to opaque green). Cornelian (translucent reddish-orange). Fire agate (translucent to opaque, reddish iridescent). Moss agate (translucent colourless chalcedony with dendritic green inclusions). Onyx (brown and white straight banded – also called sardonyx). Sard (translucent brownish-red)
Refractive index: 1.53 to 1.54 (a single vague shadow edge is visible on refractometer)
Specific gravity: 2.58 to 2.64
Hardness: 6.5
Cleavage: None
Fracture: Conchoidal
Lustre: Vitreous
Locality: Worldwide

Chrysoberyl

Composition: $BeAl_2O_4$
Crystal system: Orthorhombic
Habit: Tabular or prismatic; also trillings (repeated twinning producing pseudo-hexagonal crystal with re-entrant angles – see Figure 4.15)
Varieties: Alexandrite (transparent colour-change gem, green in daylight, red in filament light); transparent green, yellow and brown chrysoberyl; translucent yellowish chrysoberyl cat's-eye (cymophane)
Refractive index: 1.74–1.75
Double refraction and optic sign: 0.009, positive
Dispersion: Low (0.015)
Specific gravity: 3.72
Hardness: 8.5
Cleavage: Distinct (prismatic)
Fracture: Conchoidal to uneven
Lustre: Vitreous
Pleochroism: Strong in alexandrite (green, yellowish, pink – in daylight; red, orange-yellow, green – in filament light). Other colours – distinct (shades of body colour)

Luminescence: Red fluorescence in LW UV, weak red in SW UV

Absorption spectrum: Alexandrite has chromium lines in the red (doublet plus two other lines), a broad band in the yellow-green and two narrow bands in the blue (see Plate 10). Yellow chrysoberyl has a broad band at 444 nm in the blue (see Plate 10)

Inclusions: Chrysoberyl, two-phase inclusions; cat's-eye variety contains parallel tubes or needles. Alexandrite (see Chapter 16)

Localities: Alexandrite – Brazil, Myanmar, Sri Lanka, the former USSR, Zimbabwe. Other varieties – Brazil, Myanmar, the Malagasy Republic, Sri Lanka, Zimbabwe

Chrysocolla

Composition: A hydrated copper silicate
Crystal system: Monoclinic
Habit: Massive, microcrystalline
Varieties: Semi-translucent to opaque, green, blue. Eilat stone is a mixture of chryso-colla, turquoise and other copper minerals
Refractive index: 1.50
Specific gravity: 2.0–2.45 (Eilat stone 2.8–3.2)
Hardness: 2.0–4.0 (may approach 7 with quartz intergrowth)
Fracture: Even
Lustre: Vitreous
Localities: Chile, Peru, USA, USSR, Zaire (Eilat stone from near Eilat on the Gulf of Aqaba, in the Red Sea)

Corundum

Composition: Al_2O_3
Crystal system: Trigonal
Habit: Sapphire, barrel-shaped bi-pyramids with hexagonal cross-section. Ruby, tabu-lar prisms with hexagonal outline
Varieties: Ruby (pink to red) and sapphire (colourless, blue, orange – 'padparadscha', yellow, green, purple)
Refractive index: 1.76–1.77
Double refraction and optic sign: 0.008, negative
Dispersion: Low (0.018)
Specific gravity: 3.99
Hardness: 9.0
Cleavage: Poor (parting parallel to the basal and rhombohedral planes due to lamellar twinning)
Fracture: Conchoidal to uneven
Lustre: Vitreous to sub-adamantine
Pleochroism: Strong; ruby (deep red, orange-red). Medium; blue (blue, greenish blue), yellow and green sapphire (two shades of body colour), orange sapphire (orange, colourless), purple sapphire (violet and orange)
Luminescence: See Table 12.1
Absorption spectrum: Ruby (see Figure 11.1 and Plate 10), doublet in the deep red plus two further lines in the red, broad absorption band centred on 550 nm (this is broader and more intense in the ordinary ray direction), three lines in the blue (doub-let at 476/7 nm and line at 468 nm). With appropriate lighting, doublet and lines in red may be seen as emission rather than absorption lines. Sapphire, in most varieties some indication of an iron absorption band at 450 nm (see Plate 11). This is seen at

its strongest in the ordinary ray direction. In iron-rich stones, three absorption bands at 477, 460 and 450 nm (see Figure 11.1)

Inclusions: Vary widely with locality – see Chapter 16 under 'Ruby' and 'Sapphire'

Localities: Ruby from Afghanistan, Myanmar, Kampuchea (Cambodia), Kenya, Pakistan, Sri Lanka, Tanzania, Thailand. Sapphire from East Africa, Australia, Myanmar, Madagascar, Kashmir, Nigeria, Sri Lanka, Thailand, USA

Danburite

Composition: $CaB_2(SiO_4)_2$
Crystal system: Orthorhombic
Habit: Chisel-shape rhombic prism terminated by domes
Varieties: Transparent; colourless, pale yellow, pale pink
Refractive index: 1.63–1.64
Double refraction and optic sign: 0.006, negative
Dispersion: Low (0.017)
Specific gravity: 3.0
Hardness: 7.0
Cleavage: None
Fracture: Sub-conchoidal
Lustre: Vitreous
Luminescence: Sky blue fluorescence in LW and SW UV
Absorption spectrum: Some specimens show faint rare earth spectrum
Localities: Myanmar, Japan, the Malagasy Republic, Mexico, USA

Diamond

Composition: Carbon
Crystal system: Cubic
Habit: Octahedron mainly, but also occurs as cubes, dodecahedra and modifications of these (see Chapter 4, Figure 4.6). Twinned octahedra are called 'macles' and have re-entrant angles along the edges of the triangular sections (see Figure 4.15)
Varieties: Transparent; colourless and shades of yellow (cape series), brown (brown series) and green. Also 'fancy' shades of red (very rare), pink, orange, yellow, brown, blue, green. Industrial diamonds are generally of poor colour and quality, and are often microcrystalline (boart).

Type I diamonds contain nitrogen as the main impurity (Type Ia contains nitrogen atoms in clusters which do not affect the stone's colour; Type Ib contains nitrogen atoms dispersed throughout the crystal lattice which causes the yellow colour of Cape series diamonds). Natural diamonds are normally a mixture of Type Ia and Type Ib (synthetic diamonds containing nitrogen are all Type Ib)

Type Ia diamonds can be further subdivided into Type IaA (which contain nitrogen atoms in pairs) and Type IaB (which contain groups of three nitrogen atoms – the N_3 centres – and/or larger groups or aggregates of even numbers of nitrogen atoms (four or six)

Type II diamonds contain no nitrogen impurities (Type IIa contains no other impurity and is rare in nature; Type IIb diamonds are very rare and contain boron atoms which replace carbon atoms and make the stones electrically semi-conducting). While Type IIb diamonds are often blue, diamonds coloured blue by irradiation are, however, not electrically semi-conducting. Type II diamonds have significantly higher thermal conductance than Type I diamonds

Type III diamonds were discovered by Dame Kathleen Lonsdale (and named 'Lonsdaleite'). They have a hexagonal instead of a cubic crystal structure.

Refractive index: 2.417

Dispersion: High (0.044)

Specific gravity: 3.52

Hardness: 10

Cleavage: Perfect (octahedral – four directions)

Fracture: Conchoidal to irregular

Lustre: Adamantine

Luminescence: Variable fluorescence of various colours under LW UV. More consistent bluish-white, yellowish or greenish fluorescence under X-rays

Absorption spectrum: Colourless to yellow cape series stones have a band in the violet at 415.5 nm (plus other weak bands in the violet in strongly coloured stones – see Plate 10). Brown series stones have a band in the green at 504 nm (see Plate 10) sometimes with two weaker bands in this area

Inclusions: Crystal inclusions include hematite, diamond, diopside, enstatite, garnet, olivine, zircon. Black inclusions may be iron ores or graphite. Incipient cleavage and other cracks may be present

Localities: Western Australia, Brazil, Canada, China, Southern Africa (including Namibia), the former USSR

Diopside

Composition: $CaMgSi_2O_6$

Crystal system: Monoclinic

Habit: Tabular prismatic crystals with square cross-section

Varieties: Transparent to translucent; colourless, bottle-green, violet/blue, yellow and brownish; star and green chrome diopside

Refractive index: 1.67–1.70

Double refraction and optic sign: 0.027, positive

Dispersion: Very low

Specific gravity: 3.3

Hardness: 5.0

Cleavage: Good (prismatic – two directions at nearly 90°)

Fracture: Conchoidal

Lustre: Vitreous

Pleochroism: Weak to moderate

Luminescence: Variable

Absorption spectrum: Green chrome diopside has fine lines in the deep red and two narrow bands in the blue

Localities: Austria, Brazil, Myanmar, the Malagasy Republic, South Africa, Sri Lanka, USA

Dioptase

Composition: $CuSiO_3.H_2O$

Crystal system: Trigonal

Habit: Short prismatic crystals

Varieties: Transparent to translucent green
Refractive index: 1.644–1.709
Double refraction and optic sign: 0.053, positive
Dispersion: Medium (0.03)
Specific Gravity: 3.28–3.35
Hardness: 5
Cleavage: Perfect, rhombohedral
Fracture: Conchoidal to uneven
Lustre: Vitreous
Pleochroism: Weak
Luminescence: None
Absorption spectrum: Two bands in the yellow-green (560, 570 nm)
Localities: Chile, Namibia, USA, the former USSR, Zaire

Enstatite
Composition: $(Mg, Fe) SiO_3$
Crystal system: Orthorhombic
Habit: Prismatic
Varieties: Transparent to opaque, colourless, brownish-green, green, bronze; cat's-eye and star stones
Refractive index: 1.65 to 1.68
Double refraction and optic sign: 0.01, positive
Dispersion: Weak
Specific Gravity: 3.25–3.30
Hardness: 5.5
Cleavage: Distinct, prismatic (two directions at nearly 90°)
Fracture: Uneven
Lustre: Vitreous
Pleochroism: Strong in brown stones, weaker in green stones (two shades of body colour)
Luminescence: None
Absorption spectrum: Band at 506 nm in blue-green (see Plate 11) with adjacent fine lines in brown stones, and/or doublet in red in chrome-coloured stones
Localities: Austria (bronze stones), Myanmar, India (star stones), South Africa, Sri Lanka (cat's-eye stones)

Feldspar
Composition: Potassium feldspars (including the polymorphs orthoclase, microcline and sanidine), $KAlSi_3O_8$; sodium and calcium feldspars (plagioclase, an isomorphous series), $(Ca, Na) Al_2Si_2O_8$
Crystal system: Sanidine and orthoclase – monoclinic. Microcline and plagioclase – triclinic
Habit: Prismatic (see Chapter 4, Figures 4.11, 4.12); labradorite, compact aggregates
Varieties: Orthoclase (transparent to translucent pale yellow). Sanidine (rare; transparent colourless to brownish). Microcline (amazonite, opaque green). Plagioclase (labradorite, opaque iridescent sheen – Finland variety called spectrolite; also rare transparent yellowish and brownish variety). Moonstone, translucent white, cream with sheen (caused by alternate layers with orthoclase). Oligoclase, transparent yellow.

Sunstone or aventurine feldspar, opaque to translucent bronze or green spangled. Chrome albite, known as Maw-sit-sit, opaque green with black markings, containing chrome-rich jadeite

Refractive index: Orthoclase, 1.52–1.53. Plagioclase (oligoclase, moonstone, sunstone, chrome albite), 1.527 to 1.553; labradorite, 1.56–1.57

Double refraction and optic sign: Orthoclase, 0.006, negative. Plagioclase (all except labradorite), 0.007, positive or negative; labradorite, 0.009, positive

Dispersion: Low

Specific gravity: Orthoclase, 2.56. Plagioclase (all except labradorite), 2.6–2.65; labradorite, 2.7

Hardness: 6.0

Cleavage: Perfect in two near 90° directions, distinct in third direction

Fracture: Conchoidal to uneven

Lustre: Vitreous (sometimes pearly on cleavage faces)

Pleochroism: Weak

Localities: Orthoclase – the Malagasy Republic. Microcline – Brazil, India, USA, the former USSR. Plagioclase – Canada (labradorite), Finland (specularite), Myanmar (chrome albite), Norway (sunstone), USA, the former USSR

Fluorite (also called fluorspar)

Composition: CaF_2

Crystal system: Cubic

Habit: Cubes, octahedra (cleaved cubes), interpenetrant cubes (see Chapter 4, Figures 4.6, 4.15)

Varieties: Transparent to opaque, colourless, blue, violet, green, yellow, orange, pink; 'Blue John' is a banded purple/white variety

Refractive index: 1.434

Dispersion: Very low (0.007)

Specific gravity: 3.18

Hardness: 4.0

Cleavage: Perfect, octahedral (four directions)

Fracture: Sub-conchoidal

Lustre: Sub-vitreous

Luminescence: Weak to strong blue-violet in LW UV (weaker in SW UV); none in Blue John variety

Absorption spectrum: Band in yellow-red at 585 nm in green variety; rare earth spectrum sometimes present

Inclusions: Two and three phase; incipient cleavage planes

Localities: Worldwide; 'Blue John' from Derbyshire, England

Garnet

Composition: The garnet group consists of two isomorphic series. The pyralspite series consists of: almandine, $Fe_3Al_2(SiO_4)_3$; pyrope, $Mg_3Al_2(SiO_4)_3$; spessartite, $Mn_3Al_2(SiO_4)_3$. The ugrandite series consists of: andradite, $Ca_3Fe_2(SiO_4)_3$; grossular, $Ca_3Al_2(SiO_4)_3$; uvarovite, $Ca_3Cr_2(SiO_4)_3$.

As normal with an isomorphous series, the individual species listed above represent the ideal pure states. Extensive substitutions between these can occur and will affect their physical properties

Crystal system: Cubic
Habit: Dodecahedral; icositetrahedral less common (see Chapter 4, Figure 4.6)
Varieties: Almandine (transparent brownish-red to purplish-red). Andradite (transparent
   yellow; demantoid, transparent green; melanite, black). Grossular (hessonite, transparent to translucent brownish-yellow to brownish-red; hydrogrossular – composition
   includes OH – translucent jade – green, pink grey-white; tsavorite, transparent green
   coloured by vanadium and chromium). Pyrope (transparent red, purplish-red).
   Spessartite (transparent orange, yellow, red). Uvarovite (emerald-green)

Variety	Refractive index	Dispersion	Specific gravity	Hardness
Almandine	1.76 to 1.81	0.024	3.8–4.2	7.5
Andradite	1.89	0.057	3.85	6.5
Grossular	1.74 to 1.75	0.028	3.6–3.7	7.0–7.5
Pyrope	1.74 to 1.76	0.022	3.7–3.8	7.25
Spessartite	1.80 to 1.82	0.027	4.16	7.25
Uvarovite	1.87	0.030	3.77	7.5

Fracture: Sub-conchoidal
Lustre: Vitreous to sub-adamantine (andradite)
Luminescence: None except hydrogrossular (orange in X-rays), which together with
   demantoid variety of andradite may show pink through Chelsea filter
Absorption spectrum: See Plates 10 and 11 for almandine and demantoid garnet, and
   Figure 11.1 for pyrope and almandine. The pyrope spectrum is similar to that of
   almandine, but with the three bands in the green, yellow, and orange merging, and
   (in stones containing chromium) a doublet in the red. Spessartite has two very
   strong bands in the violet and may have faint almandine bands. Some grossulars
   have bands in the red and orange. Demantoid garnet has a strong iron absorption
   band in the violet which cuts off this end of the spectrum (chrome rich stones have
   a doublet in the red and may have two vague bands in the orange)
Inclusions: Almandine, crystals and associated stress cracks; acicular crystals of
   rutile. Demantoid variety of andradite – 'horsetail' inclusion consisting of radiating
   byssolite (asbestos) fibres. Grossular (particularly hessonite variety) – profusion
   of small crystals of apatite and/or zircon giving a treacly appearance. Pyrope –
   occasional needle-like crystals. Spessartite – shredded-looking feathers
Localities: Almandine – Austria, Brazil, China, Czechoslovakia, Sri Lanka. Andradite –
   Switzerland (yellow), USSR (demantoid). Grossular – Brazil, Canada, Kenya
   (tsavorite), New Zealand, South Africa (hydrogrossular), Sri Lanka (especially hessonite), Tanzania (tsavorite), USA. Pyrope – Czechoslovakia, South Africa, Sri Lanka,
   USA, the former USSR. Spessartite – Brazil, the Malagasy Republic, Myanmar,
   Sri Lanka, USA. Uvarovite – Canada, Finland, Poland, USA, the former USSR

Hematite

Composition: Fe_2O_3
Crystal system: Trigonal
Habit: Tabular or rhombohedral crystals, and massive/botryoidal
Varieties: Opaque, grey to reddish-black ('Jeweller's rouge' is powdered hematite)
Refractive index: 2.94–3.22
Double refraction and optic sign: 0.28, negative
Specific gravity: 5.0

Hardness: 5.5–6.5
Fracture: Uneven
Lustre: Metallic
Localities: Brazil, England, France, Italy (Elba), Switzerland

Idocrase (vesuvianite)

Composition: Complex calcium aluminium silicate
Crystal system: Tetragonal
Habit: Prismatic terminated by truncated pyramids, and massive
Varieties: Transparent olive green and yellowish brown; californite, translucent to opaque green (resembling jade)
Refractive index: 1.70 to 1.723
Double refraction and optic sign: 0.005, negative (lower RI values), positive (higher RI values)
Dispersion: Low (0.019)
Specific gravity: 3.32–3.42
Hardness: 6.5
Cleavage: Poor
Fracture: Sub-conchoidal
Lustre: Vitreous
Pleochroism: Distinct (shades of body colour)
Absorption spectrum: Green stones – strong band in blue at 461 nm. Brown stones – strong band in yellow at 584 nm (rare earth lines may be present)
Localities: Austria, Canada, Italy, Switzerland, USA (Californite), the former USSR

Iolite (cordierite)

Composition: Complex silicate of magnesium and aluminium
Crystal system: Orthorhombic
Habit: prisms (rare), massive
Varieties: Transparent to translucent colourless (rare) to blue
Refractive index: 1.53 to 1.55
Double refraction and optic sign: 0.009, negative
Dispersion: Low (0.017)
Specific gravity: 2.57–2.61
Hardness: 7.5
Cleavage: Distinct, basal
Fracture: Sub-conchoidal
Lustre: Vitreous
Pleochroism: Strong (blue, violet, pale yellow)
Absorption spectrum: Two bands in the yellow, a narrow band in the green (in the 'yellow' ray direction); a broader band and two narrow ones in the blue to violet (in the 'blue' ray direction)
Inclusions: Thin plates of iron oxide in Sri Lankan stones (when profuse they produce 'bloodshot' iolite)
Localities: Brazil, India, the Malagasy Republic, Myanmar, Sri Lanka

Jadeite (classified as a jade together with nephrite)

Composition: $NaAl(SiO_3)_2$
Crystal system: Monoclinic

Habit: Polycrystalline (microscopic interlocking fibrous crystals)
Varieties: Translucent to opaque white and shades of green, pink, lilac, violet, brown, black (translucent emerald green is the most valuable)
Refractive index: 1.66 to 1.68 (only single vague shadow edge visible on refractometer at 1.67 due to random orientation of crystal fibres)
Specific gravity: 3.30–3.36
Hardness: 6.5–7.0
Fracture: Splintery
Lustre: Greasy to vitreous
Absorption spectrum: Diagnostic line at 437 nm in the blue; chrome-rich green jadeite has a doublet in the red, and two bands in the red to yellow. Stained jadeite has a band in the orange and one in the yellow-green (plus the diagnostic line at 437 nm)
Localities: Myanmar, USA

Kornerupine
Composition: Complex borosilicate of aluminium, iron and magnesium
Crystal system: Orthorhombic
Habit: Prismatic
Varieties: Transparent to translucent (occasional chatoyancy and asterism) green and greenish brown
Refractive index: 1.67–1.68
Double refraction and optic sign: 0.013, negative
Dispersion: Low (0.019)
Specific gravity: 3.28–3.35
Hardness: 6.5
Cleavage: Prismatic
Fracture: Conchoidal
Lustre: Vitreous
Pleochroism: Distinct (green, yellow, brown)
Absorption spectrum: Band in the blue-green at 503 nm and a strong band at 446 nm in the violet
Localities: East Africa, the Malagasy Republic, Myanmar, South Africa, Sri Lanka

Lapis lazuli
Composition: A rock consisting of lazurite, hauynite, calcite and pyrite
Varieties: Opaque purple-blue, mottled blue and white, with flecks of pyrite
Refractive index: 1.5
Specific gravity: 2.7–2.9
Hardness: 5.5
Fracture: Uneven
Lustre: Resinous to vitreous
Luminescence: Greenish or whitish fluorescence in SW UV; orange spots (due to calcite) in LW UV
Localities: Afghanistan, Chile, the former USSR

Malachite
Composition: $Cu_2(OH)_2CO_3$
Crystal system: Monoclinic

Habit: Massive (aggregate of fine needles), mamillary
Varieties: Opaque light and dark green with concentric banding; azurmalachite is an intergrowth of malachite and azurite
Refractive index: 1.66–1.91 (usually a single shadow edge around 1.80 on refractometer)
Double refraction and optic sign: 0.26, negative
Specific gravity: 3.8
Hardness: 4.0
Fracture: Splintery, uneven
Lustre: Silky to vitreous
Localities: Australia, Chile, South Africa, USA, the former USSR, Zaire, Zimbabwe

Moonstone

Composition: Consists of alternate layers of two feldspar minerals, orthoclase ($KAlSi_3O_8$) and plagioclase ($NaAlSi_3O_8$) producing sheen
Crystal system: Monoclinic
Habit: Prismatic (cleavage fragments)
Varieties: Transparent to translucent colourless, white, pink, yellow with blue or bluish-violet sheen with varying degrees of chatoyancy
Refractive index: 1.52–1.53
Double refraction: 0.006
Dispersion: Low (0.012)
Specific gravity: 2.56
Hardness: 6.0–6.5
Cleavage: Perfect in two near 90° directions
Fracture: Uneven
Lustre: Vitreous
Inclusions: Intersecting stress cracks which look like a centipede
Localities: Mainly Sri Lanka, but also India, the Malagasy Republic, Myanmar, USA

Natural glasses

Composition: Silica glasses containing various elements (because of their high silica content they are classed as tektites)
Non-crystalline. Varieties: Moldavite (transparent to translucent green, brown) consisting of 75% silica. Obsidian (transparent to opaque black, grey, brown, green), a volcanic glass consisting of 70% silica); snowflake obsidian contains white patches on a black background; silver and golden obsidian has an iridescence caused by minute bubbles or inclusions. Basalt glass (semi-transparent to opaque black, grey-brown, dark blue, bluish-green) containing 50% silica. Libyan glass (transparent to translucent greenish-yellow) a nearly pure silica glass

Type	Refractive index (approx.)	Specific gravity	Hardness
Moldavite	1.50	2.4	5.5
Obsidian	1.50	2.3–2.5	5.0
Basalt glass	1.61	2.7–3.0	6.0
Libyan glass	1.46	2.2	6.0

Fracture: Conchoidal
Lustre: Vitreous
Inclusions: Moldavite – bubbles and swirls. Obsidian – bubbles, small crystals, dendritic crystallites and swirls
Localities: Moldavite – Czechoslovakia. Obsidian – North America, Mexico, the former USSR. Basalt glass – worldwide. Libyan glass – Libyan dessert

Nephrite (classified as a jade together with jadeite)
Composition: Complex calcium magnesium/iron silicate
Crystal system: Monoclinic
Habit: Polycrystalline (microscopic interlocking fibrous crystals)
Varieties: Translucent to opaque white, green, grey, yellowish, brown, black
Refractive index: 1.62 (only single vague shadow edge on refractometer visible due to random orientation of crystal fibres)
Specific gravity: 2.9–3.2
Hardness: 6.5
Fracture: Splintery
Lustre: Greasy to vitreous
Absorption spectrum: Indistinct, a band may be present at 509 nm in the blue/green
Localities: Canada, China, New Zealand, Taiwan, USA, the former USSR

Opal
Composition: $SiO_2.H_2O$ (a hydrated silica containing up to 10% water in chemical combination)
Non-crystalline. Varieties: Common or 'potch' opal, opaque whitish without iridescent 'play of colour'. White opal (white background with iridescence). Black opal (dark background with iridescence). Fire opal (transparent to translucent orange, occasionally with iridescence). Water opal (translucent to transparent colourless or pale brownish-yellow with iridescence). Hyalite (transparent, colourless without iridescence). Hydrophane (opaque light-coloured which becomes transparent and iridescent when soaked in water). Pink, yellow, green and blue opal (translucent but with no iridescence). Materials petrified by opal (e.g. wood), opal pseudomorphs such as shells, bones and minerals
Refractive index: 1.45 (fire opal 1.40)
Specific gravity: White and black opal 2.1. Fire opal 2.0
Hardness: 6.0
Lustre: Vitreous
Luminescence: See Table 12.1
Localities: Mainly New South Wales and Queensland in Australia, Brazil, Mexico (fire opal), USA

Peridot
Composition: $(Mg, Fe)_2SiO_4$
Crystal system: Orthorhombic
Habit: Prismatic (complete crystals are rare). See Figure 4.10
Varieties: Transparent yellow-green, olive-green, brown (rare)
Refractive index: 1.65–1.69

Double refraction and optic sign: 0.036, positive
Dispersion: Medium (0.02)
Specific gravity: 3.34
Hardness: 6.5
Cleavage: Poor
Fracture: Conchoidal
Lustre: Vitreous
Pleochroism: Weak (green, yellowish-green)
Absorption spectrum: Three bands in the blue (see Plate 11). This spectrum is strongest in the β-ray direction.
Inclusions: Chromite crystals producing flat stress cracks termed 'lily pads'; mica flakes and minute bubble-like droplets
Localities: Main sources are Zeberget (St John's Island) in the Red Sea and Upper Myanmar; also Australia, Brazil, China, Mexico, Norway, USA

Phenakite
Composition: Be_2SiO_4
Crystal system: Trigonal
Habit: Prismatic and tabular
Varieties: Transparent colourless, pink, pale yellow, greenish
Refractive index: 1.65–1.67
Double refraction and optic sign: 0.016, positive
Dispersion: Low (0.015)
Specific gravity: 2.95–2.97
Hardness: 7.5
Cleavage: Distinct, prismatic
Fracture: Conchoidal
Lustre: Vitreous
Pleochroism: Definite (colourless, orange-yellow)
Localities: Africa, Brazil, Mexico, USA, the former USSR

Pyrite
Composition: FeS_2
Crystal system: Cubic
Habit: Cube, octahedron, dodecahedron (see Figure 4.6). Striations alternate in direction on adjacent faces of cube
Varieties: Opaque brass yellow; although pyrite is called 'marcasite' when used in jewellery, the true marcasite is a polymorph of pyrite and powders when exposed to air
Refractive index: Over 1.81
Specific gravity: 4.84–5.10
Hardness: 6.5
Fracture: Conchoidal to uneven
Lustre: Metallic
Localities: Worldwide

Quartz
Composition: SiO_2
Crystal system: Trigonal

Habit: Prismatic (hexagonal, terminated in positive and negative rhombohedra – very fine growth striations at right angles to length of prism – see Figures 4.9 and 4.14); rose quartz variety massive (prisms rare)

Varieties: Rock crystal (transparent colourless). Amethyst (transparent purple to pale violet). Ametrine (bi-coloured amethyst-citrine). Citrine (transparent yellow). Rose quartz (translucent pink). Blue quartz (translucent to opaque – colour caused by bluish rutile needles. Green natural and heat-treated quartz (marketed as prasiolite). Milky quartz (translucent white). Smoky quartz (transparent brown to black). Quartz cat's-eye (translucent containing rutile). Tiger's-eye (opaque golden brown; chatoyant effect caused by the replacement of crocidolite asbestos fibres with quartz – also called pseudo-crocidolite. Blue material is called 'falcon's-eye' and a brown/blue mix is called 'hawk's-eye'). Jasper (an impure opaque polycrystalline quartz, reddish-brown, green, pink, yellow). Silicified wood (replacement of wood cell structure by quartz in the same way that quartz replaces asbestos fibres to form tiger's-eye. Microcrystalline quartz – see under chalcedony

Refractive index: 1.544–1.553

Double refraction and optic sign: 0.009, positive

Dispersion: Low (0.013)

Specific gravity: 2.65 (jasper 2.58–2.91)

Hardness: 7.0 (jasper 6.5–7.0)

Fracture: Conchoidal

Lustre: Vitreous

Pleochroism: Weak (most varieties) to definite (smoky quartz)

Inclusions: See Chapter 16

Localities: Worldwide except for tiger's-eye, falcon's-eye and hawk's-eye which come mainly from South Africa, quartz cat's-eye which comes mainly from Sri Lanka and India, and the bi-coloured ametrine which comes from Bolivia.

Rhodochrosite

Composition: $MnCO_3$

Crystal system: Trigonal

Habit: Massive (rhombohedral crystals rare)

Varieties: Transparent rose red crystals; opaque massive aggregates with light/dark red banding)

Refractive index: 1.60–1.82

Double refraction and optic sign: 0.22, negative

Specific gravity: 3.5–3.6

Hardness: 4.0

Cleavage: Perfect, rhombohedral (three directions)

Fracture: Uneven

Lustre: Vitreous

Localities: Argentina (massive), South Africa (transparent)

Rhodonite

Composition: $MnSiO_3$

Crystal system: Triclinic

Habit: Massive (well formed crystals rare)

Varieties: Transparent red (crystals) to opaque red aggregates with black dendritic inclusions of manganese oxide

Refractive index: 1.72–1.74
Double refraction and optic sign: 0.014, positive or negative
Specific gravity: 3.4–3.7
Hardness: 6.0
Cleavage: Perfect (two directions), good (one direction)
Fracture: Uneven
Lustre: Vitreous
Pleochroism: Definite (yellow-red, rose-red, red-yellow)
Localities: Australia (transparent), Mexico, South Africa, Sweden, USA, the former USSR

Rutile

Composition: TiO_2
Crystal system: Tetragonal
Habit: Prismatic and massive (twinning common – knee shaped twins)
Varieties: Transparent to translucent red, red-brown, yellowish, black
Refractive index: 2.61–2.90
Double refraction and optic sign: 0.287, positive
Dispersion: Very high (0.28)
Specific gravity: 4.2–4.3
Hardness: 6.5
Cleavage: Distinct, prismatic
Fracture: Uneven
Lustre: Adamantine
Localities: Brazil, France, Italy, the Malagasy Republic, USA, the former USSR

Scapolite

Composition: Complex sodium calcium aluminium silicate (part of an isomorphous series whose end members are marialite and meionite)
Crystal system: Tetragonal
Habit: Prismatic and massive
Varieties: Transparent colourless, pink, yellow, blue-violet
Refractive index: 1.55–1.58 (yellow); 1.545–1.560 (colourless, pink, blue-violet)
Double refraction and optic sign: 0.02 (yellow), negative; 0.016 (colourless, pink, blue-violet), negative
Dispersion: Low (0.017)
Specific gravity: 2.74 (yellow); 2.634 (colourless, pink, blue-violet)
Hardness: 6.0
Cleavage: Perfect (in two directions); distinct (in two other directions)
Fracture: Conchoidal
Lustre: Vitreous
Pleochroism: Strong in pink and violet stones (pink, blue), medium in yellow stones (yellow and colourless)
Luminescence: Yellow stones fluoresce pale yellow in LW UV and pink in SW UV
Absorption spectrum: Pink and blue-violet stones have two bands in the red and a broad band in the yellow
Localities: Brazil, the Malagasy Republic, Mozambique, Myanmar, Sri Lanka

Sinhalite
Composition: Mg(Al, Fe)BO$_4$
Crystal system: Orthorhombic
Habit: Massive (well-shaped crystals are rare)
Varieties: Transparent to translucent yellow, greenish-brown, brown
Refractive index: 1.67–1.71
Double refraction and optic sign: 0.038, negative
Dispersion: Low (0.018)
Specific gravity: 3.48
Hardness: 6.5
Fracture: Conchoidal
Lustre: Vitreous
Pleochroism: Definite (green, light brown, dark brown)
Absorption spectrum: Similar to peridot (see Plate 11), but with a fourth band in the green-blue to blue. (Sinhalite was classed as a peridot until 1952 when it was iden-tified as a separate species)
Locality: Sri Lanka

Sodalite
Composition: Complex sodium aluminium silicate with chlorine
Crystal system: Cubic
Habit: Massive
Varieties: Opaque to translucent blue, grey veined with white
Refractive index: 1.48
Specific gravity: 2.28
Hardness: 5.5–6.0
Cleavage: Distinct, dodecahedral
Fracture: Conchoidal to uneven
Lustre: Vitreous
Localities: Brazil, Canada, Namibia, USA

Sphene (titanite)
Composition: CaTiSO$_5$
Crystal system: Monoclinic
Habit: Wedge-shaped, flattened and twinned crystals with re-entrant angles, massive
Varieties: Transparent to translucent yellow, brown, green
Refractive index: 1.89–2.02
Double refraction and optic sign: 0.13, positive
Dispersion: High (0.051)
Specific gravity: 3.53
Hardness: 5.5
Cleavage: Distinct (prismatic – two directions)
Fracture: Conchoidal
Lustre: Resinous to sub-adamantine
Pleochroism: Distinct in green stones (colourless, green), strong in yellow stones (colourless, yellow, reddish)
Absorption spectrum: Bands in the orange at 582 and 586 nm; may show rare earth lines (a weak didymium spectrum)
Localities: Austria, Brazil, Canada, the Malagasy Republic, Mexico, Sri Lanka, Switzerland, USA

Spinel

Composition: MgO.Al$_2$O$_3$ (synthetic spinel – MgO.3Al$_2$O$_3$)
Crystal system: Cubic
Habit: Octahedrons and spinel twins (see Figures 4.6 and 4.15)
Varieties: Transparent wide range of colours from near-colourless (rare), through red, yellowish, green to blue-violet. Gahnospinel is a blue spinel containing a high proportion of zinc (in gahnite, zinc has entirely replaced magnesium)
Refractive index: 1.712 to 1.720 (red synthetic spinel, 1.72–1.73; other synthetic spinel colours, 1.727). 1.725 to 1.753 (gahnospinel)
Dispersion: Medium (0.02)
Specific gravity: 3.60 (and synthetic red spinel; other synthetic spinel colours, 3.64); gahnospinel 3.58–4.06 (up to 4.60 for gahnite)
Hardness: 8.0
Fracture: Conchoidal to uneven
Lustre: Vitreous
Luminescence: See Table 12.1
Absorption spectrum: Red and pink spinel show a chromium spectrum of up to eight lines in the red (described as an 'organ pipe' spectrum) which may be fluorescent (see Plate 10 and Figure 11.1). Blue spinel (Plate 11); some rare blue spinels are coloured by cobalt (like synthetic blue spinels – see Figure 11.1) and show a faint cobalt spectrum
Inclusions: Minute octahedra, iron-stained feathers and (in Sri Lankan stones) zircon haloes. (Synthetic red spinel – strong curved colour banding; see Chapter 16)
Localities: Brazil, Cambodia, Sri Lanka, Myanmar (Burma), Thailand, USA, the former USSR (pink)

Spodumene

Composition: LiAl(SiO$_3$)$_2$
Crystal system: Monoclinic
Habit: Prismatic, tabular (arrow-head surface markings on crystals)
Varieties: Transparent pink (kunzite variety), yellow, green (hiddenite variety)
Refractive index: 1.66–1.68
Double refraction and optic sign: 0.015, positive
Dispersion: Low (0.017)
Specific gravity: 3.18
Hardness: 7.0
Cleavage: Perfect (prismatic, two directions at nearly 90°)
Fracture: Uneven
Lustre: Vitreous
Pleochroism: Strong, pink stones (violet, pale red, colourless); green stones (blue-green, emerald-green, yellow-green)
Luminescence: Strong yellow-red to orange with pink stones in LW UV
Absorption spectrum: Green hiddenite variety shows a chromium spectrum with lines in the red and yellow. Yellow spodumene has two bands in the blue-violet (see Plate 11)
Localities: Kunzite – Afghanistan, Myanmar, Brazil, the Malagasy Republic. Hiddenite – USA

Steatite (soapstone)

Composition: Hydrous magnesium silicate
Crystal system: Monoclinic

Habit: Massive
Varieties: Translucent to opaque yellow, greenish, brown, reddish; basic mineral constituent is talc
Refractive index: 1.55
Specific gravity: 2.5–2.8
Hardness: 1.0 (often harder due to impurities)
Fracture: Even
Lustre: Waxy
Localities: Canada, Central Africa, India, Zimbabwe

Topaz

Composition: $Al_2(F, OH)_2SiO_4$
Crystal system: Orthorhombic
Habit: Prismatic, with dome and/or pyramidal termination
Varieties: Transparent colourless, red, pink, orange, brown, yellow, blue, green (rare)
Refractive index: 1.61–1.62* (colourless, brown, blue, yellow). 1.63–1.64* (red, pink, orange)
Double refraction and optic sign: 0.01 (colourless, brown, blue yellow), positive; 0.008 (red, pink, orange), positive
Dispersion: Low (0.014)
Specific gravity: 3.53–3.56
Hardness: 8.0
Cleavage: Perfect, basal
Fracture: Sub-conchoidal to uneven
Lustre: Vitreous
Pleochroism: Distinct to strong depending on depth of body colour
Luminescence: Weak; red, pink, light brown and yellow stones may fluoresce orange-yellow in LW UV
Absorption spectrum: In red and pink topaz there is a band in the red at 683.8 nm
Inclusions: Cavities containing two immiscible liquids, and long tube-like cavities
Localities: Africa and Asia, with Brazil as most important source

Tourmaline

Composition: Complex borosilicate of aluminium, magnesium and iron
Crystal system: Trigonal
Habit: Three-sided prism with convex faces heavily striated along prism length (see Figure 4.9)
Varieties: Transparent to translucent single and parti-coloured (lengthwise or radially); alkali-rich stones (containing sodium, lithium or potassium) are colourless, red or green; iron-rich stones are dark blue, bluish-green or black; manganese stones are colourless or yellow-brown to brownish-black
Refractive index: 1.62 to 1.66
Double refraction and optic sign: 0.014–0.021, negative
Dispersion: Low (0.018)
Specific gravity: 3.01–3.11 (black 3.15–3.26)

---

* According to Hoover, the RI of topaz is dependent upon its fluorine content and not its colour. Generalizations associating the colour and RIs of topaz should, therefore, be avoided, and its DR taken as centred around 0.009 regardless of fluorine content.

Hardness: 7.0–7.5
Fracture: Conchoidal
Lustre: Vitreous
Pleochroism: Weak to strong depending on depth of colour (two shades of body colour)
Absorption spectrum: Blue and green stones have a strong band in the green at 498 nm, and a weak one in the blue at 468 nm (see Plate 11). Pink and red stones have a broad absorption band in the green and two narrow bands in the blue at 450 nm and 458 nm (see Plate11)
Inclusions: Irregular thread-like cavities and flat films
Localities: Brazil, East Africa, the Malagasy Republic, Namibia, Myanmar, Sri Lanka, USA, the former USSR

Turquoise

Composition: $CuAl_6(PO_4)_4(OH)_8.5H_2O$
Crystal system: Triclinic
Habit: Botryoidal, massive, granular to microcrystalline
Varieties: Opaque (sometimes lined with sandstone or limonite matrix), sky-blue, blue-green, green
Refractive index: 1.61 to 1.65 (only single shadow edge at 1.62 normally seen on refractometer)
Specific gravity: 2.6–2.9 depending on source
Hardness: 5.5–6.0
Fracture: Conchoidal
Lustre: Waxy
Absorption spectrum: Lines in the blue-violet at 420 and 432 nm may be visible
Localities: Afghanistan, China (Tibet), Iran, Israel (Sinai), USA

Zircon

Composition: $ZrSiO_4$
Crystal system: Tetragonal
Habit: Prismatic; square prisms terminated by pyramids (when prism section very short, these resemble octahedra – see Figure 4.7)
Varieties: Transparent yellow, green, brown, red, orange, blue, golden brown, colourless (reddish-brown stones from Indo-China are heat treated to produce blue, golden brown and colourless). Because of naturally occurring alpha particle irradiation from trace amounts of isomorphous uranium and/or thorium, some zircons have their internal crystalline structure partially broken down into near-amorphous silica and zirconia. For this reason zircons can be arbitrarily divided into 'high', 'intermediate' and 'low' (or metamict) types, although these simply represent phases in the breakdown of the crystalline structure. Metamict zircons are virtually non-crystalline and usually brown or green stones. The degree of radioactive damage also affects the constants of zircon. Because of the presence of traces of uranium and/or thorium in most zircons, they can produce 'autoradiographs' (i.e. self-photographs) if placed on photographic paper for some hours
Refractive index: 1.93–1.99 high zircon; 1.84–1.93 intermediate zircon; 1.78–1.84 low zircon
Double refraction: 0.059, positive, high zircon. Less for intermediate and approaching zero for low type
Dispersion: High (0.039)

Specific gravity: 3.9–4.68 (low to high types)

Hardness: 6.0–7.25 (low to high types). Zircon is brittle which causes 'paper wear' damage when loose zircons are placed together in a stone paper and allowed to abrade each other

Cleavage: Poor

Lustre: Vitreous to sub-adamantine

Pleochroism: Weak except in heat-treated blue stones (blue, colourless)

Absorption spectrum: Diagnostic line at 653.5 nm, often with other lines right across spectrum (uranium spectrum – see Plate 10 and Figure 11.1)

Localities: Australia, Cambodia, France (red zircon), Indo-China, Sri Lanka, Thailand

Zoisite

Composition: $Ca_2(OH)Al_3(SiO_4)_3$

Crystal system: Orthorhombic

Habit: Prismatic (vertically striated) and massive

Varieties: Transparent blue-violet (tanzanite), opaque green (often surrounding ruby crystals) and pink (thulite)

Refractive index: 1.69–1.70

Double refraction and optic sign: 0.009, negative

Dispersion: Low (0.012)

Specific gravity: 3.35

Hardness: 6.5

Cleavage: Perfect, prismatic

Fracture: Uneven

Lustre: Vitreous

Pleochroism: Strong in blue (violet, green, blue in untreated stones; blue and violet in heat treated stones)

Absorption spectrum: A broad band in the yellow-green and two fainter bands, one in the green at 528 nm, and one in the blue at 455 nm

Localities: Tanzania (blue zoisite), Norway, Australia and South Africa (thulite)

Appendix D Man-made diamond simulants

Chemical composition, constants, characteristics, synthesis process

Cubic zirconium oxide

Composition: ZrO_2

Crystal system: Cubic

Colour: Colourless (diamond simulant), deep red (coloured by cerium), pink (erbium), yellow, orange-brown (cerium), green, lilac (neodymium), blue, purple; some colours produced by transition elements

Crystal form: Columnar with curved faces

Refractive index: 2.15 to 2.18 (depending on amount of stabilizer)

Dispersion: Very high (0.065)

Specific gravity: 5.6–6.0 (depending on amount of stabilizer)

Hardness: 8.5

Lustre: Adamantine

Luminescence: Green-yellow to yellow in SW UV, whitish in X-rays

Absorption spectrum: Strong rare earth spectrum in pink, lilac and orange-brown stones; deep red stones (coloured by cerium) have a strong band in the yellow at

575 nm, and a complete absorption of the spectrum from 560 nm to the violet end; purple stones have three bands in the orange, yellow and green
Synthesis process: Skull crucible

Gadolinium gallium garnet (GGG)

Composition: $Gd_3Ga_5O_{12}$
Crystal system: Cubic
Crystal form: 3-inch (7.5 cm) diameter columnar (originally produced as substrate for computer bubble memories)
Colour: Colourless (diamond simulant), red (manganese), yellow (praesidium), lilac (neodymium), green (chromium or cobalt), red (manganese)
Refractive index: 1.97
Dispersion: High (0.045)
Specific gravity: 7.05
Hardness: 6.0 (brittle)
Lustre: Sub-adamantine
Luminescence: Colourless stones fluoresce peach colour in SW UV (if exposed for several minutes stones turn brown, but return to colourless after several hours). Stones fluoresce lilac in X-rays
Absorption spectrum: Coloured stones show the appropriate transition element or rare earth spectrum
Synthesis process: Czochralski crystal pulling

Lithium niobate

Composition: $LiNbO_3$
Crystal system: Trigonal
Crystal form: Large columnar
Colour: Colourless (diamond simulant), red (iron), yellow (nickel or manganese), green (chromium), blue-violet (cobalt)
Refractive index: 2.21–2.30
Double refraction and optic sign: 0.09, negative
Dispersion: Very high (0.13)
Specific gravity: 4.64
Hardness: 5.5
Lustre: Adamantine
Absorption spectrum: For coloured stones, typical of dopant
Synthesis process: Czochralski crystal pulling

Synthetic moissanite

Composition: SiC
Crystal system: Hexagonal
Crystal form: Boule
Colour: Colourless, greyish green, greyish yellow, grey, greenish yellow, brownish yellow, yellow green
Refractive index: 2.65–2.69
Double refraction and optic sign: 0.043, positive
Dispersion: Very high (0.104)
Specific gravity: 3.22 (floats in di-iodomethane – diamond sinks)
Hardness: 9.25
Lustre: Sub-adamantine

Luminescence: Inert to weak orange in LW UV, inert in SW UV and X-rays
Absorption spectrum: Band from 400 to 425 nm
Inclusions: Whitish or reflective needles
Synthesis process: Sublimation from a feed powder on seed crystals

Strontium titanate

Composition: $SrTiO_3$
Crystal system: Cubic
Crystal form: Boule
Colours: Transparent (diamond simulant); red, yellow, brown, blue when doped with appropriate transition elements
Refractive index: 2.41
Dispersion: Very high (0.19)
Specific gravity: 5.13
Hardness: 5.5 (susceptible to damage in ultrasonic cleaner)
Lustre: Adamantine
Absorption spectrum: For coloured stones, typical of dopant
Synthesis process: Flame-fusion (Verneuil type furnace using special burner to provide extra oxygen necessary to produce a colourless boule)

Yttrium aluminium garnet (YAG)

Composition: $Y_3Al_5O_{12}$
Crystal system: Cubic
Crystal form: Long columnar crystals of 1.5-inch (3.75-cm) diameter; main production for use in lasers
Colour: Colourless (diamond simulant), red (manganese), pink (erbium), yellow (ytterbium), pale green (praseodymium), emerald green (chromium), blue (cobalt)
Refractive index: 1.83
Dispersion: Medium (0.028)
Specific gravity: 4.58
Hardness: 8.5
Lustre: Sub-adamantine
Luminescence: Some stones fluoresce yellowish in LW UV, and yellow or bright mauve in X-rays
Absorption spectrum: For coloured stones, typical of transition element or rare earth dopant, although some colourless stones also show a rare earth spectrum
Synthesis process: Czochralski crystal pulling, and flux-melt

Appendix E   Examination notes

Both the Foundation and Diploma examinations of the Gemmological Association of Great Britain have two theory papers. Each question in these papers carries the appropriate 'Potential' mark, which gives some indication of the depth of answer required. Each of the Foundation examination theory papers contains five questions, all of which must be attempted. Because there are no optional questions, candidates must therefore be familiar with the complete syllabus to ensure a comfortable pass mark. There is a practical element to the first Foundation paper in that answers to each of the five questions may require the examination of gem specimens (using a 10× lens and/or a spectroscope or dichroscope) and gem photographs. Three hours are allowed for each paper. A Foundation Certificate is awarded to successful candidates, but only after the use of the basic gem identification instruments has been endorsed by an approved Gem-A tutor (either before or after the exam date). These endorsements are for the use of the $10\times$ hand lens, the microscope, refractometer, spectroscope, polariscope, dichroscope, Chelsea colour filter and the use of UV light. Also included is the measurement of a gem's specific gravity. For correspondence students, this element of the Foundation course is covered by a one-day practical workshop run by the Gem-A in London (or at other recommended venues).

Each of the two Diploma theory papers contains six optional questions from which only five need be selected for answering. As with the Foundation examination, three hours are allowed for each paper.

Examination techniques

Despite feeling that it is impossible to do justice to all five questions in the time allocated, both Foundation and Diploma students are strongly advised to spend the first few minutes in reading through the question paper carefully, and understanding exactly what is being asked in each question. In the Diploma theory papers, it is important to assess which five of the six optional questions can be best answered to yield the most marks (where the candidate is equally confident with all six questions, it may help to include one which can be answered relatively quickly, perhaps by means of tabulation or diagrams).

Having checked out the questions, a reasonably strict timetable should be adopted in which 30 minutes (plus or minus 2 minutes) are allocated for each question. If 5 minutes are taken at the beginning to read the questions, this will leave around 20 minutes at the end in which to check over the answers and make sure they are complete. With both Foundation and Diploma theory papers it is most important that the candidate attempts to answer all five questions, even if one or two present problems – a poorly or partially answered question will most probably gain some marks!

One of the dangers with any written examination is the possibility of misunderstanding questions, although these are usually composed with care to avoid any ambiguity. The possibility of a misunderstood question can be reduced by becoming familiar with the contents and style of papers set in previous years. Copies of earlier papers can be obtained from the offices of the Gemmological Association (see address, e-mail address, and telephone number at the end of this appendix).

Despite the shortage of time, some effort should be made to write legibly. Diagrams should be drawn reasonably large (there is no shortage of paper!) and the vital sections clearly annotated. Each answer should aim at convincing the examiner that the subject matter is thoroughly understood – handwriting and diagrams that are difficult to follow can only serve to diminish this aim.

All answers should be written in pen or ball-point pen. Diagrams should be drawn in pencil, using a ruler where appropriate. Opaque correction fluid must *not* be used on any answer paper. The examiners realize that many students write in English as a foreign language, and they are prepared to make due allowance for this (candidates are *not* penalized for poor spelling or grammar). If relevant, students should write the following note on the front cover of all their answer books: 'My first language is not English'.

For students taking the Diploma course, the practical examination can pose more of a challenge than the theory papers. The practical examination lasts $3\frac{1}{2}$ hours and is divided into four sections. These consist of the identification and description (including crystal system) of five crystal specimens by visual/hand lens inspection alone, the precise measurement (to three places of decimals) of the refractive index and double refraction of five polished sections (these are not restricted to gem materials in the syllabus, but no identification is called for), the identification of 10 polished gemstones using the test equipment provided (describing the cut and colour, and recording the tests made – hardness tests are forbidden!) and, finally, the identification of five specimens using only the spectroscope (a record is required of the absorption lines seen, and this can be indicated by means of a sketch with the red and blue ends of the spectrum clearly marked).

As with the theory papers, it is advisable to keep a check on the time taken for individual sections. A maximum of 7.5 minutes should be allowed for each determination. The remaining 15 minutes or so at the end can be spent in checking over answers and in retesting any doubtful specimens (this is permitted if the stone is available).

As the majority of tests carried out on the examination specimens involve the use of optical equipment, the level of lighting in the examination room is usually subdued. It is therefore advisable to avoid bright sunlight just before entering the examination room, and to allow several minutes for the eyes to become adapted to this level of lighting.

A card containing a printed list of gemstone constants is provided for each candidate at the practical exam, but as it is necessary to know these figures for the theory papers, it will probably save time if this list is only referred to in the event of a mental block!

At the larger examination centres, such as London, it is possible for 'bottlenecks' to occur in the use of instruments and in the availability of specimens. The candidate must therefore be prepared to use his time to best advantage by making tests out of sequence, and by taking the next available specimen from any of the four sections, as the examination proceeds. Because of this, it is most important that the checklist provided on the rear of the examination paper is carefully ticked off as each specimen is tested in order to avoid duplication. This list should be handed to the invigilator each time a new stone is requested so a specimen not ticked off on the list can be offered.

For Diploma correspondence course students who do not have access to any gem test instruments, it is important that they attend one of the several 2- or 3-day examination preparation courses (which are run just before the Diploma exam dates) in order to become more familiar with the use of the refractometer, microscope, polariscope and heavy liquids.

Changes to the Diploma examination

These changes, which were introduced in January 2005, include the award of a Practical Certificate to all candidates who pass the practical section of the examination. Those who fail the theory section but pass the practical section will also receive a Practical Certificate. These candidates can then re-sit the theory examination (with no restriction on the number of re-sit attempts) in order to qualify for the Diploma.

The practical examination is also now available to those who do not wish to sit the Diploma theory examination, or who wish to take it at a later stage.

To obtain a Merit, Distinction, Prize or Award, a candidate must have sat and passed both Diploma theory and practical papers for the first time in the same session. No awards are given following a re-sit. In accordance with current Gem-A Diploma rules, the theory and practical examinations are only available to holders of the Foundation Certificate, or have a pass in the earlier preliminary examination, or have been exempted.

From January 2006 onwards, candidates who pass the Diploma Theory examination but need to re-sit the practical examination can take two years (four opportunities to re-sit) in order to qualify for the Diploma. After this time a candidate must re-sit both papers.

The syllabus of both the Foundation and the Diploma courses are contained in the 'Education Prospectus' which can be obtained from the Gem-A offices at

27 Greville Street, London EC1N 8TN, UK (tel. 020 7404 3334; e-mail: information@gem-a.com)

Alternatively, the appropriate syllabus can be downloaded from the Gem-A website: www.gem-a.com

Appendix F Gemstone constants (in alphabetical order of gemstone)

Gemstone	Crystal system	Approx. RI	DR and optic sign	Dispersion	SG	H
Alexandrite – see Chrysoberyl						
Almandine (garnet)	Cubic	1.77–1.81	–	0.024	3.8–4.2	7.5
Amazonite (feldspar, microcline)	Triclinic	1.52–1.54	−0.008	0.012	2.56	6.0
Amber	Non-crystalline	1.54	–	–	1.05–1.10	2.5
Andalusite	Orthorhombic	1.63–1.64	−0.01	0.016	3.18	7.5
Andradite (demantoid, garnet)	Cubic	1.89	–	0.057	3.85	6.5
Apatite	Hexagonal	1.63–1.64	−0.003	0.013	3.18–3.22	5.0
Aquamarine – see Beryl						
Benitioite	Trigonal	1.76–1.80	+0.047	0.046	3.65–3.68	6.5
Beryl (aquamarine, emerald, goshenite, heliodor, vanadium beryl)	Hexagonal	1.57–1.59	−0.005/6	0.014	2.71	7.5–8.0
Bixbite (red beryl)	Hexagonal	1.586–1.594	−0.008	0.014	2.80	7.5–8.0
Bowenite	Monoclinic	1.56	†	–	2.58–2.62	4.0–5.0
Calcite	Trigonal	1.486–1.658	−0.172	0.02	2.71	3.0
Chalcedony	Trigonal	1.53–1.54	*	–	2.58–2.64	6.5
Chrysoberyl (alexandrite, cymophane)	Orthorhombic	1.74–1.75	+0.009	0.014	3.72	8.5
Chrysocolla	Monoclinic	1.50	*	–	2.0–2.45	2.0–4.0
Coral (black)	Trigonal	–	–	–	1.34–1.46	3.5
Coral (white and pink)	Trigonal	–	†	–	2.6–2.84	3.5–4.0
Cordierite – see Iolite						
Corundum (ruby, sapphire)	Trigonal	1.76–1.77	−0.008	0.018	3.99	9.0
Cubic zirconium oxide	Cubic	2.15–2.18	–	0.065	5.6–6.0	8.5
Cymophane – see Chrysoberyl						
Danburite	Orthorhombic	1.63–1.64	±0.006	0.017	3.0	7.0
Demantoid – see Andradite						
Diamond	Cubic	2.417	–	0.044	3.52	10.0
Dichroite – see Iolite						
Diopside	Monoclinic	1.67–1.70	+0.027	–	3.3	5.0
Dioptase	Trigonal	1.644–1.709	+0.053	0.03	3.25–3.35	5.0
Emerald – see Beryl						
Enstatite	Orthorhombic	1.66–1.67	+0.01	–	3.25–3.30	5.5
Feldspar – see Amazonite, moonstone, oligoclase, orthoclase, sunstone						

(Continued)

Appendix F (*Continued*)

Gemstone	Crystal system	Approx. RI	DR and optic sign	Dispersion	SG	H
Fire opal	Non-crystalline	1.40	–	–	2.0	6.0
Fluorite	Cubic	1.434	–	0.007	3.18	4.0
Garnet – see Almandine, andradite, grossular, pyrope, spessartite, uvarovite						
GGG	Cubic	1.97	–	0.045	7.05	6.0
Goshenite – see Beryl						
Grossular (hessonite, garnet)	Cubic	1.74–1.75	–	0.028	3.6–3.7	7.0–7.5
Heliodor – see Beryl						
Hematite	Trigonal	2.94–3.22	−0.28	–	4.9–5.3	5.5–6.5
Hessonite – see Grossular						
Hiddenite – see Spodumene						
Idocrase	Tetragonal	1.70–1.723	±0.005	0.019	3.32–3.42	6.5
Iolite (cordierite, dichroite)	Orthorhombic	1.54–1.55	−0.009	0.017	2.57–2.61	7.5
Ivory (dentine)	Non-crystalline	1.54	–	–	1.70–2.00	2.0–3.0
Ivory (vegetable)	Non-crystalline	1.54	–	–	1.38–1.54	2.5
Jadeite	Monoclinic	1.66–1.68	†	–	3.30–3.36	6.5–7.0
Jet	Non-crystalline	1.64–1.68	–	–	1.3–1.4	2.5–4.0
Kornerupine	Orthorhombic	1.67–1.68	−0.013	0.019	3.28–3.35	6.5
Kunzite – see Spodumene						
Kyanite	Triclinic	1.715–1.732	−0.017	0.02	3.65–3.69	5.0 and 7.0
Labradorite (feldspar)	Triclinic	1.56–1.57	−0.009	0.012	2.69–2.72	6.0
Lapis lazuli	–	1.50	–	–	2.7–2.9	5.5
Lithium niobate	Trigonal	2.21–2.30	−0.09	0.13	4.64	5.5
Malachite	Monoclinic	1.66–1.91	−0.25	–	3.8	4.0
Maxixe and maxixe-type beryl	Hexagonal	1.584–1.592	−0.008	–	2.80	7.5
Microcline – see Amazonite						
Moissanite (synthetic)	Hexagonal	2.65–2.69	+0.043	0.104	3.22	9.25
Moldavite	Non-crystalline	1.50	–	–	2.4	5.5
Moonstone and orthoclase (feldspar)	Monoclinic	1.52–1.53	−0.006	–	2.56	6.0–6.5
Morganite (pink beryl)	Hexagonal	1.586–1.594	−0.008	0.014	2.80	7.5–8.0
Nephrite	Monoclinic	1.62	†	–	2.90–3.2	6.5
Obsidian	Non-crystalline	1.50	–	0.01	2.3–2.5	5.0
Odontolite	Non-crystalline	1.57–1.63	–	–	3.0–3.25	5.0
Oligoclase (feldspar)	Triclinic	1.53–1.54	−0.007	–	2.64	6.0
Opal (except fire Opal)	Non-crystalline	1.45	–	–	2.1	6.0
Orthoclase – see moonstone						
Pearl (native)	Orthorhombic	1.52–1.66	–	–	2.68–2.78	2.5–3.5
Peridot	Orthorhombic	1.65–1.69	+0.036	0.02	3.34	6.5
Phenakite	Trigonal	1.65–1.67	+0.016	0.015	2.95–2.97	7.5

(*Continued*)

Gemstone	Crystal system	Approx. RI	DR and optic sign	Dispersion	SG	H
Prehnite	Orthorhombic	1.61–1.64	+0.03	–	2.88–2.94	6.0
Pyrite	Cubic	–	–	–	4.84–5.10	6.5
Pyrope (garnet)	Cubic	1.74–1.77	–	0.022	3.7–3.8	7.0–7.5
Quartz	Trigonal	1.544–1.553	+0.009	0.013	2.65	7.0
Rhodochrosite	Trigonal	1.60–1.82	−0.22	–	3.5–3.6	4.0
Rhodonite	Triclinic	1.72–1.74	±0.014	–	3.4–3.7	6.0
Ruby – see Corundum						
Rutile	Tetragonal	2.61–2.90	+0.287	0.28	4.2–4.3	6.5
Sapphire – see Corundum						
Saussurite	–	1.57–1.70	†	–	3.0–3.4	6.5
Scapolite (blue, pink, colourless)	Tetragonal	1.545–1.56	−0.016	0.017	2.63	6.0
Scapolite (yellow)	Tetragonal	1.55–1.57	−0.02	0.017	2.74	6.0
Scheelite	Tetragonal	1.92–1.93	+0.016	0.038	5.9–6.1	4.5
Sinhalite	Orthorhombic	1.67–1.71	−0.038	0.018	3.48	6.5
Soapstone – see Steatite						
Sodalite	Cubic	1.48	–	–	2.28	5.5–6.0
Spessartite (garnet)	Cubic	1.80–1.82	–	0.027	4.16	7.25
Sphene – see Titanite						
Spinel (natural)	Cubic	1.712–1.725	–	0.020	3.60	8.0
Spinel (synthetic)	Cubic	1.727	–	0.020	3.64	8.0
Spodumene (hiddenite, kunzite)	Monoclinic	1.66–1.68	+0.015	0.017	3.18	7.0
Steatite (soapstone)	Monoclinic	1.55	–	–	2.5–2.8	1.0+
Strontium titanate	Cubic	2.42	–	0.19	5.13	5.5
Sunstone – see Oligoclase						
Taaffeite	Hexagonal	1.717–1.723	−0.004	–	3.60–3.61	8.0
Tanzanite – see Zoisite						
Titanite (sphene)	Monoclinic	1.89–2.02	+0.13	0.051	3.53	5.5
Topaz (red, pink, orange)	Orthorhombic	1.63–1.64§	+0.008	0.014	3.53	8.0
Topaz (white/blue/yellow, brown)	Orthorhombic	1.61–1.62§	+0.01	0.014	3.56	8.0
Tourmaline	Trigonal	1.62–1.66	−0.018	0.017	3.01–3.11	7.0–7.5
Turquoise	Triclinic	1.61–1.65	**	–	2.6–2.9	5.5–6.0
Uvarovite (garnet)	Cubic	1.87	–	–	3.77	7.5
YAG	Cubic	1.83	–	0.028	4.58	8.5
Zircon (high)	Tetragonal	1.93–1.99	+0.059	0.039	4.68	7.25
Zircon (low)	Metamict	1.78–1.84	–	–	3.9–4.1	6.0
Zirconia – see Cubic zirconium oxide						
Zoisite (tanzanite – blue zoisite)	Orthorhombic	1.69–1.70	+ 0.009	0.012	3.35	6.5

* Microcrystalline.
** Granular or microcrystalline.
† Polycrystalline.
§ According to Hoover, the RI of topaz is dependent upon its fluorine content and not its colour. Generalizations associating the colour and refractive indices of topaz should therefore be avoided, and its DR be taken as centred about 0.009 regardless of fluorine content.

Gemstone constants (in order of refractive indices)

Approx. RI	DR and optic sign	Gemstone	Crystal system	Dispersion	SG	Hardness
1.40	—	Fireopal	Non-crystalline	—	2.0	6.0
1.434	—	Fluorite	Cubic	0.007	3.18	4.0
1.45	—	Opal (except fire opal)	Non-crystalline	—	2.1	6.0
1.48	—	Sodalite	Cubic	—	2.28	5.5–6.0
1.486–1.658	−0.172	Calcite (Iceland spar)	Trigonal	0.02	2.71	3.0
1.50	*	Chrysocolla	Monoclinic	—	2.0–2.45	2.0–4.0
1.50	—	Lapis lazuli	—	—	2.7–2.9	5.5
1.50	—	Moldavite	Non-crystalline	—	2.4	5.5
1.50	—	Obsidian	Non-crystalline	0.01	2.3–2.5	5.0
1.52–1.53	−0.006	Moonstone and orthoclase (feldspar)	Monoclinic	—	2.56	6.0
1.52–1.54	−0.008	Microcline and amazonite (feldspar)	Triclinic	—	2.56	6.0
1.52–1.66	*	Pearl (native)	Orthorhombic	—	2.70–2.74	2.5–3.5
1.53–1.54		Chalcedony	Trigonal	—	2.58–2.64	6.5
1.53–1.54	−0.007	Oligoclase and sunstone (feldspar)	Triclinic	—	2.64	6.0
1.54	—	Ambet	Non-crystalline	—	1.05–1.10	2.5
1.54	—	Ivory (vegetable)	Non-crystalline	—	1.38–1.42	2.5
1.54–1.55	−0.009	Iolite (cordierite, dichroite)	Orthorhombic	0.017	2.57–2.61	7.5
1.544–1.553	+0.009	Quartz (rock crystal)	Trigonal	0.013	2.65	7.0
1.544–1.56	−0.016	Scapolite (blue, pink, colourless)	Tetragonal	0.017	2.634	6.0
1.55	—	Steatite (soapstone)	Monoclinic	—	2.5–2.8	1.0+
1.55–1.57	−0.02	Scapolite (yellow)	Tetragonal	0.017	2.74	6.0
1.56	†	Bowenite	Monoclinic	—	2.58–2.62	4.0–5.0
1.56–1.57	+0.009	Labradorite (feldspar)	Triclinic	0.012	2.69–2.72	6.0
1.57–1.63	—	Odontolite	Non-crystalline	—	3.0–3.25	5.0
1.57–1.59	−0.005/6	Beryl (aquamarine, emerald, goshenite, heliodor, vanadium beryl)	Hexagonal	0.014	2.71	7.5–8.0
1.57–1.70	†	Saussurite	—	—	3.0–3.4	6.5
1.584–1.592	−0.008	Maxixe and maxixe-type beryl	Hexagonal	0.014	2.80	7.5
1.586–1.594	−0.008	Beryl (morganite and bixbite)	Hexagonal	0.014	2.80	7.5–8.0

1.60–1.82	−0.22	Rhodochrosite	Trigonal	—	3.5–3.6	4.0
1.61	†	Nephrite	Monoclinic	—	2.9–3.2	6.5
1.61–1.62§	+0.01	Topaz (white/blue/yellow, brown)	Orthorhombic	0.014	3.56	8.0
1.61–1.64	+0.03	Prehnite	Orthorhombic	—	2.88–2.94	6.0
1.61–1.65	**	Turquoise	Triclinic	—	2.6–2.9	5.5–6.0
1.62–1.66	−0.018	Tourmaline	Trigonal	0.017	3.01–3.11	7.0–7.5
1.63–1.64	−0.01	Andalusite	Orthorhombic	0.016	3.18	7.5
1.63–1.64	−0.003	Apatite	Hexagonal	0.013	3.18–3.22	5.0
1.63–1.64	±0.006	Danburite	Orthorhombic	0.017	3.0	7.0
1.63–1.64§	+0.008	Topaz (red, pink, orange)	Orthorhombic	0.014	3.53	8.0
1.64–1.68	—	Jet	Non-crystalline	—	1.3–1.4	2.5–3.5
1.644–1.709	+0.053	Dioptase	Trigonal	0.03	3.25–3.35	5.0
1.65–1.67	+0.016	Phenakite	Trigonal	0.015	2.95–2.97	7.5
1.65–1.69	+0.036	Peridot	Orthorhombic	0.02	3.34	6.5
1.66–1.67	+0.01	Enstatite	Orthorhombic	—	3.25–3.30	5.5
1.66–1.68	†	Jadeite	Monoclinic	—	3.3–3.36	6.5–7.0
1.66–1.68	+0.015	Spodumene (kunzite, hiddenite)	Monoclinic	0.017	3.18	7.0
1.66–1.91	−0.25	Malachite	Monoclinic	—	3.8	4.0
1.67–1.68	−0.013	Kornerupine	Orthorhombic	0.019	3.28–3.35	6.5
1.67–1.70	+0.027	Diopside	Monoclinic	—	3.3	5.0
1.67–1.71	−0.038	Sinhalite	Orthorhombic	0.018	3.48	6.5
1.69–1.70	+0.009	Zoisite (tanzanite)	Orthorhombic	0.012	3.35	6.5
1.70–1.72	±0.005	Idocrase	Tetragonal	0.019	3.32–3.42	6.5
1.712–1.730	—	Spinel (natural)	Cubic	0.020	3.6	8.0
1.715–1.732	−0.017	Kyanite	Triclinic	0.02	3.65–3.69	5.0 and 7.0
1.717–1.723	−0.004	Taaffeite	Hexagonal	—	3.60–3.61	8.0
1.72–1.74	+0.014	Rhodonite	Triclinic	—	3.4–3.7	6.0
1.727	—	Spinel (synthetic)	Cubic	0.020	3.64	8.0
1.74–1.75	—	Chrysoberyl (alexandrite, cymophane)	Orthorhombic	0.014	3.72	8.5
1.74–1.75	+0.009	Grossular (hessonite garnet)	Cubic	0.028	3.6–3.7	7.0–7.5
1.74–1.77	—	Pyrope (garnet)	Cubic	0.022	3.7–3.8	7.0–7.5
1.76–1.77	−0.008	Corundum (ruby/sapphire)	Trigonal	0.018	3.99	9.0
1.76–1.80	+0.047	Benitoite	Trigonal	0.04	3.65–3.68	6.5
1.77–1.81	—	Almandine (garnet)	Cubic	0.024	3.8–4.2	7.5
1.78–1.84	—	Zircon (low)	Metamict	—	3.9–4.1	6.0

(Continued)

Gemstone constants (*Continued*)

Approx. RI	DR and optic sign	Gemstone	Crystal system	Dispersion	SG	Hardness
1.80–1.82	–	Spessartite (garnet)	Cubic	0.027	4.16	7.25
1.83	–	YAG	Cubic	0.028	4.58	8.5
1.87	–	Uvarovite	Cubic	–	3.77	7.5
1.89	–	Andradite (demantoid garnet)	Cubic	0.057	3.85	6.5
1.89–2.02	+0.13	Titanite (sphene)	Monoclinic	0.051	3.53	5.5
1.92–1.93	+0.016	Scheelite	Tetragonal	0.038	5.9–6.1	4.5
1.93–1.99	+0.059	Zircon (high)	Tetragonal	0.039	4.68	7.25
1.97	–	GGG	Cubic	0.045	7.05	6.0
2.15–2.18	–	Cubic zirconium oxide	Cubic	0.065	5.6–6.0	8.5
2.21–2.30	–0.09	Lithium niobate	Trigonal	0.130	4.64	5.5
2.417	–	Diamond	Cubic	0.044	3.52	10.0
2.42	–	Strontium titanate	Cubic	0.19	5.13	5.5
2.61–2.90	+0.287	Rutile	Tetragonal	0.28	4.2–4.3	6.5
2.65–2.69	+0.043	Moissanite (synthetic)	Hexagonal	0.104	3.22	9.25
2.94–3.22	–0.28	Hematite	Trigonal	–	4.9–5.3	5.5–6.5

* Microcrystalline.
** Granular or microcrystalline.
† Polycrystalline.
§ According to Hoover, the RI of topaz is dependent upon its flourine content and not its colour. Generalizations associating the colour and refractive indices of topaz should therefore be avoided, and its DR be taken as centred about 0.009 regardless of flourine content.

Appendix G Units of measurement

Weight

The standard international (SI) unit of weight is the kilogram (kg). The most frequently used subdivisions are the gram and the milligram.

1 kilogram = 1000 grams
    1 gram = 1000 milligrams
            = 0.03527 ounce Avoir (1 ounce Avoir = 28.3500 grams)
            = 0.03215 ounce Troy (1 ounce Troy = 31.1035 grams)

For gemstone weighing, the standard unit is the metric carat.

1 carat = 0.2 grams (1 gram = 5 carats)
       = 0.007055 ounce Avoir (1 ounce Avoir = 141.7475 carats)
       = 0.006430 ounce Troy (1ounce Troy = 155.517 carats)

For pearl weighing, the standard unit is the grain.

1 grain = 0.25 carats (1 carat = 4 grains)

Note: The weight of small rough diamonds is sometimes expressed in grains (e.g. a 1-carat stone may be called a 'four grainer').

Polished diamonds under 1.0 carat in weight are measured in points.

1 point = 0.01 carats (1 carat = 100 points)

Length

The standard international (SI) unit for the measurement of length is the metre (m). The most frequently used subdivisions are the centimetre (cm), millimetre (mm), micrometre (μm, previously called 'micron') and nanometre (nm).

$1\,\text{m} = 100\,\text{cm}$
$1\,\text{cm} = 10\,\text{mm} = 10^{-2}\,\text{m}$
$1\,\text{mm} = 1000\,\mu\text{m} = 10^{-3}\,\text{m}$
$1\,\mu\text{m} = 1000\,\text{nm} = 10^{-6}\,\text{m}$
$1\,\text{nm} = 10^{-9}\,\text{m}$

Volume

The standard international (SI) unit for the measurement of volume is the cubic metre (m^3). The most frequently used subdivisions are the litre (1) and cubic centimetre (cm^3) or millilitre (ml).

$1\,\text{m}^3 = 1000\,\text{l}$
$1\,\text{l} = 1000\,\text{cm}^3 \text{ or } 1000\,\text{ml}$

Note: For all practical purposes, 1 litre is the volume of 1 kilogram of pure water at 4°C.

Wavelength

The standard international (SI) unit for the measurement of light wavelengths is the nanometre (nm).

$1\,\text{nm} = 10^{-9}\,\text{m}$ (one-thousand-millionth of a metre)
$1\,\text{nm} = 10\,\text{Å}$ (ångström units)

Light wavelengths are also sometimes given in micrometres (μm)

$1\,\mu\text{m} = 1000\,\text{nm} = 10^{-6}\,\text{m}$

Temperature

The standard international (SI) unit for temperature is the kelvin (K), and the degree Celsius (°C), both of which span equal temperature intervals. The kelvin is used mainly for thermodynamic work and represents an absolute temperature.

$0°C = 273.16\,K$

$0\,K = -273.16°C$ (the temperature at which no more internal energy can be extracted from an object, and at which the volume of a gas is theoretically zero).

Appendix H Table of elements

Element	Symbol	Atomic number (Z)	Atomic weight	Valency	Specific gravity
Actinium	Ac	89	227.05	–	–
Aluminium	Al	13	26.98	3	2.6
Americium	Am	95	243.0†	–	–
Antimony	Sb	51	121.75	3, 5	6.6
Argon	A	18	39.944	0	–
Arsenic	As	33	74.92	3, 5	5.72
Astatine	At	85	210.0†	–	–
Barium	Ba	56	137.34	2	3.8
Berkelium	Bk	97	249.0†	–	–
Beryllium	Be	4	9.02	2	1.83
Bismuth	Bi	83	209.0	3, 5	9.8
Boron	B	5	10.82	3	2.5
Bromine	Br	35	79.904	1	3.1
Cadmium	Cd	48	112.41	2	8.64
Caesium	Cs	55	132.91	1	1.87
Calcium	Ca	20	40.08	2	1.54
Californium	Cf	98	251.0†	–	–
Carbon	C	6	12.01	4	1.9–2.3‡
Cerium*	Ce	58	140.12	3, 4	6.9
Chlorine	Cl	17	35.453	1	–
Chromium	Cr	24	51.996	3, 6	7.1
Cobalt	Co	27	58.94	2, 3	8.6
Copper	Cu	29	63.55	1, 2	8.93
Curium	Cm	96	247.0†	–	–
Dysprosium*	Dy	66	162.46	3	–
Einsteinium	Es	99	254.0†	–	–
Erbium*	Er	68	167.26	3	4.8
Europium*	Eu	63	151.96	3	–
Fermium	Fm	100	253.0†	–	–
Fluorine	F	9	18.998	1	–
Francium	Fr	87	223.0†	–	–
Gadolinium*	Gd	64	157.25	3	5.9
Gallium	Ga	31	69.72	3	5.95
Germanium	Ge	32	72.59	4	5.47
Gold	Au	79	196.967	1, 3	19.3
Hafnium	Hf	72	178.49	–	–
Helium	He	2	4.003	0	–
Holium*	Ho	67	164.93	3	–
Hydrogen	H	1	1.008	1	–
Indium	In	49	114.82	3	7.3

(Continued)

Appendix H (*Continued*)

Element	Symbol	Atomic number (Z)	Atomic weight	Valency	Specific gravity
Iodine	I	53	126.93	1	4.95
Iridium	Ir	77	192.2	4	22.4
Iron	Fe	26	55.84	2, 3	7.87
Krypton	Kr	36	83.8	0	–
Lanthanum*	La	57	138.9	3	6.12
Lawrencium	Lw	103	257.0†	–	–
Lead	Pb	82	207.2	2, 4	11.34
Lithium	Li	3	6.94	1	0.53
Lutetium*	Lu	71	174.97	3	–
Magnesium	Mg	12	24.31	2	1.74
Manganese	Mn	25	54.93	2, 3	7.4
Mendelevium	Md	101	256.0†	–	–
Mercury	Hg	80	200.59	1, 2	13.59
Molybdenum	Mo	42	95.94	4, 6	10.0
Neodymium*	Nd	60	144.24	3	6.96
Neon	Ne	10	20.17	0	–
Neptunium	Np	93	237.0†	–	–
Nickel	Ni	28	58.71	2, 3	8.8
Niobium§	Nb	41	92.91	5	8.5
Nitrogen	N	7	14.007	3, 5	–
Nobelium	No	102	254.0†	–	–
Osmium	Os	76	190.2	6	22.5
Oxygen	O	8	16.00	2	–
Palladium	Pd	46	106.4	2, 4	11.4
Phosphorus	P	15	30.97	3, 5	1.8**
Platinum	Pt	78	195.09	2, 4	21.4
Plutonium	Pu	94	244.0†	–	–
Polonium	Po	84	209.0	–	–
Potassium	K	19	39.102	1	0.86
Praseodymium*	Pr	59	140.91	3	6.48
Promethium*	Pm	61	145.0†	–	–
Protactinium	Pa	91	231.0	–	–
Radium	Ra	88	226.0	2	–
Radon	Rn	86	222.0	–	–
Rhenium	Re	75	186.2	–	21.2
Rhodium	Rh	45	102.91	3	12.44
Rubidium	Rb	37	85.48	1	1.53
Ruthenium	Ru	44	101.7	6, 8	12.3
Samarium*	Sm	62	150.35	3	7.8
Scandium	Sc	21	44.96	3	–
Selenium	Se	34	78.96	2	4.8
Silicon	Si	14	28.086	4	2.3
Silver	Ag	47	107.87	1	10.5
Sodium	Na	11	22.99	1	0.97
Strontium	Sr	38	87.63	2	2.54
Sulphur	S	16	32.064	2, 4	2.07
Tantalum	Ta	73	180.88	5	16.6
Technetium	Tc	43	99.0†	–	–
Tellurium	Te	52	127.61	2	6.25
Terbium*	Tb	65	158.93	3	–
Thallium	Tl	81	204.3	1	11.9
Thorium	Th	90	232.04	4	11.3
Thulium*	Tm	69	168.93	3	–

(*Continued*)

Appendix H (*Continued*)

Element	Symbol	Atomic number (Z)	Atomic weight	Valency	Specific gravity
Tin	Sn	50	118.69	2, 4	7.28
Titanium	Ti	22	47.9	4	4.5
Tungsten	W	74	183.85	6	19.3
Uranium	U	92	238.03	4, 6	18.7
Vanadium	V	23	50.94	3, 5	6.0
Xenon	Xe	54	131.3	0	–
Ytterbium*	Yb	70	173.04	3	5.5
Yttrium	Y	39	88.905	3	3.8
Zinc	Zn	30	65.38	2	7.1
Zirconium	Zr	40	91.22	4	6.5

* One of the rare earths in the lanthanum group.
† Isotope with the longest known half-life.
‡ Graphite.
§ Formerly called columbium (Cb).
** Yellow (red = 2.2).

Appendix I Table of principal Fraunhofer lines

Fraunhofer line	Wavelength (nm)	Element	
A	762.8 (deep red)	Oxygen	} in the Earth's
B	686.7 (red)	Oxygen	atmosphere
C	656.3 (orange)	Hydrogen	
D_1	589.6 (yellow)	Sodium	
D_2	589.0 (yellow)	Sodium	
E	527.0 (green)	Iron	
b_1	518.4	Magnesium	
b_2	517.3	Magnesium	
b_3	516.9	Iron	} elements in the Sun's
b_4	516.7	Magnesium	chromosphere
F	486.1 (blue–green)	Hydrogen	
G	430.8 (blue)	Calcium	
H	396.8 (violet)	Calcium	
K	395.3 (violet)	Calcium	

Note: The twin sodium lines D_1, D_2 (with a mean wavelength of 589.3 nm) are used as the standard source when specifying gemstone refractive indices. Dispersion is measured as the difference in refractive index at the B and G wavelengths.

Appendix J Gemstone weighing

Gemstones, including diamonds, are traditionally weighed in *carats*. The carat weight standard has a long history, its most likely origin being the seed of the middle-eastern *Carob* tree. These seeds when dried have the unusual quality of being almost identical to each other in weight, and were once used as convenient units of weight in the pearl trade.

As commerce in pearls and gemstones spread across the world, the carob seed, or carat, was adopted by many countries who modified its weight to fit in with their own national weight units. This resulted in a carat weight which varied from country to country. Eventually, early in the twentieth century, a common standard carat weight, called the *metric carat*, was agreed upon by the various countries. There are exactly 5 metric carats to the gram (see Appendix H for other weight units).

Mechanical two-pan balances are provided with a set of either gram or carat weights. A set of weights will usually span the range from 100 grams (500 carats) down to 1 gram (5 carats) in brass weights, and from 0.5 gram (2.5 carats) down to 0.001 gram (0.005 carat) in aluminium weights. When weighing a gemstone, it is tempting to use these weights in a haphazard manner when attempting to balance the scales. However, the following method is recommended as being more efficient.

First, chose a single weight which is just heavier than the object being weighed. Then substitute this weight for an equivalent value in the next subdivision of smaller weights. Starting with the smallest value, remove the weights from the pan in turn until the object is just heavier than the weights. Replace the last weight removed with the equivalent value in smaller weights, and repeat the process again, working down methodically through the weight subdivisions until a balance is reached. The weights on the balance pan can then be totalled to arrive at the weight of the specimen. When using a substitutional-type analytical balance, the handling of the internal weights is done mechanically, but the method of working systematically down through the weight ranges remains the same.

To avoid damaging the balance pivots, the balance must be brought to rest before removing or adding weights to the pan. The weights should be handled with tweezers to avoid contaminating them and thus affecting their accuracy.

With modern electronic balances the weighing is done automatically; in common with mechanical balances, the main operating requirement is a draught and vibration-free environment. In dry atmospheres it may also be necessary to introduce a humidifier to reduce the build-up of static electric charges.

Index